Molecular Biology Intelligence Unit

The Na⁺/H⁺ Exchanger

Molecular Biology Intelligence Unit

The Na⁺/H⁺ Exchanger

Larry Fliegel, Ph.D.
MRC Group in the Molecular Biology of Membranes
University of Alberta
Edmonton, Alberta, Canada

CHAPMAN & HALL
I(T)P An International Thomson Publishing Company

New York • Albany • Bonn • Boston • Cincinnati • Detroit • London • Madrid • Melbourne •
Mexico City • Pacific Grove • Paris • San Francisco • Singapore • Tokyo • Toronto • Washington

R.G. Landes Company
Austin

MOLECULAR BIOLOGY INTELLIGENCE UNIT
THE Na^+/H^+ EXCHANGER

R.G. LANDES COMPANY
Austin, Texas, U.S.A.

U.S. and Canada Copyright © 1996 R.G. Landes Company and Chapman & Hall

All rights reserved.
No part of this book may be reproduced or transmitted in any form or by any means, electronic or mechanical, including photocopy, recording, or any information storage and retrieval system, without permission in writing from the publisher.
Printed in the U.S.A.

Please address all inquiries to the Publishers:
R.G. Landes Company, 909 Pine Street, Georgetown, Texas, U.S.A. 78626
Phone: 512/ 863 7762; FAX: 512/ 863 0081

North American distributor:
Chapman & Hall, 115 Fifth Avenue, New York, New York, U.S.A. 10003

CHAPMAN & HALL

U.S. and Canada ISBN: 0-412-10471-7

While the authors, editors and publisher believe that drug selection and dosage and the specifications and usage of equipment and devices, as set forth in this book, are in accord with current recommendations and practice at the time of publication, they make no warranty, expressed or implied, with respect to material described in this book. In view of the ongoing research, equipment development, changes in governmental regulations and the rapid accumulation of information relating to the biomedical sciences, the reader is urged to carefully review and evaluate the information provided herein.

Library of Congress Cataloging-in-Publication Data

Fliegel, Larry
The Na+/H+ exchanger/ Larry Fliegel   p. cm. — (Molecular biology intelligence unit)
  Includes bibliographical references and index.
  ISBN 0-412-10471-7 (alk. paper)
  1. Sodium cotransport systems. 2. Carrier proteins. I. Series.
QP535. N2F57   1996
612' .01575--dc20   96-195
                                                          CIP

Publisher's Note

R.G. Landes Company publishes six book series: *Medical Intelligence Unit, Molecular Biology Intelligence Unit, Neuroscience Intelligence Unit, Tissue Engineering Intelligence Unit, Environmental Intelligence Unit* and *Biotechnology Intelligence Unit.* The authors of our books are acknowledged leaders in their fields and the topics are unique. Almost without exception, no other similar books exist on these topics.

Our goal is to publish books in important and rapidly changing areas of bioscience for sophisticated researchers and clinicians. To achieve this goal, we have accelerated our publishing program to conform to the fast pace in which information grows in bioscience. Most of our books are published within 90 to 120 days of receipt of the manuscript. We would like to thank our readers for their continuing interest and welcome any comments or suggestions they may have for future books.

Deborah Muir Molsberry
Publications Director
R.G. Landes Company

CONTENTS

PART I

1. **Biochemistry and Molecular Biology of the Na^+/H^+ Exchanger: An Overview** 1
   Larry Fliegel and Pavel Dibrov
   A. Introduction 1
   B. Pharmacology and Molecular Biology of the Mammalian Na^+/H^+ Exchanger 4
   C. Structure of the Transmembrane Region of the NHE1 Isoform of the Na^+/H^+ Exchanger 4
   D. The Role of the Cytoplasmic Region of the Na^+/H^+ Exchanger 8
   E. Regulation of Expression of the Na^+/H^+ Exchanger (NHE1) 10
   F. The Importance of the Na^+/H^+ Exchanger in Muscle (in Health and Disease) 10
   G. Importance of Na^+/H^+ Exchange and Intracellular pH in Tumor Cells 11
   H. Future Prospects 12

2. **Na^+/H^+ Exchange in Mammalian Kidney** 21
   Orson W. Moe and Robert J. Alpern
   A. Introduction 21
   B. Na^+/H^+ Exchange in the Mammalian Nephron and Distribution of Molecular Isoforms 22
   C. Acute Regulation of Renal Na^+/H^+ Exchanger 28
   D. Chronic Regulation of Renal Na^+/H^+ Exchanger 34

3. **Regulation of the Na^+/H^+ Exchanger in Vascular Smooth Muscle** 47
   Bradford C. Berk
   A. Introduction 47
   B. The VSMC in the Blood Vessel Wall 48
   C. Na^+/H^+ Exchange and VSMC Function In Vivo: Tone and Growth 49
   D. Regulation of the Na^+/H^+ Exchanger in VSMC 51
   E. Regulation of Na^+/H^+ Exchanger Function in VSMC 55
   F. Role of VSMC Na^+/H^+ Exchanger in Hypertension 59

4. **Molecular Studies of Intestinal Epithelial Na^+/H^+ Exchangers** 69
   C. H. Chris Yun, Chung-Ming Tse and Mark Donowitz
   A. Introduction 69
   B. Members of the Na^+/H^+ Exchanger Gene Family 69
   C. Cellular Distribution of Na^+/H^+ Exchangers 72
   D. Regulation of Na^+/H^+ Exchangers 73

5. **Regulation of Expression of the Na⁺/H⁺ Exchanger (NHE1) Promoter** 91
   Larry Fliegel and Jason R.B. Dyck
   A. Introduction 91
   B. Cloning and Analysis of the NHE1 Promoter 92
   C. Comparison of the Mouse, Human and Rabbit NHE1 Promoters 95
   D. Summary and Future Directions 96

6. **Role of the Na⁺/H⁺ Antiporter Isoforms in Cell Volume Regulation** 101
   Lamara Shrode, Ana Cabado, Greg Goss and Sergio Grinstein
   A. Introduction 101
   B. Clinical Significance and Pathophysiology of Volume Regulation 102
   C. Role of Na⁺/H⁺ Exchanger in RVI 102
   D. Mechanisms of Activation of the NHE 104
   E. Isoforms of the Na⁺/H⁺ Exchanger 111
   F. Responsiveness of the Individual Isoforms to Osmolarity ... 114
   G. Concluding Remarks 116

7. **Characteristics of the Plasma Membrane Na⁺/H⁺ Exchanger Gene Family** 123
   John Orlowski and Gary Shull
   A. Molecular Heterogeneity 124
   B. Genomic Organization 128
   C. Tissue Expression 128
   D. Membrane Topology 130
   E. Functional Characteristics 132
   F. Regulation of Na⁺/H⁺ Exchanger Activity 133
   G. Structural Components of the Na⁺/H⁺ Exchanger 137
   H. Concluding Remarks 139

8. **The Regulatory Cytoplasmic Domain of the Na⁺/H⁺ Exchanger** 149
   Bernhard M. Schmitt, Toshitaro Ikeda, Munekazu Shigekawa and Shigeo Wakabayashi
   A. Introduction 149
   B. Gross Functional Anatomy of the Cytoplasmic Domain 150
   C. Phosphorylation 152
   D. Calcium/Calmodulin 153
   E. Regulatory Cofactors and "pH Maintenance Domain" 160
   F. G-Proteins and Others 161
   G. Outlook 164

Part II

9. **Role of the Sarcolemmal Na$^+$/H$^+$ Exchanger in Arrhythmogenesis During Reperfusion of Ischemic Myocardium** 173
   Metin Avkiran and Masahiro Yasutake
   A. Introduction 173
   B. An Involvement of Free Oxygen Radicals? 174
   C. Evidence for a Key Role for the Na$^+$/H$^+$ Exchanger 175
   D. Calcium–The Common Arrhythmogenic Mediator 180
   E. Concluding Comments 181

10. **Role of Sodium-Hydrogen Exchange in Mediating Myocardial Ischemic and Reperfusion Injury. Mechanism and Therapeutic Implications** 189
    Morris Karmazyn
    A. Introduction 189
    B. Importance of Intracellular pH Regulatory Mechanisms for Maintenance of Cardiac Function 193
    C. Evidence that Na$^+$/H$^+$ Exchange Regulates Intracellular Na$^+$ and Ca^{2+} Concentrations in the Heart 193
    D. Na$^+$/H$^+$ Exchange and Mycardial Ischemic and Reperfusion Injury 195
    E. Summary, Conclusions and Future Directions 209

11. **Expression and Activity of the Sodium-Hydrogen Exchanger in Cardiac Sarcolemma in Health and Disease** 217
    Grant N. Pierce, Tracy Slotin, Larry Fliegel, James S.C. Gilchrist and Thane G. Maddaford
    A. Intracellular Hydrogen and Cell Function 217
    B. Kinetics of Na$^+$/H$^+$ Exchange in Cardiac Sarcolemma 218
    C. Response of the Na$^+$/H$^+$ Exchanger to Disease 220

PART III

12. Molecular Dissection of Bacterial Na⁺/H⁺ Antiporters 231
Shimon Schuldiner and Etana Padan
  A. Introduction 231
  B. Molecular Nature and Properties of the Na⁺/H⁺ Antiporters 233
  C. Na⁺ Excretion Machinery Other than the Na⁺/H⁺
     Antiporters in *E. Coli* 247

13. Sodium Tolerance and Export from Yeast Cells 255
*Karen M. Hahnenberger, Zhengping Jia, Larry Fliegel,
Sean Hemmingsen and Paul G. Young*
  A. Introduction 255
  B. The Fungal Sodium/Proton Antiporter 256
  C. The *ZSOD2* Sodium/Proton Antiporter
     from *Zygosaccharomyces Rouxii* 260
  D. Use of the Cytosensor Microphysiometer to Demonstrate
     Bidirectional Exchange on sod2 260
  E. *PMR2/ENA1* P-Type Sodium Export Pumps 261
  F. Heterologous Expression of the *sod2* Sodium/Proton Antiporter ... 264
  G. Concluding Remarks 264

PART IV

**14. Therapeutic Potential of Inhibitors of Na⁺/H⁺
    Exchange Activity in Tumor Selective Therapy 269**
Motoyuki Yamagata and Ian F. Tannock
  A. Introduction 269
  B. The Acidic Microenvironment of Tumors 269
  C. Regulation of Intracellular pH (pH_i) 270
  D. Role of the Na⁺/H⁺ Exchanger in Proliferation and Survival
     of Tumors Cells 274
  E. Potential Therapeutic Application of Inhibitors
     of the Na⁺/H⁺ Exchanger 275
  F. Regulation of pH_i and Hyperthermia 284
  G. Regulation of pH_i and Radiation 285
  H. Regulation of pH_i and Chemotherapy 286
  I. Summary 286

Part V

15. **The NHE Family of Na^+/H^+ Exchangers; Its Known and Putative Members, and What Can Be Learned by Comparing Them with Each Other** 295
    Otto Fröhlich
    A. Introduction 295
    B. The Members of the NHE Family of Na^+/H^+ Exchangers 296
    C. A Comparison of the Members of the NHE Family 298
    D. Are All Members of the List True Na^+/H^+ Exchangers? 301
    E. Establishing a Consensus Hydropathy Profile of NHE Proteins 303
    F. Summary 305

Index 309

ABBREVIATIONS

ANP	atrial natriuretic peptides
BCECF-AM	2',7'-bis-(2-carboxyethyl)-5-(and 6)-carboxy-fluorescein acetoxymethyl ester
βNHE	trout isoform of the Na$^+$/H$^+$ exchanger
CaM	calmodulin
CAMK	CaM-dependent protein kinase
CCCP	carbonylcyanide-3-chlorophenylhydrazone
CHO	Chinese hamster ovary cells
CREB	cAMP responsive element-binding protein
DIDS	4,4-diisothiocyanstilbene 2,2-disulfonic acid
DMA	5-(N,N-dimethyl) amiloride
DNP	dinitrophenyl
EGF	epidermal growth factor
EIPA	5-(N-ethyl-N-isopropyl) amiloride (ethyl-isopropyl amiloride)
EMA	N-ethyl-N-(1-methyl-ethyl)amino amiloride
EMT-6	experimental murine tumor cells
ERK	extracellular signal-regulated kinase
ET-1	endothelin-1
FBS	fetal bovine serum
FGF	fibroblast growth factor
G-proteins	guanine-nucleotide-binding proteins
GTPγS	Guanosine 5'-O-[3-thiotriphosphate]

HL-60	human leukemic cells
HMA	5-(N,N-hexamethylene) amiloride
HOE642	4-isopropyl-3-methylsulphonylbenzoyl-guanidine methanesulphonate
HOE694	3-methylsulphonyl-4-piperidinobenzoyl-guanidine methanesulphonate
HOG	high osmolarity glycerol
HPLC	high-performance liquid chromatography
HSPD	Na^+/H^+ exchange-deficient variant cells
I_{K1}	Inward rectifying K^+ current
IMCD-3	inner medullary collecting duct cells
IP_3	inositol-1,4,5-trisphosphate
JNK-SAPK	Jun Kinase/Stress-activated Protein kinase
K_i	half inhibition constant
MAPK	mitogen-activated protein kinase
MCP-1	monocyte chemotactic peptide-1
MDR	multiple drug resistance
MGH-U1	human bladder carcinoma cells
MIBA (MIA)	5-(N-methyl-N-isobutyl) amiloride
MIBG	m-iodobenzylguanidine
MLC	myosin light chain
MLCK	myosin light chain kinase
MPA	5-N-(methylpropyl) amiloride
MTAL	medullar thick ascending limb cells

NCE	Na$^+$/Ca^{2+} exchange
NHE	sodium proton exchanger
NHE1	sodium proton exchanger type 1 isoform
NHE2	sodium proton exchanger type 2 isoform
NHE3	sodium proton exchanger type 3 isoform
NHE3P	NHE3 pseudogene (NHE3 like gene)
NHE4	sodium proton exchanger type 4 isoform
NHE5	sodium proton exchanger type 5 isoform
NMR	nuclear magnetic resonance
NO	nitric oxide
OA	okadaic acid
ORF	open reading frame
PCR	polymerase chain reaction
PDBU	phorbol 12,13-dibutyrate
PDGF	platelet derived growth factor
pH$_e$	extracellular pH
pH$_i$	cytoplasmic (intracellular) pH
PIP$_2$	phosphoinosito-4,5-bisphosphate
PKA	protein kinase A
PKC	protein kinase C
PKG	protein kinase G
PLC	phosphoinositide-specific phosphodiesterase
PMA	phorbol 12-myristate 13-acetate

PTH	parathyroid hormone
RKPC-2	rabbit S_2 proximal tubule cells
RNHE-1	Na^+/H^+ exchange activity
RT-PCR	reverse transcription-polymerase chain reaction
RVD	regulatory volume decrease
RVI	regulatory volume increase
SHR	spontaneously hypertensive rats
TMS	transmembrane segments
Vp-NhaA	*V. parahaemolyticus* NhaA antiporter
VF	ventricular fibrillation
VSMC	vascular smooth muscle cells
VT	ventricular tachycardia
WKY	Wistar-Kyoto rats
$\Delta\tilde{\mu}_{H+}$	proton electrochemical gradient
$\Delta\tilde{\mu}_{Na+}$	sodium electrochemical gradient
ΔpH	pHi-pHo
$\Delta pNa+$	$-\log [Na^+]_i/[Na^+]_o$
$\Delta\psi$	electrical potential
1,2-DAG	1,2-diacylglycerol

EDITOR

Larry Fliegel, Ph.D.
MRC Group in the Molecular Biology of Membranes
University of Alberta
Edmonton, Alberta, Canada
chapters 1, 5, 11, 13

CONTRIBUTORS

Robert J. Alpern
University of Texas Southwestern
  Medical Center
Dallas, Texas, U.S.A.
chapter 2

Metin Avkiran
Cardiovascular Research
The Rayne Institute
St Thomas' Hospital
London, United Kingdom
chapter 9

Bradford C. Berk
Division of Cardiology
Department of Medicine
University of Washington
Seattle, Washington, U.S.A.
chapter 3

Ana G. Cabado
Division of Cell Biology
Hospital for Sick Children
Toronto, Ontario, Canada
chapter 6

Pavel Dibrov
Departments of Pediatrics
  and Biochemistry
Cardiovascular Disease
  Research Group
University of Alberta
Edmonton, Alberta, Canada
chapter 1

Mark Donowitz
Department of Medicine
The John Hopkins University
  School of Medicine
Baltimore, Maryland, U.S.A.
chapter 4

Jason R.B. Dyck
Departments of Pediatrics and
  Biochemistry
Cardiovascular Disease Research
  Group
University of Alberta
Edmonton, Alberta, Canada
chapter 5

Otto Fröhlich
Department of Physiology
Emory University School
 of Medicine
Atlanta, Georgia, U.S.A.
chapter 15

James S.C. Gilchrist
Department of Oral Biology
Faculty of Dentistry
University of Manitoba
Winnipeg, Manitoba, Canada
chapter 11

Greg G. Goss
Division of Cell Biology
Hospital for Sick Children
Toronto, Ontario, Canada
chapter 6

Sergio Grinstein
Division of Cell Biology
Hospital for Sick Children
Toronto, Ontario, Canada
chapter 6

Karen M. Hahnenberger
Molecular Devices Corporation
Sunnyvale, California, U.S.A.
chapter 13

Sean Hemmingsen
National Research Council
Plant Biotechnology Institute
Saskatoon, Saskatchewan, Canada
chapter 13

Toshitaro Ikeda
Department of Molecular
 Physiology
National Cardiovascular Center
 Research Institute
Osaka, Japan
chapter 8

Zhengping Jia
Department of Biology
 and Pathology
Queen's University
Kingston, Ontario, Canada
chapter 13

Morris Karmazyn
Department of Pharmacology
 and Toxicology
University of Western Ontario
London, Ontario, Canada
chapter 10

Thane G. Maddaford
Division of Cardiovascular Sciences
Department of Physiology
St. Boniface Hospital Research
 Centre
University of Manitoba
Winnipeg, Manitoba, Canada
chapter 11

Orson W. Moe
Department of Veterans Affairs
 Medical Center
University of Texas Southwestern
 Medical Center
Dallas, Texas, U.S.A.
chapter 2

John Orlowski
Department of Physiology
McGill University
Montréal, Québec, Canada
chapter 7

Etana Padan
Department of Microbial
  and Molecular Ecology
The Institute of Life Sciences
The Hebrew University of Jerusalem
Jerusalem, Israel
chapter 12

Grant N. Pierce
Division of Cardiovascular Sciences
Department of Physiology
St. Boniface Hospital Research
  Centre
University of Manitoba
Winnipeg, Manitoba, Canada
chapter 11

Bernhard M. Schmitt
Department of Molecular
  Physiology
National Cardiovascular Center
  Research Institute
Osaka, Japan
chapter 8

Shimon Schuldiner
Department of Microbial
  and Molecular Ecology
The Institute of Life Sciences
The Hebrew University of Jerusalem
Jerusalem, Israel
chapter 12

Munekazu Shigekawa
Department of Molecular
  Physiology
National Cardiovascular Center
  Research Institute
Osaka, Japan
chapter 8

Lamara D. Shrode
Division of Cell Biology
Hospital for Sick Children
Toronto, Ontario, Canada
chapter 6

Gary Shull
Department of Molecular
  Genetics, Biochemistry
  and Microbiology
University of Cincinnati
Cincinnati, Ohio, U.S.A.
chapter 7

Tracy Slotin
Division of Cardiovascular
  Sciences
Department of Physiology
St. Boniface Hospital Research
  Centre
University of Manitoba
Winnipeg, Manitoba, Canada
chapter 11

Ian F. Tannock
Departments of Medicine
  and Medical Biophysics
Ontario Cancer Institute
University of Toronto
Toronto, Ontario, Canada
chapter 14

Chung-Ming Tse
Department of Medicine
The John Hopkins University
  School of Medicine
Baltimore, Maryland, U.S.A.
chapter 4

Shigeo Wakabayashi
Department of Molecular
  Physiology
National Cardiovascular Center
  Research Institute
Osaka, Japan
chapter 8

Motoyuki Yamagata
Departments of Medicine
  and Medical Biophysics
Ontario Cancer Institute
University of Toronto
Toronto, Ontario, Canada
chapter 14

Masahiro Yasutake
Cardiovascular Research
The Rayne Institute
St. Thomas' Hospital
London, United Kingdom
chapter 9

Paul G. Young
Department of Biology
  and Pathology
Queen's University
Kingston, Ontario, Canada
chapter 13

C. H. Chris Yun
Department of Medicine
The John Hopkins University
  School of Medicine
Baltimore, Maryland, U.S.A.
chapter 4

EDITOR'S BIOGRAPHY

Dr. Larry Fliegel, who joined the University of Alberta in 1989, is an Associate Professor in the Departments of Pediatrics and Biochemistry. He obtained his Ph.D. with Dr. G.I. Drummond at the University of Calgary where he studied the regulation by phosphorylation of proteins in isolated heart sarcolemma. There he learned basic membrane biochemistry and isolation and characterization of membrane fractions from eukaryotic tissues. Dr. Fliegel then took a postdoctoral fellowship with Dr. David MacLennan at the University of Toronto where he studied calsequestrin, a Ca^{2+} binding protein of the sarcoplasmic reticulum. In this laboratory he isolated the cDNA clone for calsequestrin and analyzed the details of its structure and function. During a second postdoctoral period with Dr. M. Michalak he studied calreticulin, a Ca binding protein of the sarcoplasmic reticulum. For this work he was awarded the Upjohn Investigator Award. Dr. Fliegel's independent research began in 1989 and he has been continuously funded by the Medical Research Council of Canada and Heart and Stroke Foundation of Canada to study the Na^+/H^+ exchanger and pH regulation. His research program has also involved cloning and characterization of cDNA for the Na^+/H^+ exchanger. He has studied regulation of expression and activity of the Na^+/H^+ exchanger promoter. His recent studies have also examined regulation of the Na^+/H^+ exchanger by phosphorylation and the interaction of the carboxyl-terminal region with other proteins. Dr. Fliegel teaches basic biochemistry at the University of Alberta and teaches membrane biochemistry. Currently Dr. Fliegel is a Senior Scholar of the Alberta Heritage Foundation for Medical Research, and his research program is supported by the medical Research Council of Canada and the National Science and Engineering Research Council of Canada.

PREFACE

This book is about the Na^+/H^+ exchanger. In particular it concentrates on recent developments on the biochemistry, molecular biology and physiology of the protein. The study of this protein began in the late seventies, and the characterization of the activity and characteristics of the protein boomed in the early 80s. Prior to the isolation and characterization of the first cDNA clones, most work in the area concentrated on measurement and characterization of the activity of the protein. Studies often examined effects of stimulation by various rapid effectors such as phorbol esters and a variety of hormones. During this time it was suspected that there were several types of Na^+/H^+ exchangers from differences in activity and sensitivity to inhibitors such as amiloride and its derivatives. However, there was no conclusive evidence identifying different proteins. Efforts to identify the protein itself were greatly hampered by the lack of efficient labels or the absence of sequence with which to direct antibodies. The low level of expression of the protein made these efforts even more difficult.

The development of molecular biological techniques allowed the isolation and cloning of the first Na^+/H^+ exchanger in 1989. Only since that time has there has been a virtual explosion in the knowledge about the structure and function of the protein. It was shown, for example, that the protein is an integral membrane protein with a large internal cytoplasmic tail. This initial observation led to the obvious hypothesis that the internal cytoplasmic region is a site of regulation by either phosphorylation and/or other proteins. In addition a number of laboratories have now utilized low stringency hybridization techniques to clone new isoforms of the Na^+/H^+ exchanger. It is now clear that there are at least five isoforms that were appropriately labeled NHE1-NHE5. The availability of these clones has allowed the examination of the structure of the protein and a comparison of the isoforms and their characteristics. Coupled with the availability of cell lines deficient in the antiporter, a number of studies have examined distinct amino acids and the domains of the protein. In addition several laboratories including our own have isolated and characterized factors important in regulation of the gene for the first isoform (NHE1). This purpose of this book is to provide a comprehensive review of the recent developments in the field. We concentrate on the recent developments on the protein and its recently characterized isoforms. A comparison is also made with other Na^+/H^+ exchanger isoforms of nonmammalian species. In addition, we review recent evidence on the important role of the protein in the contractile failure associated with ischemia.

Although several reviews of parts of the research area have appeared, there have been few comprehensive reviews of this field of research, particularly with contributions from multiple laboratories. A great deal of progress has been made, so this volume hopes to fulfill this need. Some laboratories have been very productive in the area but usually have not been able to write comprehensive reviews. It was gratifying that they were able to take the time to make significant contributions in this instance. Textbooks in a particular area need to be written by scientists involved in the research in the field. It is their insights and up to date knowledge that can provide the concise and instructive insights. I was very fortunate and am extremely grateful that so many of the leading researchers in the field were willing to take the time to contribute such high quality articles. Unfortunately, not every leading laboratory was able to find the time to contribute to this volume. However, with all the excellent contributions the field is very well covered by the chapters within. I would like to thank those that were able to take time out of their busy schedules to make contributions to the text.

Great challenges for the future still remain. There is no information of the three-dimensional structure of the protein. This will present a great challenge for the future. In addition many questions still remain to be answered about the regulation of the protein. In particular important questions about phosphorylation and the interaction with other proteins still remain. The mechanisms involved in regulation of expression of the NHE1 and other isoforms are still largely unsolved.

The text contains four sections. Section I (chapters 1-8) contains articles on the biochemistry and molecular biology of the Na^+/H^+ exchanger. It contains contributions from a number of groups including our own. Most recently, our major contribution to the area has been our studies on regulation of NHE1 gene expression. These are summarized in this section. Section II (chapters 9-10) reviews the role in the myocardium. It has become apparent that the Na^+/H^+ exchanger plays a key role in ischemic damage to the myocardium that occurs during reperfusion. This can have important clinical applications. Section III (chapters 12 and 13) reviews the new information on other forms of the Na^+/H^+ exchanger. Recent studies have examined the Na^+/H^+ antiporter in a number of nonmammalian species. In some cases these studies have the advantages of working with more easily transformed species such as *Escherichia coli*. This provides unique opportunities for expression and mutation studies. Section IV (chapter 14) addresses a unique area, the role of the Na^+/H^+ exchanger in tumor selective therapy, and its role in the myocardium. It is apparent that the Na^+/H^+ exchanger has an important role to play in cell growth. New therapies may exploit this physiological role in tumor cells.

It is hoped that the knowledge collected in this volume will provide a valuable resource for researchers in the field and graduate students of biochemistry, molecular biology and physiology. In addition the volume should serve to stimulate interest in this burgeoning area of study, and the ideas expressed within should serve as a source of inspiration for scientists in the field.

Larry Fliegel

Acknowledgments

I am very grateful to all the contributors of chapters to this volume. It was especially gratifying that so many of the top scientists in the field were able to take time out from their busy schedule in order to write such high quality reviews. Without their efforts, it would certainly have been impossible to produce such a text. I am grateful to Marie Jose Boeglin for her secretarial assistance. I am also especially grateful to my wife Sandra for her editorial assistance during the time I worked on this book and my other related research projects. The members of my laboratory who have made my research program on the Na^+/H^+ exchanger possible deserve a special thanks. This list includes Pavel Dibrov, Jason Dyck, Robert Haworth, Norma Lucena, Dyal Singh, Huayan Wang, Weidong Yang and others. I am also grateful to the Medical Research Council of Canada and the Heart and Stroke Foundation of Canada for their continuous support of my research work concerning the Na^+/H^+ exchanger. In addition, I am thankful for the scholarship support of the Heart and Stroke Foundation of Canada and the Alberta Heritage Foundation for Medical Research.

PART I

CHAPTER 1

BIOCHEMISTRY AND MOLECULAR BIOLOGY OF THE Na^+/H^+ EXCHANGER: AN OVERVIEW

Larry Fliegel and Pavel Dibrov

A. INTRODUCTION

The Na^+/H^+ antiporter is a ubiquitous integral membrane protein present in the plasma membrane of cells. It plays a key and critical role in regulation of intracellular pH, removing excess intracellular acid from the cell in exchange for extracellular sodium ions. Since the first cloning of the prokaryotic and eukaryotic Na^+/H^+ exchangers in 1987[1] and 1989,[2] there has been a virtual explosion in the knowledge about these proteins. From that time onward, new mammalian isoforms have been cloned and isolated, and the gene for the first eukaryotic isoform has been isolated. The purpose of this book is to provide a comprehensive review of recent developments in the field. Chapter 1 provides a general introduction to the Na^+/H^+ antiporter, recent developments in the field, and an outline for the chapters contained in the text.

A.1 HISTORICAL OVERVIEW

In 1961, P. Mitchell postulated the existence of Na^+/H^+ antiport within the framework of his chemiosmotic conception. Mitchell suggested that Na^+/H^+ and Na^+/K^+ antiporters could prevent the presence of excess alkali cations in the negatively charged interior of mitochondria or bacteria. During the following three decades, progress in understanding this phenomenon has been remarkable. Na^+/H^+ exchange has been discovered to be a universal process in the eukaryotic and

Table 1.1. Brief chronology of studies on Na^+/H^+ exchange systems and developments in the field

Date	Finding	Ref.
1961	First prediction of Na^+/H^+ antiport by P. Mitchell	3
1967	First experimental demonstration of Na^+/H^+ antiport in rat liver mitochondria	4
1972	Demonstration of Na^+/H^+ exchange in the bacterium St. faecalis	5
1976	Electroneutral Na^+/H^+ exchange is documented in the plasmalemma of epithelial cells	6
1981	Demonstration of inhibition of the mammalian Na^+/H^+ antiporter by amiloride and its analogs	8
1982	First evidence that the activity of the mammalian exchanger is allosterically activated by intracellular protons	7
1981-1983	Revealing that growth factor activation of Na^+/H^+ exchange participates in mitogenesis	9-12
1987	Cloning of the major Na^+/H^+ antiporter gene of E. coli	1
1989	Cloning of the human Na^+/H^+ exchanger (NHE1 isoform)	2
1990	Elucidation that growth factors induce phosphorylation of the NHE1	13
1991	Purification and reconstitution of the functional NhaA, the electrogenic Na^+/H^+ antiporter of E. coli	14
1992	Cloning of the plasmalemmal Na^+/H^+ exchanger of the yeast Schizosaccharomyces pombe	15
1989-1995	Cloning of five distinct isoforms of mammalian Na^+/H^+ exchanger (NHE1 to NHE5); determination of their expression in different tissues	16-22
1993-1994	Experimental evidence of oligomerization of mammalian Na^+/H^+ antiporters: NHE1 and NHE3 form stable homodimers in the membrane	23-24
1994	The cytoplasmic carboxyl terminal domain of the NHE1 can bind calmodulin in a Ca^{2+}-dependent manner	25, 26
1995	The C-terminal segment of NHE1 interacts with mammalian heat-shock protein	27

the prokaryotic world. It is key, not only to cellular Na^+ and pH homeostasis, but also to diverse cell responses such as growth factor-stimulated mitogenesis. Table 1.1 presents selected works on the structure, function, and regulation of Na^+/H^+ exchange systems in different organisms.[1-27] Na^+/H^+ antiporters of plant cells, algae, and invertebrates, as well as sodium-exchanging machinery of subcellular organelles, are not included in this brief summary. These topics have been recently reviewed[28] elsewhere. Some real objectives, rather than personal preferences, underlie such a selection. The mainstream lines of research in this area seem to be studies wherein techniques of molecular cloning are applied to identify corresponding genes and manipulate them to obtain biologically significant information. In the case of bacterial exchangers, NhaA and NhaB, the pioneering works of E. Padan and colleagues yielded purification of active proteins (reviewed in chapter 12). However, this has not, to date, been accomplished for the mammalian antiporters. Nevertheless, amazing achievements have been made in the study of the mammalian Na^+/H^+ antiporters. Research has generated a number of the basic fundamental characteristics of these antiporters including, (1) similar predicted topology of the membrane-spanning regions; (2) sensitivity to amiloride and amiloride-like compounds; (3) function of the cytoplasmic domain of the antiporters and its varying role between isoforms; and (4) regulation by growth factors and different roles of the cytoplasmic domain, in this regard. In addition, the important role of these proteins in cell growth and in several diseases has been confirmed.

A.2 Physiology of Na^+/H^+ Exchange in Mammalian Cells

There is a strong hypothesis for the existence of pH homeostasis and pH regulatory proteins. Practically all mammalian cells maintain a cytoplasmic pH (pH_i) within a rather narrow physiological range of approximately 7.2, despite metabolically induced changes in pH_i. Cellular pH is under strong control during a large number of processes including cell differentiation and smooth muscle contraction.[29-31] Small increases in pH_i alter cell division and activate expression of specific genes.[32] Therefore, cells have evolved several regulative mechanisms to maintain this precise pH-homeostasis in different cell types.[33] Both Na^+ dependent and Na^+ independent Cl^-/HCO_3^- exchange as well as H^+-ATPases participate in pH regulation in various cells and organelles.[34,35]

The Na^+/H^+ exchanger is a universal device that has evolved to regulate intracellular pH.[36] Under physiological conditions, in mammalian cells, the exchanger catalyzes the net uptake of Na^+ ions coupled to efflux of cytoplasmic H^+. Thus, the Na^+ gradient drives alkalinization of the cell interior with a stoichiometry of Na^+/H^+ exchange of 1:1.[37] The exchanger is normally nearly quiescent when cytoplasmic pH is at the physiological level. Activation occurs by various stimuli including hormones (insulin, vasopressin) and growth factors (PDGF, EGF) and other stimuli such as chemotactic factors and fertilization of eggs.[36,37] The key role of Na^+/H^+ antiport in prevention of intracellular acidification is evident from experiments with mutant cell lines devoid of Na^+/H^+ antiport.[38] Na^+/H^+ exchange also participates in cell volume regulation after osmotic shrinkage[39] (chapter 6) and in a number of other different functions, depending on the cell type.[40]

B. PHARMACOLOGY AND MOLECULAR BIOLOGY OF THE MAMMALIAN Na^+/H^+ EXCHANGER

An important feature of the mammalian Na^+/H^+ exchanger is its sensitivity to the diuretic amiloride and related derivatives.[41-44] All these compounds are competitive inhibitors of Na^+; the most potent derivatives have an alkylated 5-amino group.[45] Pharmacological and kinetic studies revealed multiple isoforms of the Na^+/H^+ exchanger. Thus, amiloride and its analogs distinguish between the epithelial luminal transporter with a K_i for amiloride near 0.1 mM and the more sensitive transporter (K_i = 1-10 µM) generally present on the basolateral surfaces of polarized cells.[46-48] The latter form, NHE1 (for Na^+/H^+ exchanger type 1), appears to be present in the plasma membrane of most cells.[23,46,49] At present, five different isoforms of mammalian antiporter (NHE1-NHE5) have been identified by molecular cloning experiments, and their relative abundance in some tissues has been examined.[2,13,16,17,19,22,50-61] These isoforms vary in their tissue distribution and in their kinetic and pharmacological properties. NHE2 and NHE3 are found in gastrointestinal tissues and renal epithelial tissues with NHE2 having a more diverse distribution than NHE3. NHE4 occurs mainly in the stomach, which also contains large amounts of NHE1 and some NHE3 message.[16,52] The two epithelial isoforms, NHE2 and NHE3, are thought to reside in the apical membrane. They participate in transepithelial NaCl transport. NHE1 resides in the basolateral membrane of epithelia[57] as well as in the plasma membrane of nonpolarized cells, where it participates in pH regulation.[52,62] Newly found NHE5 seems to be very similar to NHE3; the corresponding gene is expressed in brain, testis, spleen, and skeletal muscle.[17] NHE1 has been isolated from several different species.[2,16,19,22,53-55] Chapter 2 discusses the distribution and regulation of the isoforms particularly of NHE3 and NHE1 and the mammalian kidney. Chapters 4 and 7 discuss the distribution of the exchanger isoforms in other tissues. The homology between the isoforms varies with the greatest conservation occurring in the transmembrane regions. However, the first transmembrane region is not well conserved. An analysis of the relation(s) between the isoforms is presented in chapter 15. Although it is clear that the nonmammalian antiporters display little overall homology to the mammalian exchangers, there are some critical elements that are conserved. This may include regions modulating amiloride sensitivity (chapters 12 and 15).

C. STRUCTURE OF THE TRANSMEMBRANE REGION OF THE NHE1 ISOFORM OF THE Na^+/H^+ EXCHANGER

The first known mammalian Na^+/H^+ exchanger cDNA clone was later called the NHE1 isoform.[2,13] The deduced protein contains 815 amino acids and has a predicted molecular weight of 91 kDa. The actual protein has an apparent molecular weight of 110 kDa because

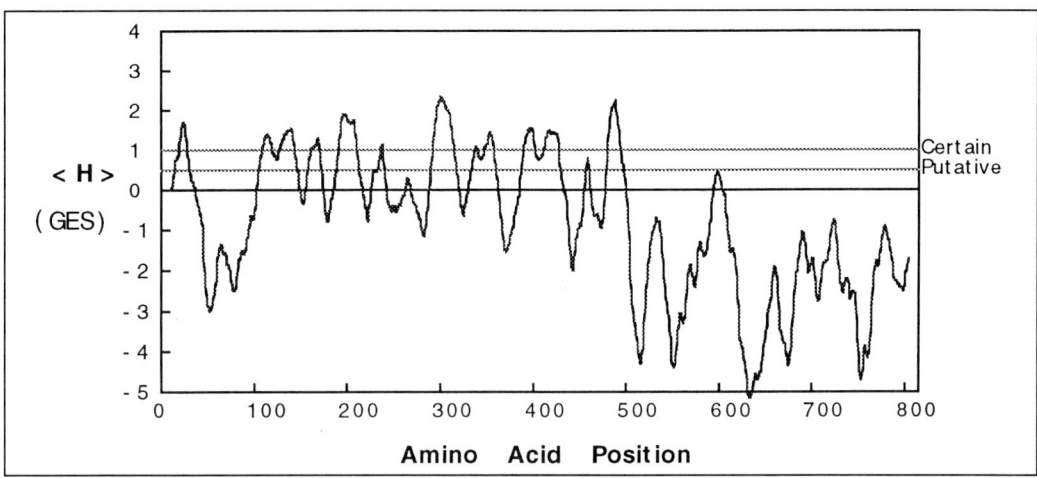

Fig. 1.1. Hydropathy plot of the human NHE1 isoform of the Na+/H+ exchanger. The plot was generated using the program TopPredII.[63] The GES parameters of Engelman et al[64] were used. The window size was 21 with a core window of 11.

it is glycosylated.[13,49] Sardet et al[2] proposed a model of the transporter that consists of two domains. The N-terminal hydrophobic domain contains 500 amino acids and could form 10 to 12 putative transmembrane helices according to the different hydropathy plots.[2,13,58] A hydrophobicity plot of the human NHE1 Na+/H+ exchanger is shown in Figure 1.1. It is based on the improved software program for prediction of protein structure TopPredII.[63] The program confirms some earlier predictions, although the model predicted varies from the standard models. Twelve transmembrane segments are suggested with 11 certain and one putative transmembrane passage of the protein (Table 1.2). There is a large hydrophilic C-terminal region. Only two proposed topologies were predicted by the program utilizing the GES parameters.[64] One of these predicted the C-terminal exterior to the cells. Because the location of the carboxyl terminal region has been demonstrated to be intracellular,[13] the appropriate model is the one with 12 transmembrane domains with a large internal cytoplasmic region. This predicted model of the transmembrane region of the Na+/H+ exchanger is presented in Figure 1.2. It should be emphasized that besides the intracellular location of the cytoplasmic tail, little is known about the actual topology of the protein. Early models predicted ten transmembrane regions,[2] and later it was suggested that this was twelve.[58] It is noteworthy that this model varies from other predictions. For example, Glu262 is often predicted to be within a transmembrane segment. Indeed, Glu262 is a highly conserved residue, and the substitution Glu262Ile results in complete inactivation of the NHE1 isoform.[24] Another more

Table 1.2. Candidate membrane-spanning segments of the human NHE1 isoform of the Na$^+$/H$^+$ exchanger

Probability of Membrane Passage (GES)	Segment Number	Amino Acids (GES)	Amino Acids (Kyte and Doolittle)
Certain	1	14-34	14-34
Certain	2	105-125	105-125
Certain	3	130-150	130-150
Certain	4	159-179	158-178
Certain	5	187-207	186-206
Certain	6	227-247	227-247
Certain	7	291-311	254-274
Certain	8	343-363	291-311
Certain	9	388-408	343-363
Certain	10	411-431	388-403
Putative	11	449-469	413-433
Certain	12	479-499	476-496

Candidate segments were identified using the program TopPredII[63] with the GES hydrophobicity scale[64] or using Kyte and Doolittle parameters.[65] The probability of membrane passage is indicated utilizing the GES parameters. The Kyte and Doolittle parameter segments shown were classified as definite transmembrane segments.

standard model of the Na$^+$/H$^+$ exchanger is shown in chapter 4 and discussed in chapter 15. The summary of its characteristics is presented in Table 1.2. We also generated this model using the same program with the standard Kyte and Doolittle[65] parameters. In this model, residues 254-274 form a transmembrane domain and 449-469 do not form one. It is clear at present, that the exact topology of the protein is not yet known and awaits proof by experimentation.

For this and other models of the Na$^+$/H$^+$ exchanger it is suggested that the first transmembrane region forms a signal sequence,[58] and this region displays lower homology between isoforms compared with the balance of the transmembrane domain (reviewed in chapter 15). To date it remains unclear if this segment is removed during the maturation of the protein or if it serves as a noncleavable anchor peptide.

Some critical elements have been identified in the protein. The fourth transmembrane domain contains residues that can confer amiloride sensitivity and resistance to antiporters. For example, replacement of Leu167 to Phe results in decreased sensitivity to amiloride analogs (reviewed in chapters 12 and 7). Glycosylation of the protein has been shown, and glycosylation is restricted to the first N-terminal domain.[66] It has also been demonstrated, in some tissues, that there are different levels of glycosylation, the function of which is not known.[49] Because

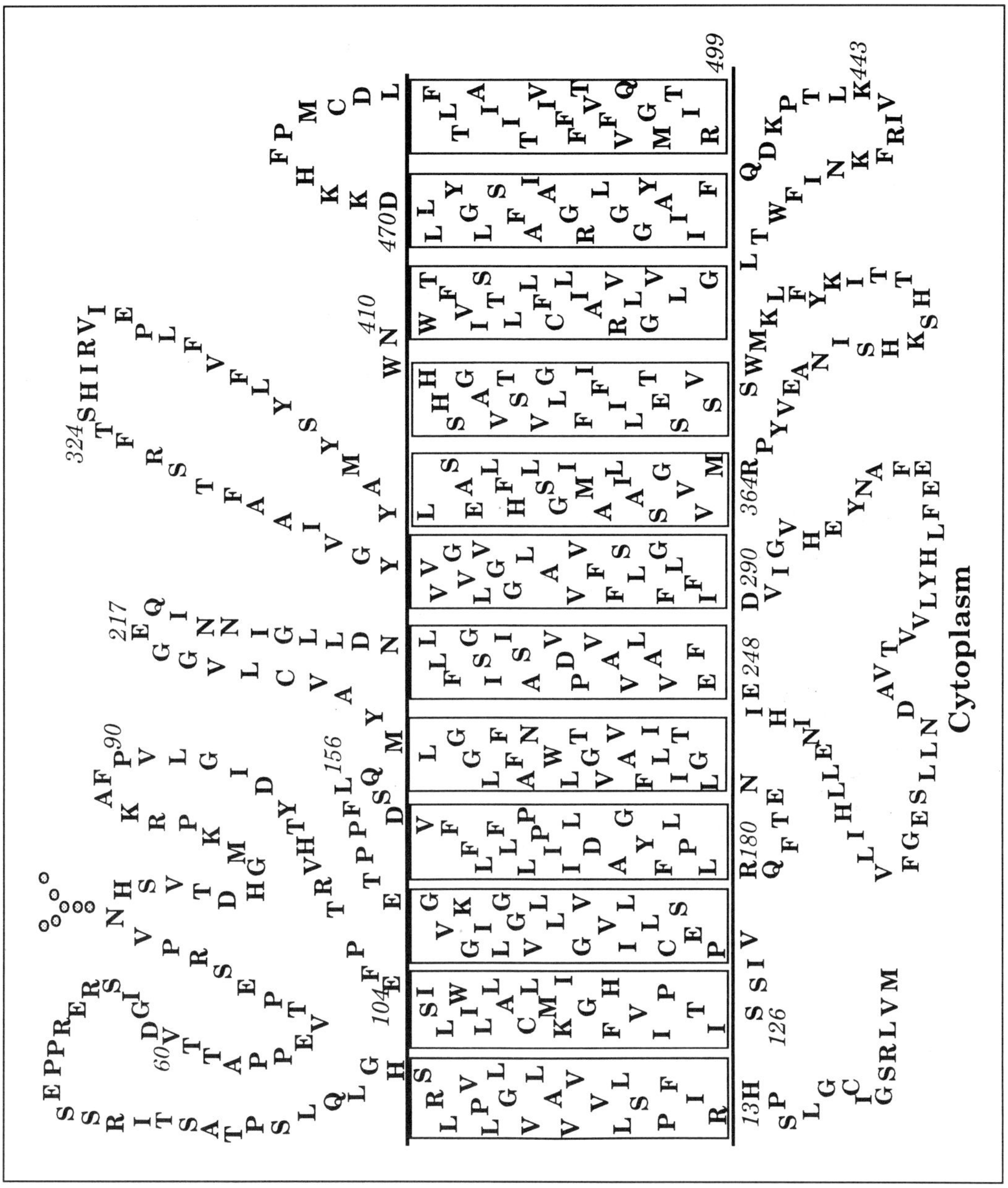

Fig. 1.2. Model of the transmembrane domain of the human NHE1 isoform of the Na+/H+ exchanger. Transmembrane domains were identified as described in Figure 1.1 and Table 1.2. Residues 1-499 of the protein are shown.

deglycosylation of right side out membrane vesicles removed the carbohydrate groups of the NHE1 isoform[49] it is confirmed this first loop is located on the extracellular face of the membrane.

D. THE ROLE OF THE CYTOPLASMIC REGION OF THE Na+/H+ EXCHANGER

The Na+/H+ exchanger is the subject of sophisticated regulation through different pathways.[36,52,58,67] Exposure of the cell to hormones and growth factors, such as platelet-derived growth factor, thrombin, angiotensin II, arginine vasopressin, and epidermal growth factor, causes rapid activation that occurs in minutes. Phorbol esters and diacylglycerol, as well as other agents such as calcium ionophores, chemotactic stimuli, fertilization, and osmotic shrinking,[36] provoke swift activation of the exchanger.[36,67-69] In general, stimulation of Na+/H+ antiport is manifested as an alkaline shift in the curve representing the pH_i dependence of activity,[36] as if the affinity of a hypothetical modifier site for protons ("pH-sensor") was increased on stimulation.

It has been firmly established that NHE1 may be activated through direct phosphorylation of the C-terminal cytoplasmic domain that is mediated directly or indirectly by protein kinase C.[2] However, phosphorylation of NHE1 cannot fully account for its activation. Indeed, removal of the NHE1 cytoplasmic tail at residue 635 eliminates all major sites of phosphorylation, but the truncated protein can still be activated[50,58] (Fig. 1.3). At the same time, a nonphosphorylatable segment containing residues 567-635 is required for the growth factor-mediated activation of the antiporter[70] (Fig. 1.3). Apparently, alternative pathways of activation of the Na+/H+ exchanger exist. One may be through intracellular Ca^{2+} acting via calmodulin (CaM) dependent protein kinase II. In some cell types the rise in intracellular Ca^{2+} precedes intracellular alkalinization induced by growth factors[71] and may be necessary to activate the exchanger.[72-74] Recent work has shown that NHE1 itself is also a Ca^{2+}/CaM-binding protein.[25,26] It was found (in these studies) that residues 637-656 of the C-terminal portion of NHE1 are involved in Ca^{2+}-dependent CaM binding;[25] moreover, it has been demonstrated for the first time that an artificial rise in intracellular $[Ca^{2+}]$ shifts the pH-dependence of Na+/H+ exchange toward alkaline pH.[26] Figure 1.3 shows a simple model of some of these elements, and chapter 8 reviews these important results in detail.

An important aspect of regulation of the Na+/H+ exchanger is its activation by hyperosmotic challenge. Exposure of cells to a solution of high osmolarity results in cell shrinkage and activation of the Na+/H+ exchanger. This results in increased cellular Na+ and subsequent compensatory water uptake.[75] An interesting aspect of the activation of the exchanger by this mechanism is that it occurs without direct phosphorylation of the protein, but is ATP dependent.[75,76] Chapter 6 reviews the mechanisms involved. They may involve a cytoskeletal com-

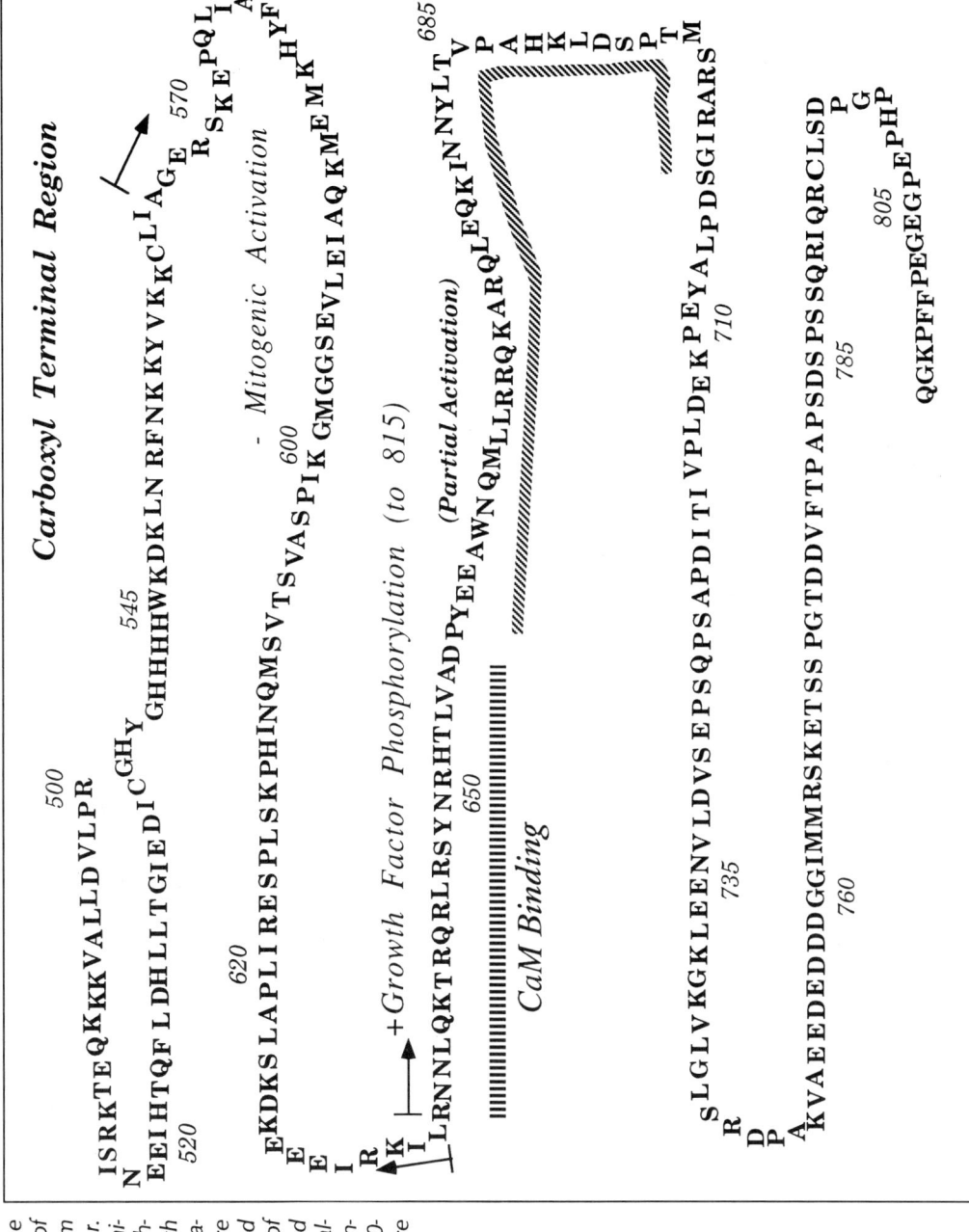

Fig. 1.3. Model of the cytoplasmic domain of the human NHE1 isoform of the Na+/H+ exchanger. Regions involved in mitogenic activation without (567-635) and with (635-815) phosphorylation of the protein are indicated and described in the text. The regions of high affinity [▨] and low affinity [▧] calmodulin binding are indicated. Residues 500-815 of the protein are shown.

ponent or other ancillary proteins, possibly acting through the cytoplasmic domain. Beside volume induced regulation, it is also postulated that some cytoskeletal and/or regulatory factors act on the cytoplasmic domain and are involved in mitogenic activation of the protein.[50,51] Chapter 8 reviews the evidence for the existence of this protein. Very recently, it has also been established (in our laboratory) that the carboxyl terminal segment of NHE1 interacts with the chaperone hsp70.[27] The resulting complex was suggested to prevent incorrect oligomerization[27] because, as it has been found in our[23] and other[24] laboratories, the exchanger exists as a dimer in the membrane.

E. REGULATION OF EXPRESSION OF THE Na^+/H^+ EXCHANGER (NHE1)

A number of stimuli are known to increase NHE1 expression at the mRNA and protein levels. In vascular smooth muscle serum, platelet-derived growth factors increase the Na^+/H^+ exchanger mRNA levels up to 25-fold.[77] With phorbol ester induced differentiation of HL60 cells, mRNA levels increase 50-fold and protein levels also increase.[78] In the kidney and vascular smooth muscle, the Na^+/H^+ exchange activity is elevated by acidosis.[79,80-87] This effect is dependent on glucocorticoids[87] and thyroid hormone.[88] It has been shown that protein kinase C is required for acid-induced increases in Na^+/H^+ exchanger expression in mouse proximal tubule cells and RKPC-2 cells.[85,89] In our laboratory, isolated hearts were subjected to varying periods of ischemia, which result in intracellular acidosis. This results in increased NHE1 expression in the myocardium.[90,91]

There have been few studies on the NHE1 gene. In the human NHE1 promoter a potential mediator of the acidosis effect is the transcription factor AP-1.[89,92] Studies in our group[93] have shown that the NHE1 gene is activated during differentiation of P19 cells and that the transcription factor AP-2 is involved. We have also shown that a 1.1 kb fragment of the NHE1 promoter is functional in primary cultures of isolated myocytes.[94] Chapter 5 reviews regulation of expression of the NHE1 isoform. Various aspects of regulation of expression are covered in chapters 2, 3 and 11.

F. THE IMPORTANCE OF THE Na^+/H^+ EXCHANGER IN MUSCLE (IN HEALTH AND DISEASE)

Several chapters of the text have been devoted to the Na^+/H^+ exchanger and its role in the myocardium and smooth muscle. In the myocardium, the regulation of intracellular pH is especially important for two major reasons. First, intracellular acidosis is important during the contractile failure associated with ischemia.[95,96] With a fall in intracellular pH, contractility can decline greatly with important and obvious consequences. The second reason why regulation of internal pH is of special importance is that the Na^+/H^+ exchanger is believed to be important in the cardiac response to reperfusion.[52,97-99] During ischemia and reperfusion of the myocardium, cell damage, necrosis, and arrhythmia can occur. Much of the damage of these episodes is associated with the reperfusion phase of the episode and not with the ischemic phase.[100,101] The mechanism by which this occurs may involve the Na$^+$/H$^+$ exchanger.[52,97-101] Ischemia results in acidification of the myocyte cytoplasm. Protons accumulate inside the myocytes until the reperfusion phase when the Na$^+$/H$^+$ exchanger becomes active and removes excess intracellular protons (see chapters 9-11). At its maximum velocity, the Na$^+$/H$^+$ exchanger has probably the highest transport capacity of all the sarcolemmal transport systems. This rate is higher than that of the Na$^+$/K$^+$ ATPase which normally removes any excess intracellular Na$^+$.[52] Because of the resulting accumulation of intracellular Na$^+$, a second antiporter, the Na$^+$/Ca^{2+} exchanger, then removes the excess intracellular Na$^+$. This, however, results in the accumulation of excess Ca^{2+} causing detrimental effects including cell necrosis, contracture and arrhythmia[52,102,103] (reviewed in chapter 10).

Amiloride and more specific amiloride analogs, such as hexamethyl amiloride, inhibit the Na$^+$/H$^+$ exchanger and improve postischemic recovery. They prevent the increases in Na$^+$ and Ca^{2+} concentrations and reduce the severity of arrhythmias after ischemia and reperfusion.[104-109] Hoe694 is a new Na$^+$/H$^+$ exchange inhibitor that is not derived from amiloride but has similar cardioprotective effects.[109] These compounds and other derivatives may have important therapeutic benefits (chapters 9 and 10).

Very recently, some evidence was reported suggesting activation of NHE1, and its gene expression is involved in molecular mechanisms of both cardiac hypertrophy and vascular smooth muscle cell proliferation.[110] Vascular smooth muscle cell growth is considered to be one of the major events occurring during repair of the vascular wall, after injury. A very low growth rate of a mutant smooth muscle cell line deficient in the amiloride-sensitive Na$^+$/H$^+$ antiporter activity was restored in a stable transformant bearing a full-length cDNA of NHE1.[110] This observation indicates that the NHE1 expression can accelerate proliferation of smooth muscle cells. These authors also examined the pathophysiological role of Na$^+$/H$^+$ exchanger on the smooth muscle cell proliferation during restenosis after angioplasty. They demonstrated that the NHE1 mRNA levels in the rabbit balloon-injured carotid arteries were elevated by as high as 80% compared with the normal state.[110] The regulation and the role of the Na$^+$/H$^+$ antiporter in smooth muscle are discussed in more detail in chapter 3.

G. IMPORTANCE OF Na$^+$/H$^+$ EXCHANGE AND INTRACELLULAR pH IN TUMOR CELLS

Studies have recently examined the Na$^+$/H$^+$ exchanger in several models of human tumors. They suggest that, in at least some cell types, the Na$^+$/H$^+$ exchanger and intracellular pH regulation associated with this protein, play an important role in tumor cell growth. Early studies showed that Na$^+$/H$^+$ exchanger deficient cells[38] lost or severely reduced their capacity to grow tumors in vivo in immune deficient mice. In addition, a variety of evidence suggested that activation of the Na$^+$/H$^+$ exchanger and the resultant increases in intracellular pH may be required for mitogenesis in some cell types. In HCO$_3$-free media, Na$^+$/H$^+$ exchanger deficient cells cannot proliferate in media of low external pH.[38] In addition, extracellular Na$^+$ is limiting in growth factor-dependent proliferation,[111] and amiloride and its analogs can block changes caused by growth factor stimulation, especially in HCO$_3$-free or low Na$^+$ medium[112] (chapter 14).

Because of the important role of the Na$^+$/H$^+$ exchanger in tumor and cell growth, amiloride and its analogs have been tested for use in tumor selective therapy. Chapter 14 reviews the use of these compounds especially for use with solid tumors. Strategies which have recently been investigated include use of a protonophore such as nigericin, in combination with other compounds, to decrease pH$_i$ of tumor cells.[113] In addition, inhibition of bicarbonate-based antiporters improves the killing effects. Overall, these relationships suggest that pH$_i$, the Na$^+$/H$^+$ exchanger, and extracellular pH all have an important relationship with tumor cell growth and initiation (see chapter 14).

H. FUTURE PROSPECTS

Clearly, tremendous progress has been made in the field. However, a number of important questions are still left unsolved. These are reviewed in the chapters and are summarized briefly (below). It is most desirable to document the actual topology of this membrane protein. Predictions of the topology are useful, however the actual structure has yet to be determined. Our own analysis (above) shows the predicted structure of the protein can vary depending on the program and parameters used. In this regard, it is also highly desirable to overexpress and purify the protein. This might lead eventually to crystallization of the protein and detailed analysis of its structure and mechanisms of transport. To date, overexpression of functional protein has been a great challenge.

As outlined (above), it is postulated that other cytoskeletal and regulatory proteins interact with the cytoplasmic region of the protein. Only initial investigations are been made in this area. There is little doubt, in the future, more aspects of mitogenic activation of the protein will be revealed. These results may be especially important, considering evidence that shows increased activity of the exchanger in hypertension. The mechanisms of the increased activity in this disease remain unsolved.

Regulation of expression of the Na$^+$/H$^+$ exchanger has yet to be completely understood. The mechanism by which acidosis increases expression of the protein is unsolved. Also, the role of increased expression of the protein and message during differentiation is not yet clear. All of these related areas pose important challenges and significant biological questions.

The Na^+/H^+ exchanger plays important roles in several diseases. There is clear evidence of its involvement in arrhythmogenesis and ischemia and reperfusion injury. In several models it has been shown that amiloride analogs help prevent the detrimental effects to which the Na^+/H^+ exchanger contributes. However, these techniques have not been used practically and await further development before use in treatments of cardiac disorders.

Perhaps the development of Na^+/H^+ exchanger inhibitors in tumor selective therapy is more advanced. Clearly, in some experimental models, Na^+/H^+ exchanger inhibitors in combination with other compounds may be of benefit. Experiments are underway to improve the efficiency of use of these compounds.

Tremendous progress has been made, but there is considerable work to be done in many areas of investigation. The next few years will be exciting and with 'enough resources' and time on our side, the promise of improving our basic understanding of this membrane protein will enhance and enable our efforts to treat the diseases with which it is involved.

ACKNOWLEDGMENTS

Research by Larry Fliegel was supported by grants from the Heart and Stroke Foundation of Canada and from the Medical Research Council program grant in the molecular biology of membranes #PG-11440. Special thanks to those in Dr. Fliegel's laboratory who helped with the research projects cited, especially Jason Dyck, Robert Haworth, Norma Lucena, Dyal Singh, Huayan Wang, and Weidong Yang.

REFERENCES

1. Goldberg B, Arbel T, Chen J, et al. Characterization of Na^+/H^+ antiporter gene of *E.coli*. Proc Natl Acad Sci USA 1987; 84:2615-2619.
2. Sardet C, Franchi A, Pouysségur J. Molecular cloning, primary structure, and expression of the human growth factor-activatable Na^+/H^+ antiporter. Cell 1989; 56:271-280.
3. Mitchel P. Coupling of phosphorylation to electron and hydrogen transfer by a chemiosmotic type of mechanism. Nature 1961; 191:144-146.
4. Mitchel P, Moyle J. Respiration-driven proton translocation in rat liver mitochondria. Biochem J 1967; 105:1147-1162.
5. Harold F, Papineau D. Cation transport and electrogenesis by *Streptococcus faecalis*. J Mem Biol 1972; 8:45-62.
6. Murer H, Hopfer U, Kinne R. Sodium/proton antiport in brush border-membrane vesicles isolated from rat small intestine and kidney. Biochem J 1976; 154:597-604.

7. Aronson PS, Nee J, Suhm MA. Modifier role of internal H$^+$ in activating the Na$^+$-H$^+$ exchanger in renal microvillus membrane vesicles. Nature 1982; 299:161-163.
8. Kinsella JL, Aronson PS. Amiloride inhibition of the Na$^+$-H$^+$ exchanger in renal microvillus membrane vesicles. Am J Physiol 1981; 241: F374-F379.
9. Schuldiner S, Rozengurt E. Na$^+$/H$^+$ antiport in Swiss 3T3 cells: mitogenic stimulation leads to cytoplasmic alkalinization. Proc Natl Acad Sci USA 1982; 79:7778-7782.
10. Frelin C, Vigne P, Lazdunski M. The amiloride-sensitive Na$^+$/H$^+$ antiport in 3T3 fibroblasts. J Biol Chem 1983; 258:6272-6276.
11. Moolenaar WH, Mummery CL, van der Saag PT, et al. Rapid ionic events and the initiation of growth in serum-stimulated neuroblastoma cells. Cell 1981; 23:789-798.
12. Pouysségur J, Chambard JC, Franchi A, et al. Growth factor activation of an amiloride-sensitive Na$^+$/H$^+$ exchange system in quiescent fibroblasts: coupling to ribosomal protein S6 phosphorylation. Proc Natl Acad Sci USA 1982; 79:3935-3939.
13. Sardet C, Counillon L, Franchi A, et al. Growth factors induce phosphorylation of the Na$^+$/H$^+$ antiporter, glycoprotein of 110 kD. Science 1990; 247:723-6.
14. Taglicht D, Padan E, Schuldiner S. Overproduction and purification of a functional Na$^+$/H$^+$ antiporter coded by nhaA (ant) from *Escherichia coli*. J Biol Chem 1991; 266:11289-11294.
15. Jia ZP, McCullough N, Martel R, et al. Gene amplification at a locus encoding a putative Na/H antiporter confers sodium and lithium tolerance in fission yeast. EMBO J 1992; 11:1631-1640.
16. Orlowski J, Kandasamy RA, Shull GE. Molecular cloning of putative members of the Na/H exchanger gene family. J Biol Chem 1992; 267:9331-9339.
17. Klanke CA, Su YR, Callen DF, et al. Molecular cloning and physical and genetic mapping of a novel human Na$^+$/H$^+$ exchanger (NHE5/SLC9A5) to chromosome 16q22.1. Genomics 1995:615-622.
18. Wang Z, Orlowski J, Shull GE. Primary structure and functional expression of a novel gastrointestinal isoform of the rat Na/H exchanger. J Biol Chem 1993; 268:11925-8.
19. Takaichi K, Wang D, Balkovetz DF, et al. Cloning, sequencing, and expression of Na-H antiporter cDNAs from human tissues. Am J Physiol:C 1992; 262:1069-C1076.
20. Tse C-M, Brant SR, Walker MS, et al. Cloning and sequencing of a rabbit cDNA encoding an intestinal and kidney-specific Na$^+$/H$^+$ exchanger isoform (NHE-3). J Biol Chem 1992; 267:9340-9346.
21. Tse C-M, Levine SA, Yun CHC, et al. Cloning and expression of a rabbit cDNA encoding a serum-activated ethylisopropylamiloride-resistant epithelial Na$^+$/H$^+$ exchanger isoform (NHE-2). J Biol Chem 1993; 268:11917-11924.

22. Tse C-M, Ma AI, Yang VW, et al. Molecular cloning and expression of a cDNA encoding the rabbit ileal villus cell basolateral membrane Na^+/H^+ exchanger. Embo J 1991; 10:1957-1967.
23. Fliegel L, Haworth RS, Dyck JRB. Characterization of the placental brush border membrane Na/H exchanger: identification of thiol-dependent transitions in apparent molecular size. Biochem J 1993; 289:101-107.
24. Fafournoux P, Noel J, Pouysségur. J. Evidence that Na/H exchanger isoforms NHE1 and NHE3 exist as stable dimers in membranes with a high degree of specificity for homodimers. J Biol Chem 1994; 269: 2589-2596.
25. Bertrand B, Wakabayashi S, Ikeda T, et al. The Na^+/H^+ exchanger isoform 1 (NHE1) is a novel member of the calmodulin-binding proteins. J Biol Chem 1994; 269:13703-13709.
26. Wakabayshi S, Bertrand B, Ikeda T, et al. Mutation of calmodulin-binding site renders the Na^+/H^+ exchanger (NHE1) highly H^+-sensitive and $Ca2^+$ regulation-defective. J Biol Chem 1994; 269:13710-13715.
27. Silva NLCL, Haworth RS, Singh D, et al. The carboxyl-terminal region of the Na/H exchanger interacts with mammalian heat shock protein. Biochemistry 1995; 34:10412-10420.
28. Padan E, Schuldiner S. Molecular physiology of Na^+/H^+ antiporters, key transporters in circulation of Na^+ and H^+ in cells. Biochem Biophys Acta 1993; 1185:129-151.
29. Nucitelli R, Webb DJ, Lagier ST, et al. 31P NMR reveals increased intracellular pH after fertilization in Xenopus eggs. Proc Natl Acad Sci USA 1981; 78:4421-4425.
30. Berk BC, Canessa M, Vallega G, et al. Agonist-mediated changes in intracellular pH: role in vascular smooth muscle cell function. J Cardiovasc Pharmacol 1988; 5:S104-114.
31. Berk BC, Aronow MS, Brock TA, et al. Thrombin-stimulated events in cultured vascular smooth-muscle cells. J Biol Chem 1987; 262:5057-5064.
32. Bingham RJ, Hall KS, Slonchewski JL. Alkaline induction of a novel gene locus, alx, in *Escherichia coli*. J Bacteriol 1990; 172:2184-2186.
33. Hoffmann E, Simonsen LO. Membrane mechanisms in volume and pH regulation in vertebrate cells. Physiol Rev 1989; 69:315-382.
34. Madshus IH. Regulation of intracellular pH in eukaryotic cells. Biochem J 1988; 250:1-8.
35. van Adelsberg J, Al-Awquati Q. Regulation of cell pH by Ca^+2-mediated exocytotic insertion of H^+-ATPases. J Cell Biol 1986; 102:1638-1645.
36. Grinstein S, Rothstein A. Mechanisms of regulation of the Na/H exchanger. J Mem Biol 1986; 90:1-12.
37. Zhuang Y-Z, Cragoe Jr EJ, Shaikewitz T, et al. Characterization of potent Na^+/H^+ exchange inhibitors from the amiloride series in A431 cells. Biochemistry 1984; 23:4481-4488.
38. Pouysségur J, Sardet C, Franchi A, et al. A specific mutation abolishing Na/H antiport activity in hamster fibroblasts precludes growth at neutral and acidic pH. Proc Natl Acad Sci USA 1984; 81:4833-4837.

39. Grinstein S, Rothestein A, Cohen S. Mechanism of osmotic activation of Na$^+$/H$^+$ exchange in rat thymic lymphocytes. J Gen Physiol 1985; 85:765-787.
40. Mahnensmith RL, Aronson PS. The plasma membrane sodium-hydrogen exchanger and its role in physiological and pathophysiological processes. Circ Res 1985; 56:773-785.
41. Seifter JL, Aronson PS. Properties and physiologic roles of the plasma membrane sodium-hydrogen exchanger. J Clin Invest 1986; 78:859-864.
42. Krulwich TA. Na$^+$/H$^+$ antiporters. Biochim Biophys Acta 1983; 726: 245-264.
43. Rocco VK, Cragoe Jr EJ, Warnock DG. N-ethoxycarbonyl-2-ethoxy-1, 2-dihydroquinoline, amiloride analogues, and renal Na$^+$/H$^+$ antiporter. Am J Physiol 1987; 252:F517-F524.
44. Vigne P, Frelin C, Cragoe Jr EJ, et al. Structure-activity relationships of amiloride and certain of its analogues in relation to the blockade of the Na$^+$/H$^+$ exchange system. Mol Pharm 1984; 25:131-136.
45. David P, Mayan H, Cragoe Jr EJ, et al. Strucutre-activity relations of amiloride derivatives, acting as antagonists of cation binding on Na$^+$/K$^+$-ATPase. Biochem Biophys Acta 1993; 1146:59-64.
46. Haggerty JG, Agarwal N, Reilly RF, et al. Pharmacologically different Na/H antiporters on the apical and basolateral surfaces of cultured porcine kidney cells (LLC-PK1). Proc Natl Acad Sci USA 1988; 85: 6797-6801.
47. Kulanthaivel P, Leibach FH, Mahesh VB, et al. The Na/H exchanger of the placental brush-border membrane is pharmacologically distinct from that of the renal brush-border membrane. J Biol Chem 1990; 265: 1249-1252.
48. Casavola V, Helmle-Kolb C, Murer H. Separate regulatory control of apical and basolateral Na$^+$/H$^+$ exchange in renal epithelial cells. Biochem Biophys Res Comm 1989; 165:833-837.
49. Haworth RS, Frohlich O, Fliegel L. Multiple carbohydrate moieties on the Na$^+$/H$^+$ exchanger. Biochem J 1993; 289:637-640.
50. Wakabayashi S, Bertrand B, Shigekawa M, et al. Growth factor activation and "H$^+$-sensing" of the Na$^+$/H$^+$ exchanger isoform 1 (NHE1). J Biol Chem 1994; 269:5583-5588.
51. Goss GG, Woodside M, Wakabayashi S, et al. ATP dependence of NHE-1, the ubiquitous isoform of the Na$^+$/H$^+$ antiporter. Analysis of phosphorylation and subcellular localization. J Biol Chem 1994; 269:8741-8.
52. Fliegel L, Frohlich O. The Na$^+$/H$^+$ exchanger: an update on structure, regulation and cardiac physiology. Biochem J 1993; 296:273-285.
53. Reilly RF, Hildebrandt F, Biemesderfer D, et al. cDNA cloning and immunolocalization of a Na$^+$/H$^+$ exchanger in LLC-PK1 renal epithelial cells. Am J Physiol 1991; 261:F1088-94.
54. Hildebrandt F, Pizzonia JH, Reilly RF, et al. Cloning, sequence and tissue distribution of a rabbit renal Na$^+$/H$^+$ exchanger transcript. Biochim Biophys Acta 1991; 1129:105-108.

55. Fliegel L, Sardet C, Pouysségur J, et al. Identification of the protein and cDNA of the cardiac Na/H exchanger. FEBS Lett 1991; 279:25-29.
56. Fliegel L, Dyck JRB, Wang H, et al. Complete nucleotide sequence of the human myocardial Na$^+$/H$^+$ exchanger. Mol Cell Biochem 1993b; 125:137-143.
57. Biemesderfer D, Reilly RF, Exner M, et al. Immunocytochemical characterization of Na-H exchanger isoform NHE-1 in rabbit kidney. Am J Physiol 1992; 263:F833-F840.
58. Noel J, Pouysségur J. Hormonal regulation, pharmacology, and membrane sorting of vertebrate Na$^+$/H$^+$ exchanger isoforms. Am J Phyiol 1995; 268:C283-C296.
59. Green RD, Frelin C, Vigne P, et al. The activity of the Na/H antiporter in cultured cardiac cells is dependent on the culture conditions used. FEBS Lett 1986; 196:163-166.
60. Little PJ, Weissberg PL, Cragoe Jr EJ, et al. Dependence of Na$^+$/H$^+$ antiport activation in cultured rat aortic smooth muscle on calmodulin, calcium, and ATP. Evidence for the involvement of calmodulin-dependent kinases. J Biol Chem 1988; 263:16780-16786.
61. Mitsuka M, Berk BC. Long-term regulation of Na$^+$-H$^+$ exchange in vascular smooth muscle cells: role of protein kinase C. Am J Physiol 1991; 260:C562-C569.
62. Bookstein C, DePaoli AM, Xie Y, et al. Na$^+$/H$^+$ exchangers, NHE-1 and NHE-3, of rat intestine. Expression and localization. J Clin Invest 1994; 93:106-113.
63. Claros MG, von Heijne G. TopPredII: an improved software for membrane protein structure predictions. Comput Applic Biosci 1994; 10:685-686.
64. Engelman DM, Steitz TA, Goldman A. Identifying nonpolar transbilayer helices in amino acid sequences of membrane proteins. Ann Rev Biophys Chem 1986; 15:321-353.
65. Kyte J, Doolittle RF. A simple method for displaying the hydropathic character of a protein. J Mol Biol 1982; 157:105-132.
66. Counillon L, Pouysségur J, Reithmeier RAF. The Na$^+$/H$^+$ exchanger NHE-1 possesses N- and O-linked glycosylation restricted to the first N-terminal extracellular domain. Biochem 1994; 33:10463-10469.
67. Grinstein S, Rotin D, Mason MJ. Na$^+$/H$^+$ exchange and growth factor-induced cytosolic pH changes. Role in cellular proliferation. Biochim Biophys Acta 1989; 988:73-97.
68. Hogue D, Michalak M, Fliegel L. The role of antiporters in the maintenance of intracellular pH in rat vascular smooth muscle. Mol Cell Biochem 1991; 102:125-137.
69. Rosoff PM. Phorbol esters and the regulation of Na$^+$/H$^+$ exchange. In: Na$^+$/H$^+$ Exchange. Grinstein S, ed. 1988;pp:243-253.
70. Wakabayashi S, Fafournoux P, Sardet C, et al. The Na$^+$/H$^+$ antiporter cytoplasmic domain mediates growth factor signals and controls "H($^+$)-sensing." Proc Natl Acad Sci USA 1992; 89:2424-8.

71. Hesketh RT, Moore JP, Morris JDH, et al. Common sequence of calcium and pH signals in the mitogenic stimulation of eukaryotic cells. Nature 1985; 313:481-484.
72. Hendey B, Mamrach MD, Putnam RW. Thrombin induces a calcium transient that mediates an activation of the Na$^+$/H$^+$ exchanger in human fibroblasts. J Biol Chem 1989; 264:18540-19547.
73. Villereal ML. Sodium fluxes in human fibroblasts: effect of serum, Ca^{2+}, and amiloride. J Cell Physiol 1981; 107:359-369.
74. Muldoon LL, Dinerstein RJ, Villereal ML. Intracellular pH in human fibroblasts: effect of mitogens, A23187, and phospholipase activation. Am J Physiol 1985; 249:C140-C148.
75. Grinstein S, Woodside M, Sardet C, et al. Activation of the Na$^+$/H$^+$ antiporter during cell volume regulation. Evidence for a phosphorylation-independent mechanism. J Biol Chem 1992; 267:23823-8.
76. Grinstein S, Cohen S, Goetz JD, et al. Osmotic and phorbol ester-induced activation of Na$^+$/H$^+$ exchange. Possible role of protein phosphorylation in lymphocyte volume regulation. J Cell Biol 1985; 101:269-276.
77. Rao GN, Sardet C, Pouyssegur J, et al. Differential regulation of Na$^+$/H$^+$ antiporter gene expression in vascular smooth muscle cells by hypertrophic and hyperplastic stimuli. J Biol Chem 1990; 265:19393-6.
78. Rao GN, de Roux N, Sardet C, et al. Na$^+$/H$^+$ antiporter gene expression during monocytic differentiation of HL60 cells. J Biol Chem 1991; 266:13485-8.
79. Bidet M, Merot J, Tauc M, et al. Na/H exchanger in proximal cells isolated from kidney. II. Short-term regulation by glucocorticoids. Am J Physiol 1987; 253:F945-F951.
80. Kinsella JL, Cujdik T, Sacktor B. Na$^+$/H$^+$ exchange in isolated renal brush-border membrane vesicles in response to metabolic acidosis. J Biol Chem 1984; 259:13224-13227.
81. Sacktor B, Kinsella J. Regulation of Na$^+$/H/up exchange activity by adaptive mechanisms. In: Na$^+$/H$^+$ Exchange. Grinstein S, ed. CRC Press, 1988; 307-324.
82. Vigne P, Jean T, Barbry P, et al. [^{3}H] Ethylpropylamiloride, a ligand to analyze the properties of the Na$^+$/H$^+$ exchange system in the membranes of normal and hypertrophied kidneys. J Biol Chem 1985; 260:14120.
83. Krapf R, Pearce D, Lynch C, et al. Expression of rat renal Na/H antiporter mRNA levels in response to respiratory and metabolic acidosis. J Clin Invest 1991; 87:747-51.
84. Moe OW, Miller RT, Horie S, et al. Differential regulation of Na$^+$/H$^+$ antiporter by acid in renal epithelial cells and fibroblasts. J Clin Invest 1991; 88:1703-8.
85. Mrkic B, Helmle-Kolib C, Krapf R, et al. Functional adaptation to high PCO2 of apically and basolaterally located Na/H exchange activities in cultured renal cell lines. Pflugers Arch 1994; H426:333-340.
86. Ruiz OS, Talor Z, Arruda JAL. Regional localization of renal Na$^+$-H$^+$

anitporter: response to respiratory acidosis. Am J Physiol 1990; 259:F512-F518.

87. Kinsella J, Cujdik T, Sacktor B. Na⁺-H⁺ exchange activity in renal brush border membrane vesicle in response to metabolic acidosis: The role of glucocorticoids. Proc Natl Acad Sci USA 984; 81:630-634.

88. Kinsella J, Sacktor B. Thyroid hormones increase Na/H exchange activity in renal brush border membranes. Proc Natl Acad Sci USA 1985; 82:3606-3610.

89. Horie S, Moe O, Yamaji Y, et al. Role of protein kinase C and transcription factor AP-1 in the acid-induced increase in Na⁺/H⁺ antiporter activity. Proc Natl Acad Sci USA 1992; 89:5236-40.

90. Dyck JRB, Lopaschuk GD, Fliegel L. Identification of a small Na⁺/H⁺ exchanger-like message in the rabbit myocardium. FEBS Lett 1992; 310:255-259.

91. Fliegel L, Maddaford TG, Pierce G, et al. Induction of expression of the Na⁺/H⁺ exchanger in the rat myocardium. Cardiovas Res 1995; 29:203-208.

92. Miller RT, Counillon L, Pages G, et al. Structure of the 5'-flanking regulatory region and gene for the human growth factor-activatable Na/H exchanger NHE-1. J Biol Chem 1991; 266:10813-10819.

93. Dyck JRB, Silva NLC L, Fliegel L. Activation of the Na⁺/H⁺ exchanger gene by the transcription factor AP-2. J Biol Chem 1995; 270:1375-1381.

94. Fliegel L, Dyck JRB. Molecular biology of the cardiac Na⁺/H⁺ Exchanger. Cardiovasc Res 1995; 29:155-159.

95. Malloy C, Matthews P, Smith M, et al. In vivo 31P NMR study of the regional metabolic responses to cardiac ischemia. Adv Myocard 1985; 6:461-464.

96. Jeffrey FMH, Mallow CR, Radda GK. Influence of intracellular acidosis on contractile function in the working rat heart. Am J Physiol 1987; 253:H1499-H1505.

97. Tani M, Neely JR. Role of intracellular Na⁺ in Ca2⁺ overload and depressed recovery of ventricular function of reperfused ischemic rat hearts: Possible involvement of H⁺-Na⁺ and Na⁺-Ca²⁺ exchange. Circ Res 1989; 65:1045-1056.

98. Karmazyn M, Moffat MP. Role of Na/H exchange in cardiac physiology and pathophysiology: mediation of myocardial reperfusion injury by the pH paradox. Cardiovasc Res 1993; 27:915-924.

99. Pierce GN, Meng H. The role of sodium-proton exchange in ischemic/reperfusion injury in the heart. Am J Cardiovasc Pathol 1992; 4:91-102.

100. Hears DJ, Humphrey SM, Bullock GR. The oxygen paradox and the calcium paradox: Two facets of the same problem? J Mol Cell Cardiol. 1978; 10:641-688.

101. Becker LC, Ambrosio G. Myocardial consequences of reperfusion. Progr Cardiovasc Dis 1987; 30:23-44.

102. Fredrich F, Sablotni J, Burckhardt G. Identification of the renal Na^+/H^+ exchanger with N,N'-dicyclohexylcarbodiimide (DCCD) and amiloride analogues. J Mem Biol 1986; 94:253-266.
103. Nakanishi T, Seguchi M, Tsuchiya T, et al. Effect of partial Na pump and Na-H exchange inhibition on [Ca]i during acidosis in cardiac cells. Am J Physiol 1991; 261:C758-C76.
104. Karmazyn M. Amiloride enhances post ischemic recovery: possible role of Na^+/H^+ exchange. Am J Physiol 1988; 255:H608-H615.
105. Meng H-P, Pierce GN. Protective effects of 5-(N,N-dimethyl) amiloride on ischemia-reperfusion injury in hearts. Am J Physiol 1990; 258: H1615-H1619.
106. Dennis SC, Coetzee WA, Cragoe Jr EJ, et al. Effects of proton buffering and of amiloride derivative on reperfusion arrhythmias in isolate hearts: Possible evidence for an arrhythmogenic role of Na^+/H^+ exchange. Circ Res 1990; 66:1156-1159.
107. Avkiran M, Ibuki C. Reperfusion-induced arrhythmias: a role for washout of extracellular protons? Circ Res 1992; 71:1429-1440.
108. Meng H-P, Maddaford TG, Pierce G. Effect of amiloride and selected analogues on postischemic recovery of cardiac contractile function. Am J Physiol 1993; 264:H1831-H1835.
109. Scholz W, Albus U, Lang HJ, et al. Hoe 694, a new Na/H exchange inhibitor and its effects in cardiac ischemia. Br J Pharmacol 1993; 109:562-568.
110. Takewaki S, Kuro O-M, Hiroi Yu, et al. Activation of $Na(^+)$-H^+ antiporter (NHE-1) gene expression during growth, hypertrophy and proliferation of the rabbit cardiovascular system. J Mol Cell Cardiol 1995; 27:729-742.
111. Burns CP, Rozengurt E. Serum, platelet-derived growth factor, vasopressin, and phorbol ester increase intracellular pH in Swiss 3T3 cells. Biochem Biophys Res Commun 1983; 116:931-938.
112. L'Allemain G, Paris S, Pouysségur J. Growth factor action and intracellular pH regulation in fibroblasts. Evidence for a major role of the Na/H antiporter. J Biol Chem 1984; 259:5809-5815.
113. Newell K, Wood P, Stratford I, et al. Effects of agents which inhibit the regulation of intracellular pH on murine solid tumors. Br J Cancer 1992; 66:311-317.

CHAPTER 2

Na^+/H^+ EXCHANGE IN MAMMALIAN KIDNEY

Orson W. Moe and Robert J. Alpern

A. INTRODUCTION

In 1947, Pitts and coworkers observed that normal human kidney excretes more acid than is present in the glomerular filtrate and concluded that acid must gain access to the urine via the tubules.[1] Although the clearance data could not support conclusions on transport mechanisms, these investigators were the first to postulate that hydrogen ions are secreted into the lumen by renal tubules in exchange for luminal Na^+.[1] When micropuncture became available, H^+ secretion into the proximal tubule lumen was found to depend on the presence of luminal Na^+ to a certain degree.[2] Although these early microperfusion data suggested the existence of Na^+/H^+ exchange, they were not definitive. The first conclusive demonstration of renal tubular Na^+/H^+ exchange was provided by Murer, Hopfer and Kinne in 1976.[3] These investigators used renal cortical brush border vesicles to show that outwardly directed pH gradients could drive counter Na^+ uptake, and inward Na^+ gradient could drive H^+ extrusion.

Since the original report of Murer et al, renal Na^+/H^+ exchange has been extensively studied in terms of its substrate specificity and pharmacokinetics, its role in transepithelial NaCl and $NaHCO_3$ transport, and its acute and chronic regulation. With the cloning of the NHE gene family, the molecular basis of these physiologic processes are now being unveiled. This chapter will highlight certain aspects of the physiology of Na^+/H^+ exchange in the kidney and some of the recent advances in the understanding of the underlying molecular mechanisms.

The Na+/H+ Exchanger, edited by Larry Fliegel. © 1996 R.G. Landes Company.

B. Na^+/H^+ EXCHANGE IN THE MAMMALIAN NEPHRON AND DISTRIBUTION OF MOLECULAR ISOFORMS

Plasma membrane Na^+/H^+ exchange is universal in the animal kingdom. Na^+/H^+ exchange has been described in prokaryotes[4] and eukaryotes ranging from yeast,[5] worms,[6] crustaceans,[7] fish[8] and mammals. Significant primary sequence homology between mammalian Na^+/H^+ exchangers and their functional counterparts in other organisms is not evident below the level of the piscine homolog.[8] In mammals, Na^+/H^+ exchange is ubiquitous. It utilizes the downhill entry of Na to energize the uphill extrusion of protons in a 1:1 stoichiometry.

In nonpolar cells, Na^+/H^+ exchange functions primarily to alkalinize the cytoplasm. This mechanism is crucial in defending the cell against acid loads. In polarized epithelia where Na^+/H^+ exchange is positioned in tandem with other H^+-equivalent transporters on the opposite plasma membrane, it constitutes a system that can affect transepithelial H^+-equivalent transport. In nonpolar cells, Na^+/H^+ exchange coupled with Cl^-/base exchange causes no net change in cell pH but rather a net increase in ion content of the cell. This is important for defending the cell against ambient hypertonicity. In polarized epithelia, coupled Na^+/H^+ exchange and Cl^-/base exchange comprises the apical membrane component of transepithelial NaCl transport. The ability of Na^+/H^+ exchangers to perform these specialized functions in epithelia rests on the differentiated phenotype of epithelial cells, their polarized architecture, and the intrinsic properties of the specific NHE isoforms that reside in transporting epithelia.

B.1 Na^+/H^+ EXCHANGE IN THE PROXIMAL TUBULE

The best studied Na^+/H^+ exchanger in the kidney is the proximal tubule apical membrane Na^+/H^+ exchanger. This transporter serves at least three important functions for the proximal tubule: (1) $NaHCO_3$ absorption, (2) NaCl absorption; and (3) NH_4^+ secretion.

B.1.1 $NaHCO_3$ Absorption

Approximately 3700 mEqs of HCO_3^- are filtered at the glomerulus daily. Fatal acidosis would ensue rapidly if the filtered HCO_3^- were not reclaimed by the remainder of the nephron. Eighty percent of HCO_3^- reabsorption (3000 mEqs) occurs in the proximal tubule.[9] This process involves predominantly H^+ secretion rather than direct HCO_3^- absorption.[10] Because of the acid luminal fluid pH and the cell voltage is approximately -70 mV relative to the lumen, apical H^+ secretion in this epithelium requires energy. Approximately one-third of this active H^+ secretion is directly coupled to ATP hydrolysis via an apical membrane H^+-ATPase with the remaining two-thirds occurring through the apical membrane Na^+/H^+ exchanger.[11] Thus, for the organism to maintain acid-base balance, approximately 2000 mEq of "net" H^+ ions must traverse the proximal tubule Na^+/H^+ exchanger everyday. However, if one takes into account that there is lumen-to-cell apical membrane H⁺ backleak and blood-to-lumen paracellular HCO_3^- backleak,[12] the required absolute translocation of H⁺ by the Na⁺/H⁺ exchanger is 3000 mEq/day or more. As this transporter turns over H⁺ ions daily of a magnitude equivalent to more than 10 times the total extracellular fluid HCO_3^- buffer capacity, it is imperative that the Na⁺/H⁺ exchanger be regulated exquisitely.

B.1.2 NaCl Absorption

In addition to HCO_3^- absorption, the proximal tubule apical membrane Na⁺/H⁺ exchanger also mediates NaCl absorption. Approximately half of filtered NaCl is absorbed in the proximal tubule where transcellular absorption contributes 30-60% with paracellular pathways accounting for the remainder.[13,14] All of transcellular NaCl absorption is mediated by the apical Na⁺/H⁺ exchanger coupled with Cl⁻/base exchange.[14-16] The apical membrane Na⁺/H⁺ exchanger thus mediates the flux of some 20-40 liters of isotonic NaCl daily, an amount equivalent to several times the extracellular fluid volume.

B.1.3 NH₄⁺ Secretion

Since the lowest urinary pH attainable is about 5.0 ($[H] = 10^{-5}$ moles/L) and the daily requirement for H⁺ excretion is on the order of about 10^{-1} moles, it follows that the majority of H⁺ in the urine is carried by buffers. Physiologically, the most important urinary buffer is NH_3 because it is the most abundant, and its production is highly regulated in response to acid.[17] The proximal tubule cell is the major site of ammoniagenesis in the kidney. The synthesized NH_3 enters the luminal fluid via nonionic diffusion across the apical membrane, and protonated NH_4^+ enters the luminal fluid via the apical membrane Na⁺/H⁺ exchanger.[18,19] As described below, NH_4^+ is accepted as a substrate on the Na⁺/H⁺ exchanger.[20]

B.1.4 The Apical Membrane Na⁺/H⁺ Exchanger

The proximal tubule apical membrane Na⁺/H⁺ exchanger has been extensively characterized in brush border membrane vesicles. Na⁺/H⁺ exchange is electroneutral with Li⁺ and NH_4^+ functioning as acceptable substrates in place of Na⁺.[20] It is competitively inhibited by the diuretic amiloride with an IC_{50} in the micromolar range.[21,22] Na⁺ follows first order kinetics with a K_{Na} of about 12 mM.[22] H⁺ kinetics are more complex in that, in addition to its role as a substrate, cellular H⁺ also allosterically activates the transporter.[23] Prior to the genetic classification of Na⁺/H⁺ exchangers by primary nucleotide and amino acid sequences,[24] Na⁺/H⁺ exchangers were classified pharmacologically according to their sensitivity to amiloride and its various analogs.[25] Apical membrane Na⁺/H⁺ exchangers in transporting epithelia in general are relatively resistant to amiloride and its derivatives.

Of the cloned molecular isoforms that have been expressed and studied in terms of their amiloride sensitivity, NHE3 exhibits a profile that bears the closest resemblance to that of the apical membrane protein.[26] Specific antibodies against NHE3 labeled only apical membranes and not basolateral membranes isolated from kidney cortex.[27,28] On tissue sections, NHE3 antibodies labeled proximal tubule brush border in both rabbit and rat kidney.[27,28] In the rat, axial heterogeneity of abundance was noted with staining most intense in the proximal convoluted tubule and diminished staining in the proximal straight tubule. This histologic pattern mirrors the heterogeneity of bicarbonate transport[29,30] and apical membrane Na^+/H^+ exchanger activity[31] in the proximal tubule. OK cells and LLC-PK cells, two proximal tubule cell lines exhibiting apical membrane ethyl-isopropyl amiloride (EIPA)-insensitive Na^+/H^+ exchange activity,[25,32,33] also express NHE3 mRNA and protein.[34,35]

Thus, the accumulated data to date support NHE3 as the isoform responsible for proximal tubule apical membrane Na^+/H^+ exchange. It remains possible that other isoforms may contribute to apical Na^+/H^+ exchange. When NHE2 is heterologously expressed in nonepithelial cells, it has an amiloride sensitivity that is similar to NHE1 with a $K_{0.5}$ of about 1 µM,[36,37] but its EIPA sensitivity is between that of NHE1 and NHE3 with a $K_{0.5}$ between 0.08-0.5 µM.[36,37] In RKPC-2 cells, a rabbit proximal tubule cell line, NHE2 is located on the apical membrane and is regulated by PTH.[38] When transfected into a colonic epithelial cell line, NHE2 is sorted to the apical membrane.[36] In a preliminary report, NHE4 protein was found in a cortical membrane preparation but it comigrated with basolateral markers in a cortical membrane preparation.[39] In situ hybridization studies have localized NHE4 mRNA to the inner medullary collecting duct.[40] Published histologic data of NHE2 and NHE4 in the kidney are currently not available.

B.1.5 The Basolateral Membrane Na^+/H^+ Exchanger

In studies with cortical membrane vesicles as well as perfused proximal convoluted tubules, Na^+/H^+ exchange activity was not demonstrated in the basolateral membrane.[41-44] However, in the S3 segment of the rabbit proximal straight tubule, a basolateral Na^+/H^+ exchanger has been described.[45] This transporter likely does not contribute to $NaHCO_3$ or NaCl absorption since it operates in a counter direction to urine acidification. A possible role for this Na^+/H^+ exchanger is that of cell volume regulation in response to changes in interstitial Na^+ concentration and tonicity. Cl^-/HCO_3^- and $Cl^-/base$ exchangers are present in the basolateral membrane in proximal tubules.[45-47] Na^+/H^+ exchange in parallel with $Cl^-/base$ exchange can acutely regulate cellular ionic content and volume in the S3 segment. This is important since S3 segments of juxtamedullary nephrons traverse into the outer medulla,[48] and interstitial tonicity in this region fluctuates and can reach as high as 500-600 mOsm under conditions of antidiuresis.[49] Basolateral Na$^+$/H$^+$ exchange has also been described in more proximal S1 and S2 segments in the extremely deep juxtamedullary nephrons.[50]

The Na$^+$ affinity of the basolateral Na$^+$/H$^+$ exchanger is lower than that of the apical transporter with a K_{Na} of 53 mM.[45] Although amiloride kinetics have not been studied, it is likely to be an amiloride sensitive Na$^+$/H$^+$ exchanger. NHE1 antibodies have localized NHE1 to the basolateral membrane of the rabbit S3 proximal tubule, although preferential distribution towards deeper nephrons was not described in that study.[51] When NHE1 transcript was analyzed in individual nephron segments by reverse transcription-polymerase chain reaction (RT-PCR), expression in proximal tubules was most consistent and prominent in juxtamedullary nephrons.[52]

B.2 Na$^+$/H$^+$ Exchange in the Descending Limb

It is controversial whether there is ongoing H$^+$ secretion or NaCl absorption in the segment immediately downstream from the proximal straight tubule. Part of this controversy stems from interspecies, internephron and intranephron variations. In the rat kidney, upper portions of long descending limbs possess apical microvilli, basolateral Na$^+$-K$^+$-ATPase, exhibit high Na$^+$ permeability and express apical NHE3 while short descending limbs cells are flat, do not have Na$^+$-K$^+$-ATPase, are relatively impermeable to Na$^+$ and have no apical NHE3.[28,48,53-55] These data suggest that upper portions of long descending limbs may participate in active Na$^+$ absorption. A parallel finding was described for carbonic anhydrase in the rat where cytoplasmic staining was obvious in long descending limbs but not short ones.[56] Apical NHE3 in the absence of luminal carbonic anhydrase in this segment can generate an acid dysequilibrium pH which can theoretically enhance NH$_4^+$ trapping in the lumen. In contrast to the rat, rabbit descending limbs have very low Na$^+$ permeability,[55,57] lack Na$^+$-K$^+$-ATPase,[53] carbonic anhydrase,[58] NHE1[51] and NHE3[27] staining suggesting that this may be a nontransporting segment in the rabbit. One puzzling finding is that apical Na$^+$/H$^+$ exchange activity has been found in this segment in the rabbit.[59] This Na$^+$/H$^+$ exchanger may be an isoform other than NHE1 or NHE3. The function of this transporter is unknown.

B.3 Na$^+$/H$^+$ Exchange in Thick Ascending Limb

B.3.1 HCO$_3^-$ Absorption

The thick ascending limb is an important site of HCO$_3^-$ absorption retrieving 30-50% of the HCO$_3^-$ exiting from the proximal tubule.[60,61] Like the proximal tubule, HCO$_3^-$ absorption in the thick ascending limb is affected by apical H$^+$ secretion which is mediated mostly by apical membrane Na$^+$/H$^+$ exchange.[61,62]

B.3.2 NaCl Absorption

In addition to $NaHCO_3$ absorption, the thick ascending limb also absorbs NaCl.[63] NaCl absorption in this segment is unlike that of the proximal tubule in that this epithelium is impermeable to water.[64] Solute transport in the thick ascending limb is important for extracellular fluid volume homeostasis as well as water homeostasis because it generates hypertonicity in the medullary interstitium as well as a hypotonic tubular fluid. NaCl absorption is mediated by the apical membrane furosemide-sensitive Na^+-K^+-$2Cl^-$ cotransporter.[64,65] In the mouse cortical thick ascending limb, an additional mechanism for net apical NaCl absorption consisting of coupled Na^+/H^+ exchange and Cl/HCO_3 exchange has been proposed.[66] The actual mechanism is likely more complicated than two coupled exchangers as the CO_2/HCO_3^--dependent component of NaCl transport is also bumetanide-sensitive and depends on luminal K suggesting involvement of the Na-K-2Cl transporter.[67] The CO_2/HCO_3^- dependent component of NaCl transport appears to be unique to the mouse cortical thick ascending limb.

B.3.3 Apical Membrane Na^+/H^+ Exchanger

Although amiloride kinetics have not been performed on this transporter, anti-NHE3 antibodies label the apical membrane of rat thick ascending limbs intensely.[28] NHE3 mRNA has also been detected in microdissected thick ascending limbs by RT-PCR.[68] As compared to rats, the rabbit thick ascending limb lacks luminal carbonic anhydrase[58] and NHE3 staining,[27] and does not absorb HCO_3.[69]

B.3.4 Basolateral Membrane Na^+/H^+ Exchanger

Basolateral membrane Na^+/H^+ exchange has been described in the rat and mouse thick ascending limb.[70,71] In addition to its role in defending cell pH,[70] inhibition of this transporter impairs the thick ascending limb cell's ability to defend cell volume against hypertonic challenges.[71] This transporter is likely NHE1 as NHE1 mRNA was detected in microdissected cortical thick ascending limbs,[52] and the basolateral membrane of thick ascending limbs were stained by anti-NHE1 antibodies.[51]

B.4 DISTAL CONVOLUTED TUBULE

The HCO_3 that eludes absorption in the proximal tubule and the thick ascending limb is recovered in the distal portions of the nephron. Significant HCO_3 absorption via apical H secretion has been described in the early distal convoluted tubule.[72,73] Although Na^+/H^+ exchange activity has not been directly measured in this segment, bicarbonate absorption is inhibited by 10^{-4} M EIPA suggesting that Na^+/H^+ exchange mediates apical membrane H^+ secretion.[74] Pharmacological characterization of this Na^+/H^+ exchanger has not been performed. Since this is an apical membrane transporter affecting transepithelial H transport, it is enticing to extrapolate from the proximal tubule and thick ascending limb data and predict this to be NHE3. However, on immunohistochemistry, the intense NHE3 staining of the apical membrane of the thick ascending limb ceases abruptly at the transition to distal convoluted tubule.[28] This finding underscores the fact that NHE3 is not the only isoform that mediates transepithelial ion transport. NHE1 mRNA was detected in the microdissected rat distal convoluted tubule. However, NHE1 antibody labeled the basolateral rather than the apical membrane in rabbit kidney.[27] Using in situ hybridization, NHE4 mRNA was localized in the collecting tubule with much higher labeling in the medulla than the cortex.[40] There is no current published immunohistochemical location of NHE4.

B.5 CORTICAL COLLECTING DUCT

The cortical collecting duct can either absorb or secrete HCO_3 depending on the acid-base status of the organism.[75-77] The cortical collecting duct is also an important site of regulated NaCl absorption and K secretion. Cell types are heterogeneous in this segment with each type specializing in specific functions. Principal cells mediate NaCl absorption and K secretion, α-intercalated cells mediate acid secretion (HCO_3 absorption), and β-intercalated cells mediate HCO_3 secretion.[17] There are no apical membrane Na^+/H^+ exchangers in this segment, and staining with anti-NHE3 antibodies is negative in both the rat and rabbit.[27,28]

In β-intercalated cells where HCO_3 is secreted into the lumen via Cl/HCO_3 exchange, H exit on the basolateral membrane is energized by an H^+-ATPase or ATPase.[17] Na^+/H^+ exchange is also present on the basolateral membrane in the β-intercalated cell[78] and can potentially function to extrude H^+ ions utilizing the low cell Na. It is unlikely that basolateral Na^+/H^+ exchange actually fulfills this role as basolateral amiloride application or Na^+ removal does not affect apical HCO_3 secretion.[77,79] When the cells are acutely loaded with acid, basolateral Na^+/H^+ exchange performs a much more conspicuous role than the H^+-ATPase in returning cell pH to normal.[78] It is likely that the basolateral Na^+/H^+ exchanger is relatively inactive at normal cell pH and that it functions mainly as a defense against acid load rather than participating in transepithelial HCO_3 secretion. One would expect this basolateral transporter to be NHE1 and indeed, NHE1 mRNA is readily detectable in microdissected rat cortical collecting tubules.[52] However, in rabbit kidney where there is ample physiologic data for basolateral Na^+/H^+ exchange, basolateral NHE1 staining of cortical collecting duct by immunohistochemistry does not colocalize with apical membrane peanut lectin binding which is a marker for β-intercalated cells.[51] Presently, it is unclear which NHE isoform is present in the basolateral membrane of β-intercalated cells.

In principal cells where NaCl absorption is affected, a basolateral membrane Na^+/H^+ exchanger has been described along with other H-equivalent transporters.[78,80] The role of these acid-base transporters is unclear since principal cells do not participate in transepithelial H^+-equivalent transport. Although principal cells themselves do not transport H^+-equivalent across the epithelium, they are intermixed with acid or base secreting cells and hence subjected to frequent changes in ambient pH. One possible role for the basolateral Na^+/H^+ exchanger in conjunction with the other basolateral acid-base transporters is cell pH regulation. In rabbit cortical collecting duct, basolateral NHE1 was detected in a majority of cells; none of which express the markers for intercalated cells, and thus these can be presumed to be principal cells.[51]

B.6 MEDULLARY COLLECTING DUCT

In this part of the nephron, H is secreted often into a bicarbonate-free tubular fluid against a steep pH gradient where most of the secreted H^+ is titrated with NH_3 to form NH_{17}^+. The segmental organization of the medullary collecting duct is complex and is discussed elsewhere.[17,48] Apical H^+ secretion in this segment does not involve Na^+/H^+ exchanger, but Na^+/H^+ exchange activity has been described on the basolateral membrane.[81-83] In the inner stripe of the outer medullary collecting duct, the basolateral Na^+/H^+ exchanger appears to be active at resting cell pH because basolateral Na^+ removal acidifies the cell and theoretically can counter the effect of apical H secretion by the H^+-ATPase.[82] Presently it is unknown whether basolateral Na^+/H^+ exchange can affect the rate of H secretion as the effect of basolateral amiloride on H flux has not been measured.

Weak staining of NHE1 was seen diffusely in basolateral membranes of rabbit medullary collecting ducts.[51] Immortalized murine inner medullary collecting duct cells from simian virus transgenic mice express both NHE1 and NHE2 transcripts but exhibit Na^+/H^+ exchange activity only on the basolateral membrane suggesting that NHE2 may be a basolateral protein.[84] NHE4 mRNA was localized by in situ hybridization to the inner and outer medullary collecting duct.[40] The identity of the isoform(s) and their relative contribution to medullary collecting duct Na^+/H^+ exchange is presently unclear.

B.7 SUMMARY

Apically located Na^+/H^+ exchangers in general mediate $NaHCO_3$ transport whereas basolaterally located Na^+/H^+ exchangers in general defend the cell against acid loads and maintain cell volume in response to changes in ambient tonicity. These different functions are subserved by different isoforms of the NHE gene family. The distribution of the isoforms are summarized in Table 2.1.

C. ACUTE REGULATION OF RENAL NA^+/H^+ EXCHANGER

There is a large body of literature on the acute and chronic regulation of Na^+/H^+ exchange in the mammalian kidney. Since regulation

Table 2.1. Distribution of NHE isoforms in the mammalian kidney

	Apical	Basolateral
Proximal tubule		
S1	NHE3	–
S2	NHE3	–
S3	NHE3	NHE1
Descending limb	NHE3	–
Thick ascending limb	NHE3	NHE1
Distal convoluted tubule	+	–
Cortical collecting duct		
Principal cell	–	NHE1
β-intercalated cell	–	+
α-intercalated cell	–	–
Medullary collecting duct	–	NHE1
		NHE4
		NHE2?

+ activity present but isoform unknown
– activity absent

of NHE1 is discussed in great detail in chapters 3-6 and 8 the present chapter will focus on recent advances in the acute and chronic regulation of NHE3.

C.1 ACUTE REGULATION OF PROXIMAL TUBULE TRANSPORT AND APICAL MEMBRANE NA$^+$/H$^+$ EXCHANGER BY HORMONES

As discussed above, the renal proximal tubule normally handles large fluxes of NaHCO$_3$ and NaCl. Thus, both acid-base and volume homeostasis rely on regulated proximal tubule NaHCO$_3$ and NaCl transport. Hormones regulate proximal tubule transport indirectly by altering systemic and glomerular hemodynamics and peritubular physical factors and directly by interacting with receptors. Regulation of NaCl and NaHCO$_3$ transport by receptor-mediated mechanisms is affected by regulation of one or more transport proteins.

For example, angiotensin II increases NaHCO$_3$ transport by stimulating both the apical membrane Na$^+$/H$^+$ exchanger and the basolateral membrane Na$^+$/H$^+$CO$_3$/CO$_3$ symporter, the major protein that mediates basolateral base exit.[85] Similarly, dopamine inhibits proximal tubule NaCl absorption by inhibiting both the apical membrane Na$^+$/H$^+$

exchanger as well as basolateral membrane Na^+-K^+-ATPase, the basolateral Na^+ exit step and the primary motor for driving Na^+ absorption.[86-88] Interestingly, all determinants of proximal tubule Na^+ transport appear to exert at least part of their effect through regulation of the apical membrane Na^+/H^+ exchanger. The hormones that acutely regulate proximal tubule transport and apical membrane Na^+/H^+ activity are summarized in Table 2.2 and are reviewed elsewhere.[17,24] Hormonal regulation of Na^+/H^+ exchanger is mediated by a number of signaling pathways. This section will review selected recent advances in acute regulation of NHE3 by four signaling systems; protein kinase A, protein kinase C, Ca-calmodulin kinase and tyrosine kinases.

C.2 ACUTE REGULATION OF NHE3 BY cAMP-DEPENDENT PROTEIN KINASE

Dopamine and PTH acutely increase cellular cAMP levels and inhibit proximal tubule apical membrane Na^+/H^+ exchange. Maneuvers that increase cAMP levels have been shown to acutely inhibit Na^+/H^+ exchange in renal cortical brush border vesicles[89-92] and in cultured proximal tubule cell lines.[33,93-95] Since the cloning and identification of NHE3 as the proximal tubule apical isoform, the effect of PKA activation on NHE3 has been examined directly.

The first step was to establish whether the NHE3 isoform is regulated by PKA. When expressed in AP-1 cells which are Na^+/H^+ exchanger null fibroblasts, NHE3 activity was inhibited by acute addition of cAMP.[96] The kinetic characteristics of the effect were not specifically addressed, but the assays were performed under V_{max} conditions for both pH_i and Na^+_o. Therefore, there is at least a V_{max} effect though changes in affinities could not be ruled out. When expressed in *Xenopus* oocytes, NHE3 activity was also inhibited by acute addition of cAMP.[96]

NHE3 conforms to the predicted topology of all the NHE isoforms with an N-terminal transmembrane domain and a C-terminal cytoplasmic domain.[97] To examine the relative roles of these two domains, full-length NHE3 and a NHE3 mutant with a cytoplasmic domain truncation were expressed in *Xenopus* oocytes.[96] Although the mutant

Table 2.2. Acute hormonal regulation of proximal tubule apical membrane Na^+/H^+ exchanger

Stimulates	Alpha adrenergic
	Angiotensin II
	Endothelin
Inhibits	Parathyroid hormone
	Dopamine

protein could sustain Na⁺ transport in oocytes to levels comparable to that of the full-length protein, activity was no longer regulated by cAMP. This modular design of transport vs. regulatory roles of the two domains is consistent with that observed for mammalian NHE1 and βNHE, the piscine homolog of NHE1.[24,98-100]

Since there are several PKA consensus phosphorylation motifs in the cytoplasmic domain, phosphorylation of this domain by PKA may mediate the acute regulation of NHE3. Since direct phosphorylation by PKA can only be tested in vitro in the absence of other kinases, purified recombinant NHE3 cytoplasmic domain was tested for phosphorylation by the purified catalytic subunit of PKA. Time dependent phosphorylation was observed exclusively on serine residues. Since false positive cryptic phosphorylation sites may be exposed in vitro, NHE3 was immunoprecipitated from AP-1 cells expressing NHE3 and was shown to be a phosphoprotein in vivo. In addition, NHE3 phosphate content was increased by acute addition of cAMP.[96] Whether the phosphorylation in vivo also occurs on the cytoplasmic domain remains to be defined.

These studies suggest that the cytoplasmic domain of NHE3 is absolutely required for PKA regulation and that inhibition of NHE3 activity is associated with phosphorylation of NHE3 protein, possibly on the cytoplasmic domain. Several points are noteworthy. First, the above data do not prove that phosphorylation and inhibition are causally related. Indeed there can be mechanisms other than phosphorylation that can regulate NHE3 activity, and there may be phosphorylated residues that do not alter NHE3 activity. Second, it is conceivable that kinases downstream from PKA, in addition to PKA itself, can also phosphorylate NHE3 in vivo. Third, other cofactors may be required for PKA to induce its effect on NHE3. When rabbit NHE3, which has a similar primary structure to rat NHE3 was expressed in a different Na⁺/H⁺ exchanger null cell line, PKA activation had no effect on NHE3 activity.[101] This discordance may be due to differences in cellular environment. A 40-55 kD cytoplasmic cofactor has been purified and cloned that functionally mediates the effect of PKA activation on Na⁺/H⁺ exchanger activity in a reconstituted proteoliposomal system.[102,103] When renal cortical brush border membranes were immunodepleted of this cofactor, cAMP addition could no longer induce inhibition of native NHE3 activity.[102,103] Restoration of regulation was achieved when the cofactor was added back into the system.[102,103]

Endocytosis remains a possible means of acute regulation of NHE3. The acute inhibition of apical membrane Na⁺/H⁺ exchange by PTH, which activates PKA, has been postulated to occur via endocytosis.[104] This was based on the concurrent decrease in Na⁺/H⁺ exchange activity in the apical membrane fraction with an increase in activity in a different microsomal fraction on a density gradient. Unfortunately, it was difficult to define the identity of these fractions and impossible then to correlate activity with antigen. Recently, the effect of acute PTH infusion on NHE3 antigenic abundance in parathyroidectomized rats was examined by immunoblots (Moe and Alpern, unpublished data). After one hour of PTH infusion apical membrane NHE3 abundance was decreased by 35% and after eight hours of PTH infusion by 75%. When total cortical membranes were examined, no significant decrease in NHE3 abundance could be detected after 1 or 8 hours of PTH. These studies demonstrate that endocytosis can be a mechanism of acute regulation of NHE3 activity. It is conceivable that acute phosphorylation of NHE3 may have a dual effect in that certain phosphorylated residues may regulate endocytosis/exocytosis, and phosphorylation of other residues may regulate the intrinsic activity of the transporter protein.

C.3 REGULATION OF NHE3 BY PROTEIN KINASE C

There has been considerable controversy regarding the effect of PKC activation on the proximal tubule apical membrane Na^+/H^+ exchanger. In the intact tubule, activation of PKC by phorbol ester either stimulates[105] or inhibits[106] volume and HCO_3 transport. One study partially resolved the discrepancy by showing a time-dependent biphasic effect in that phorbol esters stimulate transport initially, followed by inhibition after 10 minutes of application.[107] In renal brush border vesicles, Na^+/H^+ exchange activity was stimulated by direct addition of PMA to the membranes.[108] However, in LLC-PK1 cells expressing NHE3, phorbol ester inhibited apical membrane Na^+/H^+ exchanger activity.[33] Likewise, when the NHE3 gene was expressed in Na^+/H^+ exchanger null cells, NHE3 activity was inhibited by phorbol ester treatment via a decrease in its V_{max}.[101] At the present time, it is difficult to settle these disparities. Since phorbol ester probably exerts a host of effects on tubules, membranes and cells, different effects may simply reflect different interacting pathways activated by phorbol esters in these systems. Another possibility is that the systems tested have different profiles of PKC isoforms and different isoforms, may exert different effects on NHE3. Lastly, it is possible that protein kinase C is acting on a proximal tubule apical membrane Na^+/H^+ exchanger that is not encoded by NHE3.

C.4 REGULATION BY CA^{2+} CALMODULIN KINASE AND TYROSINE KINASES

Cytosolic calcium is an important second messenger for hormonal action on proximal tubule transport. The binding of Ca^{2+} to calmodulin leads to the modification of a host of downstream effectors including a number of kinases.[109] One such kinase is Ca^{2+}-calmodulin (CaM) kinase II which has broad substrate specificity and is expressed in renal cortex.[110] Activation of CaM kinase has been shown to inhibit the apical membrane Na^+/H^+ exchanger in reconstituted renal brush border proteoliposomes,[90] ileal brush border membranes[111] and LLC-PK1 cells.[112,113] It is presently unknown whether NHE3 is a substrate for CaM kinase II.

The physiologic role of this signaling pathway was demonstrated in studies with endothelin.[114,115] In the proximal tubule, low concentrations of endothelin increase proximal tubule volume absorption.[116] This effect is mediated at least in part via activation of apical membrane Na^+/H^+ exchanger activity.[117,118] The signaling mechanisms for endothelin were recently examined in OKP cells overexpressing either the ET_A or ET_B endothelin receptors. Endothelin-1 (ET-1) stimulated Na^+/H^+ exchanger activity in cells overexpressing ET_B (OKP/ET_B cells) but not ET_A receptors. ET-1 also caused an increase in intracellular Ca^{2+} in OKP/ET_B cells. The increase in Na^+/H^+ exchanger activity was inhibited 50% by either clamping intracellular Ca^{2+} low with BAPTA or inhibiting CaM kinase with KN62, suggesting that CaM kinase is responsible for half the stimulation of the apical membrane Na^+/H^+ exchanger activity by endothelin.

In the same study, the role of tyrosine kinases in mediating endothelin stimulation of Na^+/H^+ exchanger activity were also examined. Inhibition of tyrosine kinases with either herbimycin A or tyrphostin eliminated 50% of the endothelin-induced stimulation of Na^+/H^+ exchanger activity. The identies of the responsible tyrosine kinases are presently unknown. In OKP/ET_B cells, ET-1 induced tyrosine phosphorylation of five proteins of 68, 110, 125, 130 and 210 kD. The identities of the 68 and 125 kD proteins have been revealed to be paxillin and focal adhesion kinase respectively. In the presence of cytochalasin D, the ET-1-induced tyrosine phosphorylation of all the above proteins was blocked except for the 210 kD protein. Interestingly, cytochalasin D did not block the effect of ET-1 on Na^+/H^+ exchanger activity leaving the 210 kD tyrosine phosphoprotein as a possible mediator for the effect of ET-1 on the exchanger.

When both BAPTA and herbimycin A were added, the endothelin-induced stimulation was completely blocked. These findings suggest that endothelin acutely activates the Na^+/H^+ exchanger by dual pathways involving CaM kinase and tyrosine kinases.

Bicarbonate absorption in the rat medullary thick ascending limb is inhibited by acute hyperosmolarity.[119] Since thick ascending limb apical membrane NHE3 mediates almost all the HCO_3 absorption, it is reasonable to presume that NHE3 is regulated. Seventy to eighty percent of this inhibition could be blocked by tyrosine kinase inhibitors, genistein and herbimycin A while PKC inhibition by staurosporine or activation of PKA by cAMP did not affect the hyperosmolarity effect. Under baseline conditions, tyrosine kinase inhibitors increased HCO_3 absorption by 30% while tyrosine phosphatase inhibitors inhibited it by 50%.

Table 2.3. Chronic conditions associated with increased proximal tubule apical membrane Na⁺/H⁺ exchanger activity

Metabolic acidosis
K deficiency
Hyperfiltration
Respiratory acidosis
Uncontrolled diabetes
ECF volume contraction
Glucocorticoids
Thyroid hormone

D. CHRONIC REGULATION OF RENAL Na⁺/H⁺ EXCHANGER

In a number of chronic conditions, increased proximal tubule $NaHCO_3$ and/or NaCl absorption is observed in association with increased apical membrane Na^+/H^+ exchanger activity. Table 2.3 catalogs these conditions. Here, we review several advances made in defining the molecular mechanisms of NHE3 adaptation in chronic metabolic acidosis and chronic glucocorticoid administration.

D.1 Chronic Metabolic Acidosis

Chronic metabolic acidosis induces a host of adaptive changes in the mammalian proximal tubule including increased ammoniagenesis, increased citrate reabsorption and increased HCO_3 absorptive capacity.[120] Increased bicarbonate reabsorptive capacity is secondary to parallel increases in apical membrane Na^+/H^+ exchanger activity and basolateral $Na^+/H^+CO_3/CO_3$ symporter activity.[121,122] This phenotypic change of the proximal tubule cell in response to low pH requires sustained, complex and coordinated sensing and effector mechanisms. This review will focus only on the molecular mechanisms of increased Na^+/H^+ exchanger activities and the proximal tubule cellular response to low ambient pH.

Systemic acidosis can alter hemodynamics, renal nerve activity, and a variety of endocrine and paracrine responses in addition to pH. Cultured epithelial cells have the advantage that the effect of pH per se can be studied and that large sustained changes in pH can be instituted to amplify a response for detection. When rabbit proximal tubule cells and MCT cells were exposed to acid pH (pH 7.0) for 48 hours, an increase in Na^+/H^+ exchanger activity was noted that was dependent on protein synthesis. In MCT cells and LLC-PK1 cells, proximal tubule cell lines that also express NHE1, incubation in acid increased Na^+/H^+ exchange activity and NHE1 mRNA abundance.[123,124] A similar increase in renal cortical NHE1 mRNA was also found in rats fed an acid diet.[124,125] In contrast to epithelial cells, acidosis suppressed gene expression and activity of NHE1 in fibroblasts. These results underscore the importance of NHE1 as a defender of cell pH particularly in epithelial cells.

However, these studies did not address the mechanism by which acidosis increases proximal tubule apical membrane Na^+/H^+ exchange and H transport since NHE1 likely plays no role in mediating proximal tubule H secretion. To study apical membrane EIPA-insensitive Na^+/H^+ activity, OKP cells and LLK-PK cells were used. EIPA-insensitive Na^+/H^+ exchanger activity was increased by chronic incubation in low pH medium in both LLC-PK1 cells and OKP cells.[34,123,124] In OKP cells, increases in EIPA-insensitive Na^+/H^+ exchange activity were accompanied by increases in NHE3 mRNA abundance[34] and in LLC-PK cells, by increase in NHE3 protein.[35] In the thick ascending limb, chronic metabolic acidosis also causes increased HCO_3^- absorptive capacity.[126] A preliminary report detected increases in NHE3 mRNA abundance in microdissected thick ascending limbs by quantitative RT-PCT in rats fed an acid diet.[68] In a similar model of acidosis in rats, brush border NHE3 protein abundance was increased but NHE3 mRNA abundance was not.[127] It is possible that small increases in NHE3 mRNA, at levels that elude detection, are sufficient to sustain increased protein abundance in the kidney. Alternatively, regulatory mechanisms at the translational or post-translational level may be operational. Interestingly, in the OKP cells, the acid-induced increase in NHE3 activity preceded the increase in NHE3 mRNA abundance and was not blocked by cycloheximide.[34] Studies in proximal tubule suspensions have shown that the acute adaptation of apical membrane Na^+/H^+ exchange to acid load is blocked by colchicine but not cycloheximide.[122] These studies suggest that membrane trafficking may play a role in the acute regulation of NHE3 by acid.

Recently, a number of studies were directed at examining the signaling mechanisms by which acid activates Na^+/H^+ exchangers. In MCT cells, the increase in NHE1 activity is mediated by chronic activation of protein kinase C as chronic downregulation of PKC can block the effect of acid on NHE1 activity.[128] For NHE3, the situation is different. PKC and PKA inhibitors did not block the effect of acid on NHE3 activity in OKP cells, but inhibition of tyrosine kinases blocked the effect.[129] Thus chronic activation of NHE1 proceeds via a pathway involving PKC, while chronic activation of NHE3 undertakes a pathway involving tyrosine kinases.

Acutely, acid invokes a series of responses in epithelial cells. In MCT cells, acid increases the expression of a number of primary response genes or early response genes including c-fos, c-jun, junB and egr-1.[130] Acid incubation was found to increase the rate of transcription of these genes. Primary response genes themselves do not confer a specific cellular response, but they frequently activate secondary and tertiary response genes that are cell specific and may lead to a change in phenotype that is appropriate to the environmental stimuli. The transcriptional factor AP-1 consists of dimers of various leucine zipper proteins including primary response gene products such as c-fos and c-jun. When AP-1 activity was assayed in MCT cells, acid acutely increased its transcriptional activity.[128]

Both PKC and tyrosine kinase pathways can lead to activation of early response genes. Since PKC activation can induce expression of early response genes and PKC has been shown to mediate the response of NHE1 to chronic acid, the effect of PKC on acid-induced early response gene expression was examined. When PKC was downregulated in MCT cells, acid still induced immediate early genes suggesting acid activated early response genes via a non-PKC pathway. Congruent with this observation was the fact that clamping intracellular Ca^{2+} low with BAPTA did not prevent the induction of early response genes by acid.[130] In contrast, when tyrosine kinase pathways were inhibited, induction of early response genes was blocked. These studies established a link between acid and early response genes via tyrosine kinases.

To examine the possible substrates of tyrosine phosphorylation, immunoblots with antiphosphotyrosine antibodies were performed on extracts from acid-treated MCT cells. Acid incubation caused an increase in the phosphotyrosine content of several proteins of 60-70 kD and 120 kD.[131] All proteins were in the cytosolic fraction raising the possibility that proteins of the nonreceptor tyrosine kinase family such as c-src may be involved. c-src is a 60 kD proto-oncogene that belongs to a family of structurally related proteins that perform a variety of functions in different tissues. c-src is abundantly expressed in the proximal tubule, but its function is currently unknown. In the basal state, the C-terminal tyrosine (Y527) is phosphorylated and binds to a region called the src homology 2 or SH2 domain, which conceals its tyrosine kinase catalytic site. Upon activation, Y527 is dephosphorylated, releasing it from the SH2 domain and the catalytic site is rendered active.

To examine the effect of acid on c-src, c-src was immunoprecipitated from acid-treated MCT cells, and its activity assayed in vitro. Decrements in intracellular pH as small as 0.07 pH units increased c-src kinase activity up to 2-fold as early as 0.5 mins.[131] As mentioned above, the increases in NHE3 activity and mRNA are dependent on tyrosine kinase pathways. To study the relevance of acid-induced c-src activation on NHE3 adaptation, c-src activity was inhibited in OKP cells. To achieve this, c-src kinase or csk was expressed in OKP cells.[129] csk phosphorylates c-src on Y527 and maintains c-src in an inactive configuration. In clones with the highest expression of csk, the effect of acid on c-src activity and NHE3 activity and mRNA abundance was blocked. These results are highly suggestive of c-src playing a major role in mediating the effect of acid on NHE3.

D.2 GLUCOCORTICOIDS

Glucocorticoid administration in whole animals increases proximal tubule H secretion[132] and apical membrane Na+/H+ exchanger.[133] Glucocorticoids can stimulate the Na+/H+ exchanger by increasing GFR as well as by a direct action on the proximal tubule, as this effect could be demonstrated in suspended tubules as well as cultured cells.[134,135] The effect of glucocorticoid on OKP cell NHE3 activity was dose-dependent, apparent within four hours and dependent on protein synthesis.[135] In OKP cells, dexamethasone increases NHE3 mRNA and activity with a similar time course. The increase in NHE3 mRNA in OKP cells was due to increased transcription with no change in transcript stability.[136] Thus in OKP cells, stimulation of NHE3 by glucocorticoids can all be accounted for by increased NHE3 gene transcription. Similarly, increased activity was described in conjunction with increased steady-state mRNA in whole animals, although a detailed time course was not performed.[137,138] In suspended tubules, after three hours of incubation with dexamethasone, Na+/H+ exchange activity was increased, although there was no change in NHE3 mRNA.[138] This finding contrasts with what was observed in OKP cells and suggests that there may be additional levels of regulation by glucocorticoids.

The effect of glucocorticoids on NHE3 expression is important in maturation of proximal tubule function. Neonates have lower HCO_3^- thresholds due to impaired proximal tubule acidification and are thus prone to acidosis.[139,140] The profile of maturation of proximal tubule HCO_3^- transport capacity exactly parallels that of NHE3 activity, mRNA and protein expression.[140-142] The expression of NHE1 appears to be constitutive from newborn to adult.[142] However, the appearance of NHE3 expression and maturation of proximal acidification function coincides with a surge of endogenous glucocorticoids.[143] When neonates were given exogenous glucocorticoid, proximal tubule acidification, renal cortical NHE3 mRNA and protein expression were induced to levels approaching that of adults.[142-144]

SUMMARY

Na+/H+ exchangers in renal epithelia are vital for both cellular and whole organism homeostasis. The role of Na+/H+ exchangers in cell pH and volume defense assumes paramount importance since the nephron is subjected to fluctuations in ambient extracellular pH and tonicity in magnatudes far in excess of any other cell types. As transepithelial Na+ and H+- equivalent transporters, Na+/H+ exchangers are critical for maintaining extracellular fluid volume and systemic acid-base balance. These transport functions are designed towards defense of the systemic milieu rather than the renal epithelial cell and are tightly regulated in response to acute as well as chronic perturbations. Within a single nephron segment, one encounters diverse roles performed by various Na+/H+ exchanger isoforms. The house-keeping and transepithelilal transporting functions of Na+/H+ exchangers often reside in one and the same cell. The kidney provides a unique model for the study of differential gene expression, membrane targeting and regulation of the various Na^+/H^+ exchanger isoforms.

REFERENCES

1. Pitts RF, Lotspeich WD, Scheiss WA, et al. The renal regulation of acid-base balance in man. I. The nature of mechanism of acidifying the urine. J Clin Invest 1947; 27:46-56.
2. Ullrich KJ, Radtke HW, Rumrich G. The role of bicarbonate and other buffers on isotonic fluid absorption in the proximal convolutions of the rat kidney. Pflugers Arch 1971; 330:149-161.
3. Murer H, Hopfer U, Kinne R. Sodium/proton antiport in brush-border membrane vesicles isolated from rat small intestine and kidney. Biochem J 1976; 154:597-604.
4. Padan E, Schuldiner S. Molecular physiology of the Na^+/H^+ antiporter in *Escherichia coli*. J Exp Biol 1994; 196:443-456.
5. Jia ZP, McCullough N, Martel R, et al. Gene amplification at a locus encoding a putative Na^+/H^+ antiporter confers sodium and lithium tolerance in fission yeast. EMBO J 1992; 11:1631-1640.
6. Marra MA, Prasad SS, Baillie DL. Molecular analysis of two genes between let-56 in the unc-22(IV) region of *Caenorhabditis elegans*. Mol Gen Genet 1993; 236:289-298.
7. Ahearn GA, Clay LP. Kinetic analysis of electrogenic 2 Na^+-1 H^+ antiport in crustacean hepatopancreas. Amer J Physiol 1989; 257:R484-R493.
8. Morais R, Borgese F, Fievet B, et al. Regulation of Na^+/H^+ exchange and pH in erythrocytes of fish. Comp Biochem Physiol 1992; 102A:597-602.
9. Alpern RJ. Cell mechanisms of proximal tubule acidification. Physiol Rev 1990; 70:79-114.
10. DuBose Jr TD, Pucacco LR, Seldin DW, et al. Direct determination of PCO_2 in the rat renal cortex. J Clin Invest 1978; 62:338-348.
11. Preisig PA, Ives HE, Cragoe EJ, et al. Role of the Na^+/H^+ antiporter in rat proximal tubule bicarbonate absorption. J Clin Invest 1987; 80:970-978.
12. Alpern RJ, Cogan MG, Rector Jr FC. Effect of luminal bicarbonate concentration on proximal acidification in the rat. Amer J Physiol 1982; 243:F53-F59.
13. Alpern RJ, Howlin KJ, Preisig PA. Active and passive components of chloride transport. J Clin Invest 1995; 76:1360-1366.
14. Preisig PA, Rector FC, Jr. Role of Na-H antiport in rat proximal tubule NaCl absorption. Amer J Physiol 1988; 255:F461-F465.
15. Baum M. Evidence that parallel Na-H and $Cl-HCO_3(OH)$ antiporters transport NaCl in the proximal tubule. Amer J Physiol 1987; 252:F338-F345.
16. Schild L, Giebisch G, Karniski LP, et al. Effect of formate on volume reabsorption in the rabbit proximal tubule. J Clin Invest 1987; 79:32-38.
17. Alpern RJ, Stone DK, Rector Jr FC. Renal acidification mechanisms. In Brenner BM, Rector Jr FC, eds. The Kidney. Philadelphia, W.B. Saunders Co. 1991:318-379.

18. Nagami GT. Luminal secretion of ammonia in the mouse proximal tubule perfused in vitro. J Clin Invest 1988; 81:159-164.
19. Preisig PA, Alpern RJ. Pathways for apical and basolateral membrane NH_3 and NH_4 movement in rat proximal tubule. Amer J Physiol 1990; 259:F587-F593.
20. Kinsella JL, Aronson PS. Interaction of NH_4 and Li with the renal microvillus membrane Na-H exchanger. Amer J Physiol 1981; 241: C220-C226.
21. Kinsella JL, Aronson PS. Amiloride inhibition of the Na-H exchanger in renal microfillus membrane vesicles. Amer J Physiol 1981; 241:F374-F379.
22. Warnock DG, Reenstra WW, Yee VJ. Na^+/H^+ antiporter of brush border vesicles. studies with acridine orange uptake. Amer J Physiol 1982; 242:F733-F739.
23. Aronson PS, Nee J, Suhm MA. Modifier role of internal H in activating the Na-H exchanger in renal microvillus membrane vesicles. Nature 1982; 299:161-163.
24. Noel J, Pouysségur J. Hormonal regulation, pharmacology, and membrane sorting of vertebrate Na^+/H^+ exchanger isoforms. Amer J Physiol 1995; 268:C283-C296.
25. Haggerty JG, Agarwal N, Reilly RF, et al. Pharmacologically different Na^+/H^+ antiporters on the apical and basolateral surfaces of cultured porcine kidney cells (LLC-PK_1). Proc Natl Acad Sci USA 1988; 85:6797-6801.
26. Orlowski J. Heterologous expression and functional properties of amiloride high affinity (NHE-1) and low affinity (NHE-3) isoforms of the rat Na^+/H^+ exchanger. J Biol Chem 1993; 268:16369-16377.
27. Biemesderfer D, Pizzonia J, Exner M, et al. NHE3: a Na^+/H^+ exchanger isoform of the renal brush border. Amer J Physiol 1993; 265:F736-F742.
28. Amemiya M, Loffing J, Lotscher M, et al. Expression of NHE-3 in the apical membrane of rat renal proximal tubule and thick ascending limb. Kidney Int 1995; 48:1206-1215.
29. McKinney TD, Burg MB. Bicarbonate and fluid absorption by renal proximal straight tubules. Kidney Int 1977; 21:1-8.
30. Burg MB, Green N. Bicarbonate transport by isolated perfused rabbit proximal convoluted tubules. Amer J Physiol 1977; 233:F307-F314.
31. Baum M. Axial heterogeneity of rabbit proximal tubule luminal H and basolateral HCO_3 transport. Amer J Physiol 1989; 256:F335-F341.
32. Montrose MH, Murer H. Polarity and kinetics of Na-H exchange in cultured opossum kidney cells. Amer J Physiol 1990; 259:C121-C133.
33. Casavola V, Helmle-Kolb C, Murer H. Separate regulatory control of apical and basolateral Na^+/H^+ exchange in renal epithelial cells. Biochem Biophys Res Comm 1989; 165:833-837.
34. Amemiya M, Yamaji Y, Cano A, et al. Acid incubation increases NHE-3 mRNA abundance in OKP cells. Amer J Physiol 1995; 269:C126-C133.
35. Soleimani M, Bookstein C, McAteer JA, et al. Effect of high osmolality on Na^+/H^+ exchange in renal proximal tubule cells. J Biol Chem 1994; 269:15613-15618.

36. Tse C-M, Levine SA, Yun CHC, et al. Cloning and expression of a rabbit cDNA encoding a serum-activated ethylisopropylamiloride-resistant epithelial Na$^+$/H$^+$ exchanger isoform (NHE-2). J Biol Chem 1993; 268:11917-11924.
37. Yu FH, Shull GE, Orlowski J. Functional properties of the rat Na$^+$/H$^+$ exchanger NHE-2 isoform expressed in Na$^+$/H$^+$ exchanger-deficient Chinese hamster ovary cells. J Biol Chem 1993; 268:11925-11928.
38. Mrkic BC, Tse C-M, Forgo J, et al. Identification of PTH-responsive Na$^+$/H$^+$ exchanger isoforms in a rabbit proximal tubule cell line (RKPC-2). Pflugers Arch 1995; (in press).
39. Soleimani M, Bizal GL, Hattabaugh Y, et al. Expression and subcellular localization of Na$^+$/H$^+$ exchanger isoforms NHE-3 and NHE-4 in rabbit kidney. J Amer Soc Nephrol 1993; 4:847(Abstract).
40. Bookstein C, Musch MW, DePaoli A, et al. A unique sodium-hydrogen exchange isoform (NHE-4) of the inner medulla of the rat kidney is induced by hyperosmolarity. J Biol Chem 1994; 269:29704-29709.
41. Ives HE, Yee VJ, Warnock DG. Asymmetric distribution of the Na$^+$/H$^+$ antiporter in the renal proximal tubule epithelial cell. J Biol Chem 1983; 258:13513-13516.
42. Grassl SM, Aronson PS. Na$^+$/H$^+$CO$_3$ co-transport in basolateral membrane vesicles isolated from rabbit renal cortex. J Biol Chem 1986; 261:8778-8783.
43. Alpern RJ. Mechanism of basolateral membrane H/OH/HCO$_3$ transport in the rat proximal convoluted tubule. J Gen Physiol 1985; 86:613-636.
44. Krapf R, Alpern RJ, Rector Jr FC, et al. Basolateral membrane Na/base cotransport is dependent on CO$_2$/HCO$_3$ in the proximal convoluted tubule. J Gen Physiol 1987; 90:833-853.
45. Kurtz I. Basolateral membrane Na$^+$/H$^+$ antiport, Na/base cotransport, and Na-independent Cl/base exchange in the rabbit S$_3$ proximal tubule. J Clin Invest 1989; 83:616-622.
46. Alpern RJ, Chambers M. Basolateral membrane Cl/HCO$_3$ exchange in the rat proximal convoluted tubule. J Gen Physiol 1987; 89:581-598.
47. Sasaki S, Yoshiyama N. Interaction of chloride and bicarbonate transport across the basolateral membrane of rabbit proximal straight tubule. J Clin Invest 1988; 81:1004-1011.
48. Kriz W, Kaissling B. Structural organization of the mammalian kidney. In: Seldin DW, Giebisch G, eds. The kidney: physiology & pathophysiology. New York, Raven Press 1992:707-778.
49. Roy DR, Layton HE, Jamison RL. Countercurrent mechanism and its regulation. In: Seldin DW, Giebisch G, eds. The kidney: physiology & pathophysiology. New York, Raven Press 1992:1649-1692.
50. Geibel J, Giebisch G, Boron WF. Effects of acetate on luminal acidification processes in the S3 segment of the rabbit proximal tubule. Amer J Physiol 1989; 257:F586-F594.
51. Biemesderfer D, Reilly RF, Exner M, et al. Immunocytochemical characterization of Na-H exchanger isoform NHE-1 in rabbit kidney. Amer J Physiol 1992; 263:F833-F840.

52. Krapf R, Solioz M. Na$^+$/H$^+$ antiporter mRNA expression in single nephron segments of rat kidney cortex. J Clin Invest 1991; 88:783-788.
53. Ernst SA, Schreibner JH. Ultrastructural localization of NaKATPase in rat and rabbit kidney medulla. J Cell Biol 1981; 91:803-813.
54. Imai M. Functional heterogeneity of the descending limb of Henle's loop. I. Internephron heterogeneity in the hamster. Pflugers Arch 1984; 402:385-392.
55. Imai M. Functional heterogeneity of the descending limb of Henle's loop. II. Interspecies differences among rabbit, rat, and hamster. Pflugers Arch 1984; 402:393-401.
56. Lonnerholm G, Ridderstrale Y. Intracellular distribution of carbonic anhydrase in the rat kidney. Kidney Int 1980; 17:162-174.
57. Kokko JP. Sodium chloride and water transport in the descending limb of Henle. J Clin Invest 1970; 49:1838-1846.
58. Dobyan DC, Magill LS, Friedman PA, et al. Carbonic anhydrase histochemistry in rabbit and mouse kidneys. Anat Rec 1982; 204:185-197.
59. Kurtz I. Apical and basolateral Na$^+$/H$^+$ exchanger in the rabbit outer medullary thin descending limb of Henle: role in intercellular pH regulation. J Membr Biol 1988; 106:253-260.
60. DuBose Jr TD, Pucacco LR, Lucci MS, et al. Micropuncture determination of pH, PCO$_2$, and total CO$_2$ concentration in accessible structures of the rat renal cortex. J Clin Invest 1979; 64:476-482.
61. Good DW. Sodium-dependent bicarbonate absorption by cortical thick ascending limb of rat kidney. Amer J Physiol 1985; 248:F821-F829.
62. Kikeri D, Azar S, Sun A, et al. Na-H antiporter and Na-(HCO$_3$)$_n$ symporter regulate intracellular pH in mouse medullary thick limbs of Henle. Amer J Physiol 1990; 258:F445-F456.
63. Rocha AS, Kokko JP. Sodium chloride and water transport in the medullary thick ascending limb of Henle. Evidence for active chloride transport. J Clin Invest 1973; 52:612-623.
64. Hebert SC, Culpepper RM, Andreoli TE. NaCl transport in mouse medullary thick ascending limbs. I. Functional heterogeneity and ADH-stimulated NaCl cotransport. Amer J Physiol 1981; 241:F412-F431.
65. Koenig B, Ricapito S, Kinne R. Chloride transport in the thick ascending limb of Henle's loop: Potassium dependence and stoichiometry of the NaCl cotransport system in plasma membrane vesicles. Pflugers Arch 1983; 399:173-179.
66. Friedman PA, Andreoli TA. CO$_2$-stimulated NaCl absorption in the mouse renal cortical thick ascending limb of Henle: Evidence for synchronous Na$^+$/H$^+$ and Cl/HCO$_3$ exchange in apical plasma membranes. J Gen Physiol 1982; 80:683-711.
67. Friedman PA, Andreoli TE. Effects of (CO$_2$+HCO$_3$-) on electrical conductance in cortical thick ascending limbs. Kidney Int 1986; 30:325-331.
68. Borensztein P, Laghman K, Froissart M, et al. RT-PCR analysis of Na$^+$/H$^+$ exchanger isoforms mRNAs in the rat outer medulla: adaptation to in vivo chronic metabolic acidosis. J Amer Soc Nephrol 1994; 5:248 (Abstract).
69. Iino Y, Burg MB. Effect of acid-base status in vivo on bicarbonate transport by rabbit renal tubules in vitro. Jpn J Physiol 1981; 31:99-107.

70. Krapf R. Basolateral membrane H/OH/HCO₃ transport in the rat cortical thick ascending limb. J Clin Invest 1988; 82:234-241.
71. Hebert SC. Hypertonic cell volume regulation in mouse thick limbs. II. Na-H and Cl-HCO₃ exchange in basolateral membranes. Amer J Physiol 1986; 250:C920-C931.
72. Lucci MS, Pucacco LR, Carter NW and DuBose Jr T. Evaluation of bicarbonate transport in the rat distal tubule: effect of acid-base status. Am J Phisio 1982; 243:F335-F339.
73. Wang T, Malnic G, Giebisch G, et al. Renal bicarbonate reabsorption in the rat. IV. Bicarbonate transport mechanisms in the early and late distal tubule. J Clin Invest 1993; 91:2776-2784.
74. Good DW, Knepper MA, Burg MB. Ammonia and bicarbonate transport by thick ascending limb of rat kidney. Amer J Physiol 1984; 247:F35-F44.
75. McKinney TD, Burg MB. Bicarbonate transport by rabbit cortical collecting tubules: effect of acid and alkali loads in vivo on transport in vitro. J Clin Invest 1977; 60:766-768.
76. McKinney TD, Burg MB. Bicarbonate absorption by rabbit cortical collecting tubules in vitro. Amer J Physiol 1978; 234:F141-F145.
77. McKinney TD, Burg MB. Bicarbonate secretion by rabbit cortical collecting tubules in vitro. J Clin Invest 1978; 61:1421-1427.
78. Weiner ID, Hamm LL. Regulation of intracellular pH in the rabbit cortical collecting tubule. J Clin Invest 1990; 85:274-281.
79. Schuster VL. Cyclic adenosine monophosphate-stimulated bicarbonate secretion in rabbit cortical collecting tubules. J Clin Invest 1985; 75:2056-2064.
80. Chaillet JR, Lopes AG, Boron WF. Basolateral Na-H exchange in the rabbit cortical collecting tubule. J Gen Physiol 1985; 86:795-812.
81. Breyer MD, Jacobson HR. Regulation of rabbit medullary collecting duct cell pH by basolateral Na⁺/H⁺ and Cl/base exchange. J Clin Invest 1989; 84:996-1004.
82. Hays SR, Alpern RJ. Inhibition of Na-independent H pump by Na-induced changes in cell Ca^{2+}. J Gen Physiol 1991; 98:791-813.
83. Hays SR, Alpern RJ. Apical and basolateral membrane H extrusion mechanisms in inner stripe of rabbit outer medullary collecting duct. Amer J Physiol 1990; 259:F628-F635.
84. Soleimani M, Singh G, Bizal GL, et al. Na⁺/H⁺ exchanger isoforms NHE-2 and NHE-1 in inner medullary collecting duct cells. J Biol Chem 1994; 269:27973-27978.
85. Geibel J, Giebisch G, Boron WF. Angiotensin II stimulates both Na-H exchange and Na⁺/H⁺CO₃ cotransport in the rabbit proximal tubule. Proc Natl Acad Sci USA 1990; 87:7917-7920.
86. Felder CC, Campbell T, Albrecht F, et al. Dopamine inhibits Na-H exchanger activity in renal BBMV by stimulation of adenylate cyclase. Amer J Physiol 1990; 259:F297-F303.
87. Aperia A, Bertorello A, Seri I. Dopamine causes inhibition of Na-K-ATPase activity in rat proximal convoluted tubule segments. Amer J Physiol 1987; 252:F39-F45.

88. Bello-Reuss E, Higashi Y, Kaneda Y. Dopamine decreases fluid absorption in rabbit proximal tubule. Min Electrolyte Metab 1983; 9:147-150.
89. Kahn AM, Dolson GM, Hise MK, et al. Parathyroid hormone and dibutyryl cAMP inhibit Na$^+$/H$^+$ exchange in renal brush border vesicles. Amer J Physiol 1985; 248:F212-F218.
90. Weinman EJ, Dubinsky WP, Fisher K, et al. Regulation of reconstituted renal Na$^+$/H$^+$ exchanger by calcium-dependent protein kinases. J Membr Biol 1988; 103:237-244.
91. Weinman EJ, Dubinsky WP, Shenolikar S. Reconstitution of cAMP-dependent protein kinase regulated renal Na-H exchanger. J Membr Biol 1988; 101:11-18.
92. Weinman EJ, Dubinsky WP, Dinh Q, et al. Effect of limited trypsin digestion on the renal Na-H exchanger and its regulation by cAMP-dependent protein kinase. J Membr Biol 1989; 109:233-241.
93. Pollock AS, Warnock DG, Strewler GJ. Parathyroid hormone inhibition of Na-H antiporter activity in a cultured renal cell line. Amer J Physiol 1986; 250:F217-F225.
94. Miller RT, Pollock AS. Modification of the internal pH sensitivity of the Na$^+$/H$^+$ antiporter by parathyroid hormone in a cultured renal cell line. J Biol Chem 1987; 262:9115-9120.
95. Cano A, Preisig P, Alpern RJ. Cyclic adenosine monophosphate acutely inhibits and chronically stimulates Na$^+$/H$^+$ antiporter in OKP cells. J Clin Invest 1993; 92:1632-1638.
96. Moe OW, Amemiya M, Yamaji Y. Activation of protein kinase A acutely inhibits and phosphorylates Na$^+$/H$^+$ exchanger NHE-3. J Clin Invest 1995; 96:2187-2194.
97. Orlowski J, Kandasamy RA, Shull GE. Molecular cloning of putative members of the Na$^+$/H$^+$ exchanger gene family. J Biol Chem 1992; 267:9331-9339.
98. Borgese F, Sardet C, Cappadoro M, et al. Cloning and expression of a cAMP-activated Na$^+$/H$^+$ exchanger. evidence that the cytoplasmic domain mediates hormonal regulation. Proc Natl Acad Sci USA 1992; 89:6765-6769.
99. Guizouarn H, Borgese F, Pellissier B, et al. Role of protein phosphorylation and dephosphorylation in activation and desensitization of the cAMP-dependent Na$^+$/H$^+$ antiport. J Biol Chem 1993; 268:8692-8699.
100. Borgese F, Malapert M, Fievet B, et al. The cytoplasmic domain of the Na$^+$/H$^+$ exchangers dictates the nature of the hormonal response: behavior of a chimeric human NHE-1/trout 6-NHE antiporter. Proc Natl Acad Sci USA 1994; 91:5431-5435.
101. Tse C-M, Levine SA, Yun CHC, et al. Functional characteristics of a cloned epithelial Na$^+$/H$^+$ exchanger (NHE3): resistance to amiloride and inhibition by protein kinase C. Proc Natl Acad Sci USA 1993; 90:9110-9114.
102. Weinman EJ, Steplock D, Schenolikar S. cAMP-mediated inhibition of the renal brush border membrane Na-H exchanger requires a dissociable phosphoprotein cofactor. J Clin Invest 1993; 92:1781-1786.

103. Morell G, Steplock D, Shenolikar S, et al. Identification of a putative Na$^+$/H$^+$ exchanger regulatory cofactor in rabbit renal BBM. Amer J Physiol 1990; 259:F867-F871.
104. Hensley CB, Bradley ME, Mircheff AK. Parathyroid hormone induced translocation of Na-H antiporters in rat proximal tubules. Amer J Physiol 1989; 257:C637-C645.
105. Liu FY, Cogan MG. Role of protein kinase C in proximal bicarbonate absorption and angiotensin signalling. Amer J Physiol 1990; 258: F927-F933.
106. Baum M, Hays SR. Phorbol myristate acetate and dioctanoylglycerol inhibit transport in rabbit proximal convoluted tubule. Amer J Physiol 1988; 254:F9-F14.
107. Wang T, Chan YL. Time- and dose-dependent effects of protein kinase C on proximal bicarbonate transport. J Membr Biol 1990; 117:131-139.
108. Weinman EJ, Shenolikar S. Protein kinase C activates the renal apical membrane Na$^+$/H$^+$ exchanger. J Membr Biol 1986; 93:133-139.
109. Means AR, Tash JS, Chafouleas JG. Physiological implications of the presence, distribution and regulation of calmodulin in eucaryotic cells. Physiol Rev 1992; 62:1-39.
110. Hanley RM, Shenolikar S, Pollack J, et al. Identification of calcium-calmodulin multifunctional protein kinase II in rabbit kidney. Kidney Int 1990; 38:63-66.
111. Rood RP, Emmer E, Wesolek J, et al. Regulation of the rabbit ileal brush-border Na$^+$/H$^+$ exchanger by ATP-requiring Ca^{++}/calmodulin-mediated process. J Clin Invest 1988; 72:1091-1097.
112. Burns KD, Homma T, Harris RC. Regulation of Na$^+$-H$^+$ exchange by ATP depletion and calmodulin antagonism in renal epithelial cells. Amer J Physiol 1991; 261:F607-F616.
113. Burns KD, Homma T, Breyer MD, et al. Cytosolic acidification stimulates a calcium influx that activates Na$^+$-H$^+$ exchange in LLC-PK$_1$. Amer J Physiol 1991; 261:F617-F625.
114. Chu TS, Cano A, Yanagisawa M, et al. Endothelin activates the Na$^+$/H$^+$ antiporter in OKP cells stably transfected with the ET$_B$ receptor cDNA. J Amer Soc Nephrol 1994; 5:250 (Abstract).
115. Chu TS, Cano A, Yanagisawa M, et al. ET$_B$ receptor activates NHE-3 by a Ca^{2+}-dependent pathway in OKP cells. 1995.(in press).
116. Garcia NH, Garvin JL. Endothelin's biphasic effect on fluid absorption in the proximal straight tubule and its inhibitory cascade. J Clin Invest 1994; 93:2572-2577.
117. Eiam-Ong S, Hilden SA, King AJ, et al. Endothelin-1 stimulates the apical Na$^+$/H$^+$ and Na$^+$/H$^+$CO$_3$ transporters in rabbit renal cortex. Kidney Int 1992; 42:18-24.
118. Guntupalli J, DuBose TD. Effects of endothelin on rat renal proximal tubule Na-P$_i$ cotransport and Na-H exchange. Amer J Physiol 1994; 266:F658-F666.
119. Good DW. Hyperosmolality inhibits bicarbonate absorption in rat medullary thick ascending limb via a protein-tyrosine kinase-dependent pathway. J Biol Chem 1995; 270:9883-9889.

120. Alpern RJ, Yamaji Y, Amemiya M, et al. The renal response to acid. News in Physiol Sci 1995; 10:77-81.
121. Preisig PA, Alpern RJ. Chronic metabolic acidosis causes an adaptation in the apical membrane Na^+/H^+ antiporter and basolateral membrane $Na(HCO_3)_3$ symporter in the rat proximal convoluted tubule. J Clin Invest 1988; 82:1445-1453.
122. Soleimani M, Bizal GW, McKinney TD, et al. Effect of in vitro metabolic acidosis on luminal Na^+/H^+ exchange and basolateral $Na:HCO_3$ cotransport in rabbit kidney proximal tubules. J Clin Invest 1992; 90:211-218.
123. Igarashi P, Freed MI, Ganz MB, et al. Effects of chronic metabolic acidosis on Na-H exchangers in LLC-PK_1 renal epithelial cells. Amer J Physiol 1992; 263:F83-F88.
124. Moe OW, Miller RT, Horie S, et al. Differential regulation of Na^+/H^+ antiporter by acid in renal epithelial cells and fibroblasts. J Clin Invest 1991; 88:1703-1708.
125. Krapf R, Pearce D, Lynch C, et al. Expression of rat renal Na^+/H^+ antiporter mRNA levels in response to respiratory and metabolic acidosis. J Clin Invest 1991; 87:747-751.
126. Good DW, Kurtz I. Effects of chronic metabolic acidosis (CMA) and mineralocorticoid on H/HCO_3 transport in rat medullary thick ascending limb (MAL). Kidney Int 1989; 35:454.
127. Ambuehl PM, Amemiya M, Lotscher M, et al. Chronic metabolic acidosis increases NHE-3 protein abundance. J Amer Soc Nephrol 1995; (in press).
128. Horie S, Moe O, Yamaji Y, et al. Role of protein kinase C and transcription factor AP-1 in the acid-induced increase in Na^+/H^+ antiporter activity. Proc Natl Acad Sci USA 1992; 89:5236-5240.
129. Yamaji Y, Cano A, Miller RT, et al. c-src is activated by cell acidification and plays a key role in acid activation of the Na^+/H^+ antiporter. J Amer Soc Nephrol 1994; 5:266 (Abstract).
130. Yamaji Y, Moe OW, Miller RT, et al. Acid activation of immediate early genes in renal epithelial cells. J Clin Invest 1994; 94:1297-1303.
131. Yamaji Y, Moe OW, Miller RT, et al. Acidosis activates a src-related tyrosine kinase in renal cells. J Amer Soc Nephrol 1993; 4:506.
132. Baum M, Quigley R. Glucocorticoids stimulate rabbit proximal convoluted tubule acidification. J Clin Invest 1993; 91:110-114.
133. Kinsella JL, Cujkit T, Sactor B. Na-H exchange activity in renal brush border membrane vesicles in response to metabolic acidosis: the role of glucocorticoids. Proc Natl Acad Sci USA 1984; 81:630-634.
134. Bidet M, Merot J, Tauc M, et al. Na-H exchanger in proximal cells isolated from kidney. II. Short-term regulation by glucocorticoids. Amer J Physiol 1987; 253:F945-F951.
135. Baum M, Cano A, Alpern RJ. Glucocorticoids stimulate Na^+/H^+ antiporter in OKP cells. Amer J Physiol 1993; 264:F1027-F1031.
136. Baum M, Alpern RJ, Moe OW. Glucocorticoids regulate NHE-3 transcription in OKP cells. Amer J Physiol 1995; (in press).

137. Kinsella JL, Freiberg JM, Sacktor B. Glucocorticoid activation of Na^+/H^+ exchange in renal brush border vesicles: kinetic effects. Amer J Physiol 1985; 248:F233-F239.
138. Baum M, Moe OW, Gentry DL, et al. Effect of glucocorticoids on renal cortical NHE-3 and NHE-1 mRNA. Amer J Physiol 1994; 267: F437-F442.
139. Edelman Jr CM, Soriano JR, Boichis H, et al. Renal bicarbonate reabsorption and hydrogen ion excretion in normal infants. J Clin Invest 1967; 46:1309-1317.
140. Schwartz GJ, Evan AP. Development of solute transport in rabbit proximal tubule. I. HCO_3- and glucose absorption. Amer J Physiol 1983; 245:F382-F390.
141. Quigley R, Baum M. Developmental changes in rabbit juxtamedullary proximal convoluted tubule bicarbonate permeability. Pediatr Res 1990; 28:663-666.
142. Baum M, Biemesderfer D, Gentry D, et al. Ontogeny of rabbit renal cortical NHE3 and NHE1: effect of glucocorticoids. Amer J Physiol 1995; 268:F815-F820.
143. Ballard PL. Glucocorticoids and differentiation. In: Baxter JD, Rousseau GG, eds. Monographs on endocrinology, glucocorticoid hormone action. New York, Springer-Verlag 1979:493-515.
144. Baum M, Quigley R. Prenatal glucocorticoids stimulate neonatal juxtamedullary proximal convoluted tubule acidification. Amer J Physiol 1991; 261:F746-F752.

CHAPTER 3

REGULATION OF THE Na⁺/H⁺ EXCHANGER IN VASCULAR SMOOTH MUSCLE

Bradford C. Berk

The Na⁺/H⁺ exchanger in vascular smooth muscle cells (VSMC) represents a major mechanism for Na⁺ influx and is also one of the principal mechanisms responsible for the regulation of intracellular pH (pH_i). In this chapter, the relationship between pH_i and vascular smooth muscle cell growth, the regulation of the NHE1 isoform of the Na⁺/H⁺ exchanger by vasoactive agents and growth factors, and the second messenger pathways that may be involved in activation of the exchanger will be discussed. Abnormalities in the regulation of the exchanger as exemplified by studies in hypertension are discussed. The increased activity in cells and tissues from hypertensive patients and genetically hypertensive rats is shown to be due to alterations in post-translational regulation rather than alterations in the exchanger DNA sequence, mRNA or protein expression. Analysis of regulatory mechanisms suggests that signal transduction pathways involving hormone-receptor coupling and/or phosphorylation of critical regulatory proteins are likely to be abnormal in hypertension. These altered signal transduction pathways may then influence several VSMC functions including growth, contractile tone, migration and inflammatory responses. It is proposed that the alterations in regulation of the Na⁺/H⁺ exchanger are a feature of human disease such as hypertension and that understanding the mechanisms responsible for abnormal regulation should provide insight into disease pathogenesis.

A. INTRODUCTION

The amiloride-sensitive Na⁺/H⁺ exchanger is an electroneutral transporter that mediates the 1:1 exchange of extracellular Na⁺ for intracellular H⁺. Rapid progress in understanding the functional regulation of the Na^+/H^+ exchanger has been made since the exchanger was cloned in Dr. Pouysségur's laboratory in 1989.[1] Four isoforms of the Na^+/H^+ exchanger, termed NHE1 through NHE4 have been identified in mammalian tissue.[1-3] The isoforms are structurally and functionally related but exhibit different tissue distribution, pharmacological properties and regulation by hormones and growth factors.[4,5] To date the only isoform of the exchanger that has been identified in vascular smooth muscle cells (VSMC) is the NHE1 isoform,[6] so the remainder of this discussion will focus on regulation of NHE1. The protein encoded by NHE1 is a ~110 kD glycoprotein consisting of 10-12 putative transmembrane spanning regions followed by a hydrophilic cytoplasmic domain.[1] The transmembrane domain alone is sufficient for transport while the cytoplasmic domain contains several consensus sequences for phosphorylation and appears to play a regulatory role.

In VSMC, the Na^+/H^+ exchanger is involved in three processes that are critical to normal cell physiology: regulation of intracellular Na^+, regulation of intracellular H^+ [and hence intracellular pH, (pH_i)], and regulation of cell volume (reviewed in refs. 7-9). In its normal transport mode the Na^+/H^+ exchanger is activated by an acid load, extrudes H^+ in exchange for Na^+ and returns pH_i to resting levels. Thus, the Na^+/H^+ exchanger, in association with the acid-activated Na^+-dependent Cl^-/HCO_3^- exchanger and the alkaline-activated Cl^-/HCO_3^- exchanger, regulates pH_i and indirectly intracellular Na^+. The Na^+/H^+ exchanger is also powerfully stimulated by cell shrinkage and appears to be an important regulator of cell volume in association with the $Na^+/K^+/2Cl^-$ cotransporter. In addition to these roles in normal VSMC physiology, the Na^+/H^+ exchanger has been suggested to play an essential role in gene expression and growth. In this chapter, the role of the Na^+/H^+ exchanger in normal VSMC physiology in vivo, the regulation of Na^+/H^+ exchange by vasoactive agents and growth factors, the second messenger pathways that mediate changes in Na^+/H^+ exchange, and the role of the exchanger in the pathogenesis of hypertension will be discussed.

B. THE VSMC IN THE BLOOD VESSEL WALL

The normal blood vessel is composed of three morphologically distinct layers. The inner layer adjacent to the vessel lumen is comprised of a single cell layer made solely of endothelial cells. Beneath the endothelium is the internal elastic lamina which separates the endothelium from the medial layer, which is composed of VSMC and fibroblasts. Finally, the outermost portion of the vessel is termed the adventitia, which is composed of fibroblasts, adipocytes, and in larger vessels, nerves and vasa vasorum.

VSMC perform multiple functions in the blood vessel. Although they play a passive, structural role in maintaining vessel integrity, the dynamic role of VSMC in vessel tone, growth and inflammation is critical. In resistance arteries, VSMC are under constant tension and by regulating vessel lumen diameter maintain blood pressure and modulate blood flow. In response to transient physiological stimuli such as endothelial derived relaxing factors (e.g., NO and prostacyclin), VSMC relax and the vessel dilates. In response to chronic stimuli such as increased blood pressure, VSMC are capable of undergoing several responses that alter vessel architecture. During atherosclerogenesis or after balloon injury, VSMC undergo a proliferative growth response that is characterized by increases in DNA synthesis and cell number.[10,11] In response to an increase in work, as in essential hypertension, or in response to vasoactive agents, VSMC undergo a hypertrophic response that is characterized by increases in protein synthesis and cell size, without cell proliferation.[10-12] In addition, VSMC may rearrange their orientation within the vessel wall creating a smaller lumen with a thicker wall, a response termed remodeling.[13] After arterial injury VSMC migrate from the media into the intima layer normally occupied only by endothelial cells. Over time this gives rise to a "neointima" which is one of the pathologic features of atherosclerosis. Finally, it has recently been appreciated that VSMC may play a role in the vascular inflammation present in atherosclerosis via their ability to express class I MHC antigens, to secrete cytokines such as monocyte chemotactic peptide-1 (MCP-1), and to express matrix metalloproteases.[14]

C. Na^+/H^+ EXCHANGE AND VSMC FUNCTION IN VIVO: TONE AND GROWTH

The Na^+/H^+ exchanger has been studied in intact vessels because alterations in pH_i may be important in regulating the force generated by actin-myosin crossbridges, and growth of VSMC in hypertension and atherosclerosis may require Na^+/H^+ exchange. Aalkjaer and colleagues[15] were among the first to demonstrate the presence of a pH regulatory mechanism in intact vessels. Foster et al[16] demonstrated the presence of Na^+/H^+ exchange in resistance vessels such as mesenteric arteries. More recently, Aickin[17] showed that recovery from an acid load in guinea pig femoral artery was mediated by an amiloride-sensitive pathway consistent with Na^+/H^+ exchange. Based on studies performed in the presence or absence of HCO_3^- and CO_2 he concluded that the exchanger was largely responsible for extrusion of acid equivalents in these vessels. Carr et al[18] studied human subcutaneous small arteries and found that pH_i was regulated by both Na^+/H^+ exchange and by HCO_3^- dependent mechanisms. In response to hormonal stimuli (norepinephrine, angiotensin II, vasopressin) there was a fall in pH_i in the absence of HCO_3^- despite significant activation of the exchanger. Thus these investigators concluded that an increase in pH_i was not likely to accompany hormonal stimulation. In summary, although the exchanger is present in multiple vascular beds, it appears that the relative contribution of the exchanger to recovery from an acid load may differ in different arteries. In addition, it appears unlikely that the exchanger mediates a net alkalinization in vessels under any normal physiologic condition.

A critical role for the Na^+/H^+ exchanger in cell growth has been proposed based on the fact that intracellular alkalinization is one of the earliest signals stimulated by growth factors[19] and that Na^+/H^+ exchange inhibitors such as amiloride block cell growth. However, it has become clear that in the presence of CO_2 and HCO_3^-, growth factors rarely stimulate alkalinization,[20] and may cause acidification. In addition, growth inhibition caused by the amiloride-like inhibitors may be unrelated to inhibition of Na^+/H^+ exchange.[21-23] Nonetheless, it is possible that Na^+/H^+ exchange is required for cell growth due to its effects on intracellular sodium and cell volume. Two recent studies report that inhibitors of Na^+/H^+ exchange decrease neointima formation in carotid arteries following balloon injury.[24,25] Kranzhofer et al[25] reported that the specific Na^+/H^+ exchange blockers, HOE694 (3-methylsulfonyl-4-piperidino-benzoyl guanidine mesylate) and ethylisopropyl amiloride (EIPA) decreased the cross-sectional area of the neointima in a concentration dependent manner (67% reduction at maximal dose). This decrease in neointima formation was associated with a 24% decrease in DNA content, a measure of cell replication in this model. Since two structurally unrelated Na^+/H^+ exchange blockers were used, it is likely that these results were due to exchanger inhibition. Mitsuka et al[24] from our laboratory also reported that EIPA decreased neointima formation following balloon injury. To determine the mechanisms by which Na^+/H^+ exchange blockers inhibited neointima formation, several experiments were carried out in vitro, using cultured VSMC and a clonal VSMC cell line that was devoid of Na^+/H^+ exchange activity (RNHE(-)). We found that EIPA blocked DNA synthesis and VSMC growth and that RNHE(-) cells failed to grow in the absence of bicarbonate at extracellular pH ≤ 6.8.[24] However, RNHE(-) cells were able to grow in the presence of bicarbonate at pH ≥ 7.5. It is clear that caution is required in interpreting results of EIPA experiments, as suggested by the finding that EIPA inhibited cell migration with equal effectiveness in control cells and RNHE(-) cells. Thus, mechanisms other than inhibition of Na^+/H^+ exchange may be involved, such as inhibition of tyrosine kinases.[21-23]

In summary, it is clear that Na^+/H^+ exchange is present in blood vessels and activated by intracellular acidosis. It appears unlikely that activation of the exchanger mediates a net alkalinization in vivo. However, Na^+/H^+ exchanger function may be required for VSMC growth in response to injury. Future studies, perhaps involving transgenic mice deficient in the NHE1 isoform, will be required to define the range of physiological functions mediated by the exchanger in blood vessels.

D. REGULATION OF THE Na^+/H^+ EXCHANGER IN VSMC

The mechanisms by which the exchanger is regulated have been reviewed in several recent articles,[4,5,26,27] so this discussion will focus primarily on regulation pertinent to VSMC. VSMC in culture have pleiotropic phenotypes and respond to appropriate stimuli with migration, expression of inflammatory molecules, proliferation (in response to mitogens such as PDGF, serum, and EGF) and hypertrophy (in response to vasoconstrictors such as angiotensin II and thrombin). Thus, cultured VSMC represent a good model for studying the responses that VSMC exhibit in the vessel wall. The signals that mediate these different types of VSMC responses have not been clearly defined, although alterations in ion transport (intracellular Ca^{2+} and pH_i), phospholipases, and kinases have been implicated.[11]

D.1 ROLE OF THE Na^+/H^+ EXCHANGER IN VSMC MIGRATION AND INFLAMMATION

There is very little data regarding the role of the exchanger in VSMC migration or expression of inflammatory cytokines and antigens. Mitsuka et al[24] demonstrated that EIPA inhibited VSMC migration, suggesting a role for the exchanger in VSMC migration. However, migration of a VSMC-derived cell line, RNHE(-), devoid of the exchanger, was normal suggesting that Na^+/H^+ exchange was not required for migration. Taubman and colleagues[28] showed that induction of MCP-1 mRNA by PDGF did not require Na^+/H^+ exchange as there was no decrease in MCP-1 expression when cells were stimulated in the absence of extracellular sodium. Similar results were obtained with other early response genes such as c-fos,[29] suggesting that the ability of early response genes to be induced without Na^+/H^+ exchange is a general property of VSMC. Thus, there appears to be little role for the exchanger in VSMC migration and inflammation.

D.2 Na^+/H^+ EXCHANGE AND VSMC GROWTH

In contrast to the lack of a role for the exchanger in migration and inflammation, data from several groups indicate that there is an important relationship between Na^+/H^+ exchange and VSMC growth. We and others have shown that actively proliferating VSMC have a more alkaline pH_i than cells maintained in a growth-arrested state.[12,30,31] In addition, hypertrophic agents such as angiotensin II, endothelin, and thrombin also lead to cellular alkalinization.[12,32-37] This suggests that although hypertrophic and hyperplastic agents have different effects on VSMC growth they share common signaling mechanisms. One such mechanism is the activation of Na^+/H^+ exchange, since both hyperplastic and hypertrophic agents raise pH_i by stimulating exchanger activity. These data raise the possibility that activation of Na^+/H^+ exchange in VSMC is responsible for the cellular alkalinization and/or

Na⁺ entry that is necessary for cell proliferation or growth. To establish that Na⁺/H⁺ exchange plays a role in VSMC growth several criteria must be met, as outlined by Grinstein et al.[19] These include: (1) activation of Na⁺/H⁺ exchange by hypertrophic and hyperplastic agents, (2) VSMC proliferation induced by cytoplasmic alkalinization in the absence of mitogens, (3) demonstration that the proliferative response is dependent on extracellular Na⁺; and (4) demonstration that inhibitors of Na⁺/H⁺ exchange block cell growth. Each criterion will be considered below.

D.2.1 Activation of Na⁺/H⁺ Exchange by Hyperplastic and Hypertrophic Stimuli

Data from several laboratories indicate that VSMC mitogens increase Na⁺/H⁺ exchanger activity (Fig. 3.1). Serum, PDGF, EGF and tumor promoting phorbol esters cause a proliferative response in VSMC that is associated with an increase in pH_i and activation of Na⁺/H⁺ exchange.[38] We and others have shown that the cellular alkalinization induced by EGF and PDGF is dependent on external Na⁺ and inhibited by amiloride derivatives, indicating the involvement of Na⁺/H⁺ exchange.[12,38] Data from our laboratory demonstrated that addition of PDGF to cultured VSMC caused a rapid increase (within 10 min.) in pH_i that was sustained for at least 24 hours after treatment with the mitogen.[12] This increase in pH_i correlated well with the PDGF-induced increase in [³H]thymidine incorporation. Data from other cell types suggest that, in general, activation of the Na⁺/H⁺ exchanger by growth factors causes a shift in the sensitivity of the exchanger for intracellular pH such that it is activated at higher pH_i values.[39-41]

Fig. 3.1. (Facing page) Proposed mechanisms of regulation of Na⁺/H⁺ exchange activity in VSMC by vasoconstrictors, growth factors, and hypertension. In the basal state, the Na⁺/H⁺ exchanger (large gray circle) is phosphorylated (P) but not active and an activating regulatory factor(s) (R) is not associated. Vasoconstrictors, such as angiotensin II, stimulate phosphorylation (P) of the exchanger on different residues than those phosphorylated basally. This involves PKC-dependent, PKC-independent and Ca^{2+}-dependent pathways. The phosphorylated exchanger may undergo a conformational change that allows it to associate with a regulatory factor (R) that may also be phosphorylated. In addition, interaction with Ca^{2+}-calmodulin may remove an "autoinhibitory" domain that permits activation. Exposure to growth factors may cause phosphorylation of the exchanger on a different set of site(s) through a tyrosine kinase signaling pathway. Alternatively, the tyrosine kinase pathway may work through similar pathways as vasoconstrictors such as PLC. The regulatory factor also appears to be regulated by phosphorylation and a conformational change is likely to occur (0 to Δ). There are also important interactions with cytoskeletal components. In the hypertensive state, there appears to be an increase in phosphorylation. This may be due to phosphorylation at a novel site (P), or increased phosphorylation at sites utilized by vasoconstrictors and growth factors. The mechanism for the increased phosphorylation and activation in hypertension appears to be due to alterations in hormone-receptor coupling.*

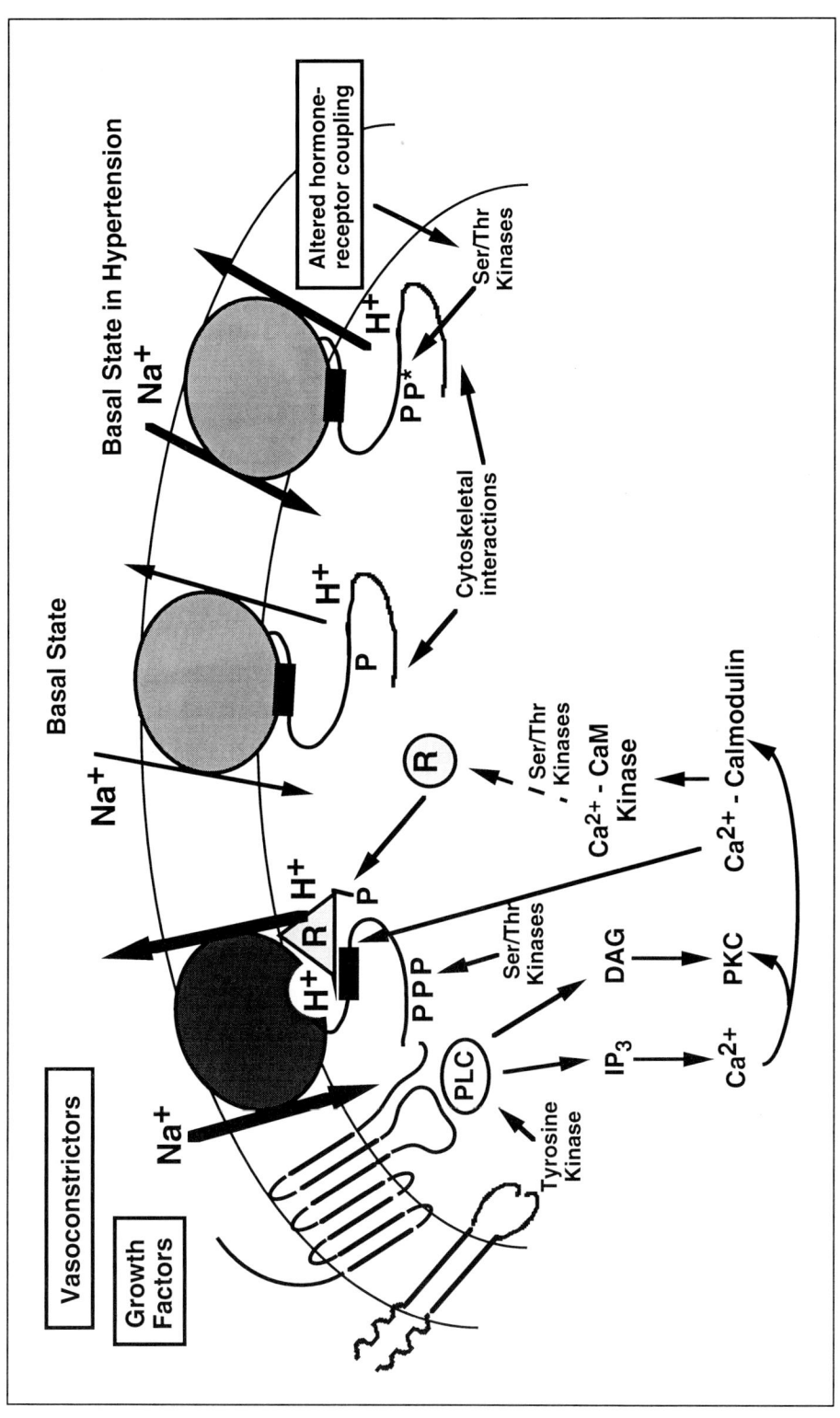

In cultured VSMC, vasoactive agents such as angiotensin II elicit a hypertrophic growth response and stimulate many of the same early signals as do growth factors. These signals include stimulation of phospholipase C with inositol trisphosphate formation and mobilization of intracellular Ca^{2+}, activation of protein kinase C, induction of c-fos mRNA, and alterations in pH_i.[34,42-45] Endothelin,[37] thrombin and angiotensin II all have been shown to regulate Na^+/H^+ exchanger activity and pH_i.[12,32-37] These agonists induce a biphasic pH_i change with a transient acidification followed by a more sustained alkalinization. The transient acidification step is independent of Na^+/H^+ exchanger activity and is thought to be associated with an increase in intracellular Ca^{2+}.[32] The later alkalinization is due to activation of the Na^+/H^+ exchanger since it can be inhibited by EIPA and is dependent on external Na^+. Thus, both hyperplastic and hypertrophic VSMC agonists rapidly stimulate Na^+/H^+ exchange, but have different long-term effects on pH_i.

D.2.2 VSMC Proliferation is Influenced by Cellular Alkalinization

Evidence that links increased pH_i to VSMC growth comes from observations that DNA synthesis in VSMC can be induced by cellular alkalinization in the absence of mitogens. Bobik et al reported that an elevation in pH_i from 7.25 to 7.35 was associated with a 30% increase in DNA synthesis, as measured by [^{3}H]thymidine incorporation.[30] In contrast, cell acidification resulted in a 40% decrease in DNA synthesis.

D.2.3 The Growth Response of VSMC is Dependent on Extracellular Na^+

Two types of studies suggest that growth responses of VSMC are influenced by extracellular Na^+. First, stimulation of VSMC protein synthesis by angiotensin II is more strongly dependent on the presence of extracellular Na^+ than extracellular Ca^{2+}.[46] Second, as shown in Table 3.1, VSMC derived from spontaneously hypertensive rats (SHR) proliferate more rapidly than those derived from normotensive Wistar-Kyoto (WKY) rats, and this enhanced rate of proliferation appears to be closely associated with an increased rate of Na^+ influx and exchanger activity.[42,47-50]

D.2.4 Inhibitors of Na^+/H^+ Exchange Block VSMC Growth

The importance of cell alkalinization in modulating cell proliferation has been demonstrated by the close correlation between the ability of amiloride derivatives to inhibit Na^+/H^+ exchange and their ability to inhibit cell proliferation. Bobik et al demonstrated that the proliferative response of primary cultures of VSMC in HCO_3^--containing media was blocked by amiloride and its derivatives: increases in both cell number and DNA synthesis were inhibited in a concentration-dependent

Table 3.1. Comparison of Na+/H+ exchanger function in hypertensive versus normotensive cells and tissues

Characteristic	Rats: SHR vs WKY	Humans: Hypertensive vs Normotensive
Activity in cells	Increased[42,47-50,83]	Increased[80-82]
Activity in tissues	Increased[87]	Not determined
NHE1 DNA sequence	Not determined	No change[89,90]
mRNA expression	No change[6,50]	Not determined
Protein expression	No change[85,86]	No change[80]
Phosphorylation state	Increased[86]	Increased[80]
Cell growth	Increased[42,50]	Increased[81,82]

manner by these inhibitors.[30] Most important was the fact that the potency for growth inhibition correlated with the potency for inhibition of Na+/H+ exchange activity (EIPA>DMA>>amiloride). The exact mechanisms by which Na+/H+ exchange influences VSMC growth are poorly understood. Bobik et al reported that EIPA had differing effects on growth factor stimulated protein synthesis; early increases (2-3 hr) were unaffected, while later increases (6-8 hr) were strongly inhibited.[30] More recently, Vairo and colleagues showed that EIPA blocked the expression of M1 and M2 ribonucleotide reductases in mouse bone marrow macrophages induced by colony-stimulating factor-1.[51] These ribonucleotide reductases are key enzymes involved in the control of S phase progression. This observation suggests that Na+/H+ exchange inhibition prevents progression of VSMC from the G1 phase of the cell cycle.

E. REGULATION OF Na+/H+ EXCHANGER FUNCTION IN VSMC

E.1 REGULATION BY mRNA AND PROTEIN EXPRESSION

NHE1 mRNA is relatively rare and can only be detected by Northern blot analysis using poly A+ purified RNA.[6,52] Several studies have shown that NHE1 mRNA is dynamically regulated in VSMC. Berk and colleagues[52] were the first to demonstrate that growth factors may regulate the level of exchanger gene expression by showing that serum, PDGF and FGF stimulated an ~15-fold increase in the steady-state levels of Na+/H+ exchanger mRNA that reached a maximum 24 hours after growth factor addition. This increase in steady-state mRNA levels was associated with only a 1.4-fold increase in V_{max}, suggesting that post-translational regulatory mechanisms such as phosphorylation, protein synthesis or protein degradation were involved in mediating the response to these mitogens. Similar increases in NHE1 mRNA were demonstrated by Lucchesi et al[6] who failed to demonstrate growth factor mediated induction of mRNAs for NHE2, NHE3 or NHE4. More recently Takewaki et al[53] observed developmentally regulated expression of NHE1 in rabbit VSMC with higher expression in embryonic and neonatal vessels than adult vessels. They also demonstrated increased NHE1 mRNA in rapidly growing, serum-stimulated VSMC in culture, and in arteries following balloon injury. In summary, there appears to be significant regulation of NHE1 mRNA in VSMC which correlates positively with cell growth state.

Few studies to date have examined the regulatory mechanisms for induction of NHE1 mRNA expression in VSMC. An important role has been suggested for PKC based on findings of Mitsuka et al[31] and Williams and Howard.[54] The strong induction of NHE1 mRNA by growth factors and phorbol esters suggests that Sp1 and AP-1 elements may be important. There are several putative regulatory sequences present in the NHE1 gene.[55] To date, the only analysis of the NHE1 promoter in VSMC was performed in the A7r5 cell line.[56] Kolyada et al[56] found that a positive regulatory element was present between -252 and +16 of the promoter. By footprinting analysis (in rat liver extracts) four protected regions were identified of which the sequences in -239 to -215 appeared most critical. This region has homology with the consensus sequence of C/EBP suggesting that the C/EBP family of proteins may be important in NHE1 gene expression.[56] Other investigators have suggested that an AP-2 binding site located at -105 to -95 may also be involved in NHE1 regulation.[57]

There is only limited information regarding the relationship between NHE1 mRNA expression and NHE1 protein expression. Rao et al[52] as well as Mitsuka et al[31] found a dissociation between changes in NHE1 mRNA levels and the V_{max} of Na$^+$/H$^+$ exchange in cultured VSMC. Similar discrepancies were also observed for HL60 cells induced to differentiate into either monocytes[58] or granulocytes.[59] As discussed in detail below, expression of NHE1 protein as measured by Western blot analysis correlates well with mRNA expression in SHR and WKY VSMC. Thus, it appears that there may be a close correlation between mRNA expression and protein expression, but not between protein expression and activity. These findings suggest that posttranslational mechanisms are important for regulation of Na$^+$/H$^+$ exchanger activity.

E.2 Regulation By Kinases and Phosphatases

A requirement for phosphorylation-dependent pathways in the activation of the exchanger is suggested by a large number of studies. Data from Sardet et al[60,61] indicate that the rapid activation of the exchanger by mitogens is associated with increases in its phosphorylation. It appears that modulation of the exchanger by phosphorylation shifts the pH range over which the exchanger's "pH$_i$ sensor" regulates ion exchange. Phosphorylation of the exchanger is associated with a shift in pH$_i$ dependence towards more alkaline pH values and an increase in maximum activity at acidic pH values.[60] More recent work with NHE1[4,62,63] has established that there are both stimulatory and inhibitory phosphorylation sites. As shown in Pouysségur's laboratory for the human NHE1, progressive deletions of the carboxyl portion of the exchanger first cause an increase in activity and then a decrease in activity. Critical serine residues essential for activity are present between amino acids 567 and 638. However, direct phosphorylation of serines present between 567 and 638 in response to growth factors is not the mechanism by which growth factors stimulate cytoplasmic alkalinization and alter pH$_i$ sensitivity.[62] In addition, the regulatory sites stimulated by growth factors appear to differ from those stimulated by protein kinase C and hyperosmolarity based on the findings of Winkel.[64] Individual replacement of the serine resides between amino acids 567 and 635 with alanine does not alter growth factor activation of the exchanger. Thus, the amino acids that are phosphorylated in response to growth factors are present between amino acids 636 and 815.[62] In summary, these findings support the concept that shifts in pH$_i$ sensitivity are mediated by a regulatory factor(s) that binds to NHE1 by interacting with the cytoplasmic tail between 567 and 635.

Many growth factor receptors that possess tyrosine kinase activity or stimulate intracellular tyrosine kinases have been shown to increase phosphorylation of the exchanger. However, the exchanger is phosphorylated only on serine after growth factor stimulation suggesting that whatever receptor type was initially activated, signal transduction events converge on a common serine/threonine kinase.[60] Thus, whether the tyrosine kinase coupled-EGF receptor, or the G protein-coupled thrombin receptor stimulated Na$^+$/H$^+$ exchange only phosphoserine was observed. The NHE1 isoform, which is the only isoform present in rat VSMC[6], has consensus sequences for phosphorylation by protein kinase C (PKC), MAP kinase and Ca^{2+}/calmodulin-dependent protein kinase.[3] In fact, Fliegel et al[65] reported that NHE1, cloned from rabbit myocardium, can be phosphorylated by Ca^{2+}/calmodulin-dependent kinase in vitro. More recently, Wakabayashi et al[63] demonstrated that a novel calmodulin binding domain was present in NHE1 at amino acids 636-656. This region normally functions as an "auto-inhibitory domain;" increases in intracellular Ca^{2+} cause binding of Ca^{2+}/calmodulin to this region and remove the negative regulation. Thus, the effect of Ca^{2+}/calmodulin appears unrelated to direct phosphorylation by Ca^{2+}/calmodulin-dependent kinase. The MAP kinase signaling pathway is an attractive mechanism for growth factor-mediated NHE1 activation because it is activated by both growth factor tyrosine kinase-coupled receptors as well as G protein-coupled receptors, such as the angiotensin II receptor.[66] Recent studies in a cell made deficient in

MAP kinase by overexpression of a dominant-negative, catalytically inactive MAP kinase, showed approximately a 50% reduction in Na+/H+ exchange activity suggesting an important role for MAP kinase.[67]

There is controversy as to whether activation of the exchanger by kinases is due to phosphorylation of the exchanger itself, or of a regulatory cofactor. Recent data suggest that regulation of exchanger activity may occur through phosphorylation of a regulatory factor that affects both the affinity of the "pH$_i$ sensor" of the Na+/H+ exchanger as well as the maximal capacity for transport.[62,63,68-70] This concept is supported by the finding that deletion of all potential phosphorylation sites COOH to 635 had no effect on pH$_i$ sensing, but caused a 50% decrease in V$_{max}$. In addition it has been shown that okadaic acid, a serine/threonine phosphatase inhibitor, activates Na+/H+ exchange,[60] and this activation is preserved when all the phosphorylation sites are deleted.[62] Similarly, phorbol ester and acidification stimulate Na+/H+ exchange in HL60 cells, yet there is no change in the phosphorylation state of the exchanger.[71] Also, ATP depletion inhibits activity of the exchanger, but there is no change in its phosphorylation.[68] A putative 44 kD regulatory factor has been identified in the proximal tubule that is regulated by protein kinase A.[72] Based on these findings it is proposed that phosphorylation of a regulatory factor(s) increases the affinity of the exchanger for intracellular H+ while direct phosphorylation of the exchanger may alter maximal transport capacity.

E.2.1 PKC-Dependent Pathways for Activation of Na+/H+ Exchange in VSMC

There is compelling evidence that several signaling pathways play a role in the activation of Na+/H+ exchange in VSMC. Data from numerous laboratories[33,36,37,47,54,73-75] indicate that both PKC-dependent and -independent pathways are involved (Fig. 3.1). Both vasoactive agents and mitogens rapidly activate PKC in VSMC. An important role for PKC in the activation of Na+/H+ exchange activity by these agents has been suggested by several findings. 1) Phorbol esters such as phorbol 12-myristate 13-acetate (PMA) and phorbol 12,13-dibutyrate (PDBU), which activate PKC without mobilizing calcium in VSMC, stimulate Na+/H+ exchange; 2) PKC inhibitors prevent agonist-mediated activation of Na+/H+ exchange; 3) PKC downregulated cells show diminished Na+/H+ exchange;[37] and 4) sphingosine, a PKC antagonist, blocks both PMA- and diacylglycerol-induced increases in pH$_i$ and Na+/H+ exchange in cultured VSMC.[76] In a similar manner, increased exchange activity observed in VSMC grown in high glucose also appears to be mediated by PKC.[54] However, PKC inhibition only partially blocks Na+/H+ exchange activation by most vasoactive agents and growth factors. For example, activation of Na+/H+ exchange by phorbol esters was significantly inhibited in actively proliferating VSMC, yet the stimulation by angiotensin II was the same as in growth arrested cells.[12,32]

PKC downregulation also partially inhibited cellular alkalinization and Na$^+$/H$^+$ exchange activation in response to thrombin[33,34,36,76] and PDGF,[76] but had no effect on the activation by serum, EGF and PDGF.[12,73]

It is not known which of the PKC isozymes are involved in Na$^+$/H$^+$ exchange regulation. Both Ca^{2+}-dependent and Ca^{2+}-independent PKC isozymes have been identified in VSMC.[77] Recent evidence suggests that Ca^{2+}-independent PKC isozymes play a role in the MAP kinase pathway.[79] Because the MAPK pathway is also activated by angiotensin II in VSMC,[66] it is possible that these Ca^{2+}-independent PKC isozymes may represent the PKC-dependent pathway for Na$^+$/H$^+$ exchange activation in VSMC.

E.2.2 Role of Protein Kinase C-Independent Pathways in VSMC

The failure of PKC down regulation and pharmacological PKC inhibition to inhibit completely activation of the Na$^+$/H$^+$ exchanger indicates the involvement of an additional signaling pathway. Many studies suggest that this PKC-independent pathway in VSMC involves a Ca^{2+}-dependent signaling pathway. For example, exposure of VSMC to the Ca^{2+} ionophore, ionomycin, transiently increased pH$_i$ and EIPA-sensitive ^{22}Na flux, which was abolished in the presence of EGTA.[63] Huang et al[36] demonstrated that activation of the Na$^+$/H$^+$ exchanger by thrombin, in PKC downregulated neonatal rat aortic VSMC, was sensitive to reductions in intracellular Ca^{2+}.[36] In VSMC from adult rats, chelation of intracellular Ca^{2+} was shown by Berk et al[33] to inhibit thrombin-induced activation of the exchanger, similar to the findings of Huang et al.[34,36] These findings suggest that activation of the Na$^+$/H$^+$ exchange in VSMC by thrombin involves a Ca^{2+} dependent pathway. Little et al[78] reported that activation of Na$^+$/H$^+$ exchanger by exposure to serum or by acid loading could be markedly attenuated by calmodulin antagonists.[78] These results suggest that activation of the Na$^+$/H$^+$ exchanger in VSMC is dependent on Ca^{2+}/calmodulin-dependent processes. This would agree well with recent findings by Wakabayashi[63] that the exchanger contains a novel binding motif for Ca^{2+}/calmodulin as discussed above.

It should be noted that the role for these kinases in activating Na$^+$/H$^+$ exchange is likely to be indirect. Specifically, it appears likely that PKC, MAP kinase and even Ca^{2+}/calmodulin-dependent kinase are "upstream" of the kinases that actually phosphorylate the exchanger. In addition, if an as yet unidentified regulatory factor requires phosphorylation to modulate exchanger activity, there may be multiple kinases required for activation of the exchanger.

F. ROLE OF VSMC Na$^+$/H$^+$ EXCHANGER IN HYPERTENSION

An important etiologic role for the exchanger has been proposed in genetic models of animal hypertension and human hypertensive patients (see ref. 8). In the following discussion potential insights into exchanger function obtained from these pathophysiologic conditions will be analyzed. Due to the activation of the Na⁺/H⁺ exchanger by both hyperplastic and hypertrophic agents, it has been proposed that abnormal function of this protein may be involved in the pathophysiology of hypertension.[8,9,26,79] Evidence for dysfunction of the Na⁺/H⁺ exchanger in hypertension is provided by observations that its activity is increased in neutrophils, lymphocytes and platelets from SHR[8] animals and in platelets and immortalized lymphoblasts from hypertensive patients.[80-82] Data from our laboratory as well as others[8,42,47-50,80,83-86] indicate that both cultured VSMC and intact mesenteric arteries from SHR[87] have a greater capacity for Na⁺/H⁺ exchange and altered kinetics. Further support for an etiologic role for altered exchanger function in genetic hypertension comes from the finding that there is no increase in Na⁺/H⁺ exchange in WKY rats made hypertensive by treatment with deoxycorticosterone acetate.[88]

Alterations in Na⁺/H⁺ exchange regulation leading to increased activity in hypertension can be theoretically divided into three categories: mutation in the gene, increased expression of the gene product and altered post-translational regulation of existing exchangers. By RFLP analysis, there is no linkage between the human Na⁺/H⁺ exchanger gene and essential hypertension.[89,90] As shown in Table 3.1 there also does not appear to be any alteration in NHE1 mRNA[6,50] or protein expression in the SHR.[85] However, there is clearly an increase in both Na⁺/H⁺ exchanger phosphorylation and cell growth in cells derived from the SHR and human hypertensive patients. Specifically, Siczkowski et al[86] showed that the phosphorylation state of the exchanger in growth arrested SHR VSMC was 2.2-fold greater than in WKY VSMC. Of interest, stimulation by serum failed to increase phosphorylation in SHR VSMC (and decreased NHE1 phosphorylation in WKY VSMC), despite stimulating Na⁺/H⁺ exchanger activity. These findings support the simple concept that increased activity of an NHE1 kinase (or decreased activity of an NHE1 phosphatase) is responsible for increased basal activity of the exchanger in SHR VSMC. However, the failure of serum to increase exchanger phosphorylation in the VSMC (as well as the discussion above regarding the role of phosphorylation in growth factor-mediated activation) suggests that alterations in a regulatory protein may be responsible for the increased exchanger activity and cell growth in response to mitogens and vasoconstrictors. Thus increased activity of the NHE1 protein appears to be a common pathogenic feature of essential hypertension.

The mechanisms leading to increased exchanger phosphorylation and activity in hypertension remain undefined. However, the findings discussed above suggest that an abnormality in the signal transduction pathway(s) leading to Na⁺/H⁺ exchanger activation is present in hypertension. This abnormality could be in the signaling molecules that couple hormone receptors to intracellular events (e.g., G-proteins, phospholipases, cytoskeletal proteins), in the kinases and phosphatases that regulate the NHE1 kinases, or in the kinases that mediate phosphorylation of the exchanger itself. Future studies utilizing both biochemical and genetic approaches should provide insight into the both regulation of the Na^+/H^+ exchanger and its role in the pathogenesis of hypertension.

ACKNOWLEDGMENTS

I wish to acknowledge the excellent work of my colleagues from my laboratory including E. Elder, M. Kusuhara, P. Lucchesi, M. Mituska, A. Muslin, G. Rao, and H. Tseng as well as my collaborators M. Canessa, J. Pouysségur and G. Vallega.

REFERENCES

1. Sardet C, Franchi A, Pouysségur J. Molecular cloning, primary structure and expression of the human growth factor-activatable Na^+/H^+ antiporter. Cell 1989; 56:271-280.
2. Levine SA, Montrose MH, Tse C-M, et al. Kinetics and regulation of three cloned mammalian Na^+/H^+ exchangers stably expressed in a fibroblast cell line. J Biol Chem 1993; 268:25527-25535.
3. Orlowski J, Kandasamy RA, Shull GE. Molecular cloning of putative members of the Na/H exchanger gene family–cDNA cloning, deduced amino acid sequence, and messenger RNA tissue expression of the rat Na/H exchanger NHE1 and 2 structurally related proteins. J Biol Chem 1992; 267:9331-9339.
4. Noel J, Pouysségur J. Hormonal regulation, pharmacology, and membrane sorting of vertebrate Na^+/H^+ exchanger isoforms. Am J Physiol 1995; 268:C283-C296.
5. Fliegel L, Frohlich O. The Na^+/H^+ exchanger: an update on structure, regulation and cardiac physiology. Biochem J 1993; 296:273-285.
6. Lucchesi PA, DeRoux N, Berk BC. Na^+/H^+ exchanger expression in vascular smooth muscle of spontaneously hypertensive and Wistar-Kyoto rats. Hypertension 1994; 24:734-738.
7. Wray S. Smooth muscle intracellular pH: measurement, regulation and function. Am J Physiol 1988; 254:C213-C225.
8. Rosskopf D, Dusing R, Siffert W. Membrane sodium-proton exchange and primary hypertension. Hypertension 1993; 21:607-617.
9. Lucchesi PA, Berk BC. Regulation of sodium-hydrogen exchange in vascular smooth muscle. Cardiovasc Res 1995; 29:172-177.
10. Owens G, Schwartz S. Alterations in vascular smooth muscle mass in the spontaneously hypertensive rat. Role of cellular hypertrophy, hyperploidy, and hyperplasia. Circ Res 1982; 51:280-289.
11. Owens GK. Control of hypertrophic versus hyperplastic growth of vascular smooth muscle cells. Am J Physiol 1989; 257:H1755-H1765.
12. Berk BC, Elder E, Mitsuka M. Hypertrophy and hyperplasia cause differing effects on vascular smooth muscle cell Na^+/H^+ exchange and intracellular pH. J Biol Chem 1990; 265:19632-19637.
13. Mulvany MJ. Resistance vessel structure in hypertension: growth or remodeling? J Cardiovasc Pharmacol 1993; 22:S44-S47.

14. Libby P. Molecular bases of the acute coronary syndromes. Circulation 1995; 91:2844-2850.
15. Aalkjaer C, Cragoe Jr EJ. Intracellular pH regulation in resting and contracting segments of rat mesenteric resistance vessels. J Physiol Lond 1988; 402:391-410.
16. Foster CD, Hill WAG, Honeyman TW, et al. Characterization of Na+/H+ exchange in segments of rat mesenteric artery. Am J Physiol 1992; 262:H1651-H1656.
17. Aickin CC. Regulation of intracellular pH in smooth muscle cells of the guinea-pig femoral artery. J Physiol Lond 1994; 479:331-340.
18. Carr P, McKinnon W, Poston L. Mechanisms of pHi control and relationships between tension and pHi in human subcutaneous small arteries. Am J Physiol 1995; 268:C580-C589.
19. Grinstein S, Rotin D, Mason J. Na+/H+ exchange and growth factor-induced cytosolic pH changes. Role in cellular proliferation. Biochim Biophys Acta 1989; 988:73-97.
20. Ganz MB, Boyarsky G, Sterzel RB, et al. Arginine vasopressin enhances pHi regulation in the presence of HCO3- by stimulating three acid-base transport systems. Nature 1989; 337:648-651.
21. Eriksson J, Lonnroth P, Wesslau C, et al. Amiloride inhibits insulin sensitivity and responsiveness in rat adipocytes through different mechanisms. Biochem Biophys Res Commun 1991; 176:1277-1284.
22. Presek P, Reuter C. Amiloride inhibits the protein tyrosine kinases associated with the cellular and the transforming src-gene products. Biochem Pharmacol 1987; 36:2821-2826.
23. Woll E, Ritter M, Offner F, et al. Effects of HOE 694—a novel inhibitor of Na+/H+ exchange—on NIH 3T3 fibroblasts expressing the RAS oncogene. Eur J Pharmacol 1993; 246:269-273.
24. Mitsuka M, Nagae M, Berk BC. Na+/H+ exchange inhibitors decrease neointimal formation after rat carotid injury. Effects on smooth muscle cell migration and proliferation. Circ Res 1993; 73:269-275.
25. Kranzhofer R, Schirmer J, Schomig A, et al. Supression of neointimal thickening and smooth muscle cell proliferation after arterial injury in the rat by inhibitors of Na+-H+ exchange. Circ Res 1993; 73:264-268.
26. Hogue D, Michalak M, Fliegel L. The role of ion antiporters in the maintenance of intracellular pH in rat vascular smooth muscle cells. Mol Cell Biochem 1991; 102:125-137.
27. Fliegel L, Dyck JR. Molecular biology of the cardiac sodium/hydrogen exchanger. Cardiovasc Res 1995; 29:155-159.
28. Taubman MB, Rollins BJ, Poon M, et al. mRNA accumulates rapidly in aortic injury and in platelet-derived growth factor-stimulated vascular smooth muscle cells. Circ Res 1992; 70:314-25.
29. Taubman MB, Berk BC, Izumo S, et al. Angiotensin II induces c-fos mRNA in aortic smooth muscle. Role of Ca^{2+} mobilization and protein kinase C activation. J Biol Chem 1989; 264:526-530.

30. Bobik A, Grooms A, Little PJ, et al. Ethylisopropylamiloride-sensitive pH control mechanisms mediate vascular smooth muscle cell growth. Am J Physiol 1991; 260:C581-C588.
31. Mitsuka M, Berk BC. Long-term regulation of Na$^+$-H$^+$ exchange in vascular smooth muscle cells - Role of protein kinase-C. Am J Physiol 1991; 260:C562-C569.
32. Berk BC, Brock TA, Gimbrone Jr MA et al. Early agonist-mediated ionic events in cultured vascular smooth muscle cells. calcium mobilization is associated with intracellular acidification. J Biol Chem 1987; 262:5065-72.
33. Berk BC, Taubman MT, Cragoe Jr EJ et al. Thrombin signal transduction mechanisms in rat vascular smooth muscle cells: Calcium and protein kinase C-dependent and -independent pathways. J Biol Chem 1990; 265:17334-17340.
34. Berk BC, Taubman MB, Griendling KK, et al. Thrombin-stimulated events in cultured vascular smooth-muscle cells. Biochem J 1991; 274:799-805.
35. Berk BC, Rao GN. Angiotensin II-induced vascular smooth muscle cell hypertrophy: PDGF A-chain mediates the increase in cell size. J Cell Physiol 1993; 154:368-380.
36. Huang CL, Cogan MG, Cragoe Jr EJ et al. Thrombin activation of the Na$^+$/H$^+$ exchanger in vascular smooth muscle cells. Evidence for a kinase C-independent pathway which is Ca^{2+}-dependent and pertussis toxin-sensitive. J Biol Chem 1987; 262:14134-14140.
37. Lonchampt MO, Pinelis S, Goulin J, et al. Proliferation and Na$^+$/H$^+$ exchange activation by endothelin in vascular smooth muscle cells. Am J Hypertens 1991; 4:776-779.
38. Little PJ, Weissberg PL, Cragoe Jr EJ et al. Sodium-hydrogen exchange in vascular smooth muscle: regulation by growth factors. J Hypertens 1987; 5:S277-S279.
39. Grinstein S, Cohen S, Goetz JD, et al. Characterization of the activation of Na$^+$/H$^+$ exchange in lymphocytes by phorbol esters: change in cytoplasmic pH dependence of the antiport. Proc Natl Acad Sci USA 1985; 82:1429-1433.
40. Paris S, Pouysségur J. Growth factors activate the Na$^+$/H$^+$ antiporter in quiescent fibroblasts by increasing its affinity for intracellular H$^+$. J Biol Chem 1984; 259:10989-10994.
41. Moolenaar WH, Tsien RY, van der Saag PT, et al. Na$^+$/H$^+$ exchange and cytoplasmic pH in the action of growth factors in human fibroblasts. Nature 1983; 304:645-648.
42. Berk BC, Vallega G, Muslin AJ, et al. Spontaneously hypertensive rat vascular smooth muscle cells in culture exhibit increased growth and Na$^+$/H$^+$ exchange. J Clin Invest 1989; 83:822-829.
43. Griendling KK, Berk BC, Socorro L, et al. Secondary signalling mechanisms in angiotensin II-stimulated vascular smooth muscle cells. Clin Exp Pharmacol Physiol 1988; 15:105-112.
44. Griendling KK, Berk BC, Alexander RW. Evidence that Na$^+$/H$^+$ exchange regulates angiotensin II-stimulated diacylglycerol accumulation in vascular smooth muscle cells. J Biol Chem 1988; 263:10620-10624.

45. Naftilan A, Pratt R, Dzau V. Induction of c-fos, c-myc and PDGF-A chain gene expressions by angiotensin II in cultured vascular smooth muscle cells. J Clin Invest 1989; 83:1419-1424.
46. Berk BC, Vekshtein V, Gordon HM, et al. Angiotensin II-stimulated protein synthesis in cultured vascular smooth muscle cells. Hypertension 1989; 13:305-314.
47. Inariba H, Kanayama Y, Takaori K, et al. Increased Na^+/H^+ exchange activity in vascular smooth muscle cells of spontaneously hypertensive rats and possible involvement of protein kinase C. Clin Exp Pharmacol Physiol 1992; 19:171-176.
48. Okada K, Ishikawa S, Saito T. Enhancement of intracellular sodium by vasopressin in spontaneously hypertensive rats. Hypertension 1993; 22:300-305.
49. Hoffmann G, Ko Y, Sachinidis A, et al. Kinetics of Na^+/H^+ exchange in vascular smooth muscle cells from WKY and SHR: effects of phorbol ester. Am J Physiol 1995; 268:C14-C20.
50. LaPointe MS, Ye M, Moe OW, et al. Na^+/H^+ antiporter (NHE1 isoform) in cultured vascular smooth muscle from the spontaneously hypertensive rat. Kidney Int 1995; 47:78-87.
51. Vairo G, Cocks BG, Cragoe Jr EJ, et al. Selective suppression of growth factor-indcued cell cycle gene expression by Na^+/H^+ antiport inhibitors. J Biol Chem 1992; 267:19043-19046.
52. Rao GN, Sardet C, Pouysségur J, et al. Differential regulation of Na^+/H^+ antiporter gene expression in vascular smooth muscle cells by hypertrophic and hyperplastic stimuli. J Biol Chem 1990; 265:19393-19396.
53. Takewaki S, Kuro OM, Hiroi Y, et al. Activation of Na^+/H^+ antiporter (NHE1) gene expression during growth, hypertrophy and proliferation of the rabbit cardiovascular system. J Mol Cell Cardiol 1995; 27:729-742.
54. Williams B, Howard RL. Glucose-induced changes in Na^+/H^+ antiport activity and gene expression in cultured vascular smooth muscle cells. Role of protein kinase C. J Clin Invest 1994; 93:2623-31.
55. Miller RT, Counillon L, Pages G, et al. Structure of the 5'-flanking regulatory region and gene for the human growth factor-activatable Na/H exchanger NHE1. J Biol Chem 1991; 266:10813-10819.
56. Kolyada AY, Lebedeva TV, Johns CA, et al. Proximal regulatory elements and nuclear activities required for transcription of the human Na^+/H^+ exchanger (NHE1) gene. Biochim Biophys Acta 1994; 1217:54-64.
57. Dyck JR, Silva NL, Fliegel L. Activation of the Na^+/H^+ exchanger gene by the transcription factor AP-2. J Biol Chem 1995; 270:1375-1381.
58. Rao GN, de Roux N, Sardet C, et al. Na^+/H^+ antiporter gene expression during monocytic differentiation of HL60 cells. J Biol Chem 1991; 266:13485-13488.
59. Rao GN, Sardet C, Pouysségur J, et al. Na^+/H^+ antiporter gene expression increases during retinoic acid-induced granulocytic differentiation of HL60 cells. J Cell Physiol 1992; 151:361-366.

60. Sardet C, Fafournoux P, Pouysségur J. a-Thrombin, epidermal growth factor, and okadaic acid activate the Na$^+$/H$^+$ exchanger, NHE1, by phosphorylating a set of common sites. J Biol Chem 1991; 266:19166-19171.
61. Sardet C, Counillon L, Franchi A, et al. Growth factors induce phosphorylation of the Na$^+$/H$^+$ antiporter, glycoprotein of 110 kD. Science 1990; 247:723-726.
62. Wakabayashi S, Bertrand B, Shigekawa M, et al. Growth factor activation and "H$^+$-sensing of the Na$^+$/H$^+$ exchanger isoform 1 (NHE1). J Biol Chem 1994; 269:5583-5588.
63. Wakabayashi S, Bertrand B, Ikeda T, et al. Mutation of calmodulin-binding site renders the Na$^+$/H$^+$ exchanger (NHE1) highly H$^+$-sensitive and Ca^{2+} regulation-defective. J Biol Chem 1994; 269:13710-13715.
64. Winkel GK, Sardet C, Pouysségur J, et al. Role of cytoplasmic domain of the Na$^+$/H$^+$ exchanger in hormonal activation. J Biol Chem 1993; 268:3396-3400.
65. Fliegel L, Walsh MP, Singh D, et al. Phosphorylation of the C-terminal domain of the Na$^+$/H$^+$ exchanger by Ca^{2+}/calmodulin-dependent protein kinase II. Biochem J 1992; 282:139-145.
66. Duff JL, Berk BC, Corson MA. Angiotensin II stimulates the pp44 and pp42 mitogen-activated protein kinases in cultured rat aortic smooth muscle cells. Biochem Biophys Res Commun 1992; 188:257-264.
67. Pages G, Lenormand P, L'Allemain G, et al. Mitogen-activated protein kinases p42mapk and p44mapk are required for fibroblast proliferation. Proc Natl Acad Sci USA 1993; 90:8319-8323.
68. Goss GG, Woodside M, Wakabayashi S, et al. ATP dependence of NHE1, the ubiquitous isoform of the Na$^+$/H$^+$ antiporter. Analysis of phosphorylation and subcellular localization. J Biol Chem 1994; 269:8741-8748.
69. Bianchini L, Pouysségur J. Molecular structure and regulation of vertebrate Na$^+$/H$^+$ exchangers. J Exp Biol 1994; 196:337-345.
70. Wakabayashi S, Fafournoux P, Sardet C, et al. The Na$^+$/H$^+$ antiporter cytoplasmic domain mediates growth factor signals and controls "H$^+$-sensing." Proc Natl Acad Sci USA 1992; 89:2424-2428.
71. Rao GN, Sardet C, Pouysségur J, et al. Phosphorylation of Na$^+$-H$^+$ antiporter is not stimulated by phorbol ester and acidification in granulocytic HL-60 cells. Am J Physiol 1993; 264:C1278-C1284.
72. Weinman EJ, Steplock D, Wang Y, et al. Characterization of a protein cofactor that mediates protein kinase A regulation of the renal brush border membrane Na$^+$/H$^+$ exchanger. J Clin Invest 1995; 95:2143-2149.
73. Berk BC, Aronow MS, Brock TA, et al. Angiotensin II-stimulated Na$^+$/H$^+$ exchange in cultured vascular smooth muscle cells. Evidence for protein kinase C-dependent and -independent pathways. J Biol Chem 1987; 262:5065-5072.
74. Livne AA, Aharonovitz O, Fridman H, et al. Modulation of Na$^+$/H$^+$ exchange and intracellular pH by protein kinase C and protein phosphatase in blood platelets. Biochim Biophys Acta 1991; 1068:161-6.

75. Ogawa A, Ishikawa Y, Sasakawa S. Na⁺/H⁺ exchange activity induced by thrombin is not inhibited by protein kinase inhibitors, staurosporine, K/252a, H-7 and sphingosine, in human platelets. Thromb Res 1993; 70:139-149.
76. Lowe J-N, Huang C-L, Ives HE. Sphingosine differentially inhibits actviation of the Na⁺/H⁺ exchanger by phorbol esters and growth factors. J Biol Chem 1990; 265:7188-7194.
77. Khalil RA, Morgan KG. PKC-mediated redistribution of mitogen-activated protein kinase C during smooth muscle cell activation. Am J Physiol 1993; 265:C406-C411.
78. Little PJ, Weissberg PL, Cragoe Jr EJ et al. Dependence of Na⁺/H⁺ antiport activation in cultured rat aortic smooth muscle on calmodulin, calcium, and ATP. Evidence for the involvement of calmodulin-dependent kinases. J Biol Chem 1988; 263:16780-16786.
79. Frelin C, Vigne P, Lazdunski M. The Na⁺/H⁺ exchanger in vascular smooth muscle cells. Acta Nephrol 1990; 19:17-30.
80. Ng LL, Sweeney FP, Siczkowski M, et al. Na⁺-H⁺ antiporter phenotype, abundance, and phosphorylation of immortalized lymphoblasts from humans with hypertension. Hypertension 1995; 25:971-977.
81. Rosskopf D, Fromter E, Siffert W. Hypertensive sodium-proton exchanger phenotype persists in immortalized lymphoblasts from essential hypertensive patients. A cell culture model for human hypertension. J Clin Invest 1993; 92:2553-2559.
82. Rosskopf D, Schroder KJ, Siffert W. Role of sodium-hydrogen exchange in the proliferation of immortalised lymphoblasts from patients with essential hypertension and normotensive subjects. Cardiovasc Res 1995; 29:254-259.
83. Kobayashi A, Nara Y, Nishio T, et al. Increased Na⁺-H⁺ exchange activity in cultured vascular smooth muscle cells from stroke-prone spontaneously hypertensive rats. J Hypertens 1990; 8:153-157.
84. Livne AA, Aharonovitz O, Paran E. Higher Na⁺-H⁺ exchange rate and more alkaline intracellular pH set-point in essential hypertension: effects of protein kinase modulation in platelets. J Hypertens 1991; 9:1013-1019.
85. Siczkowski M, Davies JE, Ng LL. Sodium-hydrogen antiporter protein in normotensive Wistar-Kyoto rats and spontaneously hypertensive rats. J Hypertens 1994; 12:775-781.
86. Siczkowski M, Davies JE, Ng LL. Na⁺/H⁺ exchanger isoform 1 phosphorylation in normal Wistar-Kyoto and spontaneously hypertensive rats. Circ Res 1995; 76:825-831.
87. Foster CD, Honeyman TW, Scheid CR. Alterations in Na⁺-H⁺ exchange in mesenteric arteries from spontaneously hypertensive rats. Am J Physiol 1992; 262:H1657-H1662.
88. Ellstrom DR, Honeyman TW, Scheid CR. Role of blood pressure in regulating Na⁺/H⁺ exchange in vascular smooth muscle. Am J Hypertens 1994; 7:340-345.

89. Lifton RP, Hunt SC, Williams RR, et al. Exclusion of the Na^+-H^+ antiporter as a candidate gene in human essential hypertension. Hypertension 1991; 17:8-14.

90. Dudley CR, Giuffra LA, Raine AE, et al. Assessing the role of APNH, a gene encoding for a human amiloride-sensitive Na^+/H^+ antiporter, on the interindividual variation in red cell Na^+/Li^+ countertransport. J Am Soc Nephrol 1991; 2:937-943.

CHAPTER 4

MOLECULAR STUDIES OF INTESTINAL EPITHELIAL Na$^+$/H$^+$ EXCHANGERS

C. H. Chris Yun, Chung-Ming Tse and Mark Donowitz

A. INTRODUCTION

The existence of multiple isoforms of mammalian Na$^+$/H$^+$ exchangers had long been predicted largely on the basis of widely different sensitivity to inhibition by amiloride and differential regulation by protein kinases depending on cell type and cellular location.[1,2] Since the cloning of NHE1 by Sardet et al in 1989,[3] at least three additional members of the mammalian gene family have been fully cloned and characterized by our group and by Orlowski and Shull.[4-8] In this chapter, we present the molecular and biochemical features of the cloned Na$^+$/H$^+$ exchanger isoforms. We focus on characteristics of epithelial Na$^+$/H$^+$ exchanger isoforms, NHE2 and NHE3, and discuss differences in the mode of protein kinase regulation compared to the ubiquitously expressed NHE1. A special attempt has been made to relate the findings on these exchangers to the biochemical aspects of intestinal Na$^+$/H$^+$ exchange.

B. MEMBERS OF THE Na$^+$/H$^+$ EXCHANGER GENE FAMILY

NHE1 is the Na$^+$/H$^+$ exchanger isoform initially cloned by Sardet et al[3] from human genomic DNA. NHE1 has been cloned from human,[3] rabbit,[4] rat,[7] pig,[9] and Chinese hamster.[10]

NHE2 and NHE3 were initially identified by our group by screening rabbit ileal villus cell cDNA libraries under differential hybridization conditions of high and low stringency.[6] The presence of NHE2 and NHE3 was subsequently demonstrated in rat[8] and recently in humans.[11,12]

The Na$^+$/H$^+$ Exchanger, edited by Larry Fliegel. © 1996 R.G. Landes Company.

The diuretic amiloride and its 5'-amino substituted analogs are potent inhibitors of the Na$^+$/H$^+$ exchangers and block the exchanger by competing with Na$^+$ for the external Na$^+$ binding site[13] although noncompetitive inhibition by amiloride and its analogs has also been reported.[14,15] Functional expression of NHE1-3 in the PS120 fibroblast cell line, which lacks all endogenous Na$^+$/H$^+$ exchange, has shown that Na$^+$/H$^+$ exchanger isoforms vary widely in their sensitivity to inhibition by amiloride and its analogs. K_i's for amiloride for both NHE1 and NHE2 are 1-3 μM. However, NHE2 is 50-fold less sensitive to ethyl-isopropyl amiloride (EIPA) than NHE1. NHE3 is the most resistant isoform being about 39 and 400-fold more resistant to amiloride and EIPA, respectively, than NHE1.

Recently, the benzoyguanidine derivative HOE694, (3-methylsulphonyl-4-piperidino-benzoyl) guanidine methanesulphonate has been shown to block Na$^+$/H$^+$ exchange activity in erythrocytes, platelets, and endothelial cells.[16,65] A study using the PS120 fibroblast cell line revealed that the sensitivity of NHEs to HOE694 decreases in the order of NHE1>NHE2>>NHE3.

Nonamiloride compounds such as cimetidine, clonidine, and harmaline also show different sensitivity to different NHEs, with NHE1 being the most sensitive exchanger. An exception is clonidine which shows a 5-fold higher sensitivity to NHE2 than NHE1.[17]

All the mammalian Na$^+$/H$^+$ exchanger isoforms are similar in the primary structure (NHE1: 815-820 amino acids depending on species; NHE2: 809-813 aa; NHE3: 831-832 aa; and NHE4: 717 aa) and the predicted secondary structure. The hydropathy profiles and sequence comparisons have led to the proposal that all the Na$^+$/H$^+$ exchangers have two structurally distinct domains: a membrane bound N-terminal domain and a long cytoplasmic C-terminal domain. Conservation of amino acid residues among the membrane associated N-terminal domains is much higher (50-60%) than that among the cytoplasmic domains (20-23%). Using the algorithms of Kyte and Doolittle,[18] and Engelman-Goldman-Steitz,[19] the N-terminal domain is predicted to contain 10-12 putative transmembrane α-helices (Fig. 4.1).[20]

Only limited information so far is available on the topology of the Na$^+$/H$^+$ exchangers. Antibody studies confirm the topology that the C-termini for NHE1, NHE2, and NHE3 are intracellular,[4,21] based on the requirement for membrane permeabilization for labeling by antibody. The putative fourth transmembrane domain M4 has been identified as a binding site for amiloride and its 5'-amino substituted analogs since mutations in this region led to changes in the sensitivity of NHE1 and NHE2 to amiloride and its analogs.[22,23] Extracellular loop a is extracellular based on N-glycosylation of NHE1.[24] The N/terminus and the first transmembrane helix are predicted to be a signal peptide. Whether this putative signal sequence is removed in the mature protein is not known. Among the putative transmembrane helices, M5A and M5B are most conserved among the Na$^+$/H$^+$ ex-

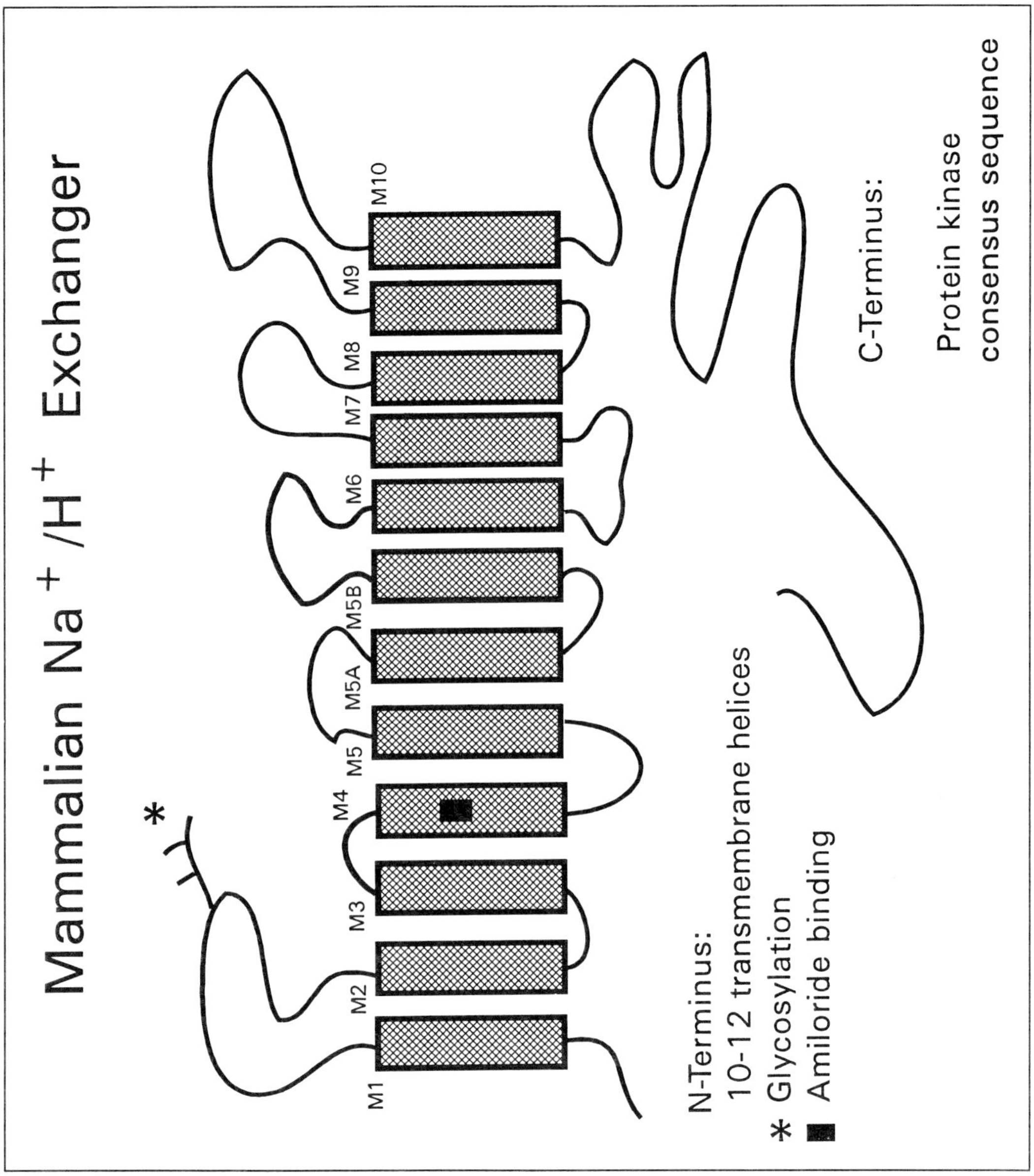

Molecular Studies of Intestinal Epithelial Na⁺/H⁺ Exchangers

Mammalian Na$^+$/H$^+$ Exchanger

M1, M2, M3, M4, M5, M5A, M5B, M6, M7, M8, M9, M10

N-Terminus:
10-12 transmembrane helices

* Glycosylation
■ Amiloride binding

C-Terminus:
Protein kinase consensus sequence

Fig. 4.1. Topological model for Na$^+$/H$^+$ exchanger isoforms. Model is derived based on hydropathy profiles and biochemical studies.

changer isoforms, suggesting that these regions may be of importance in either binding or transport of ions. Of note, replacement of a conserved amino acid E262 in putative M5B of NHE1 resulted in nonfunctional Na$^+$/H$^+$ exchanger without affecting expression of the protein.[25]

One of the characteristic features of mammalian Na$^+$/H$^+$ exchangers is the allosteric nature of the Na$^+$/H$^+$ exchange with respect to intracellular H$^+$. This allosteric behavior is attributed to the presence of an "H$^+$ modifier site" or an "H$^+$ sensor." This H$^+$ modifier site is believed to be located within the membrane bound N-terminal domain since truncation of the entire cytoplasmic C-terminus preserved the allosteric nature of the exchanger. It is anticipated that an amino acid residue involved in the modifier site undergoes protonation at physiologic pH. Of note, mutation of histidine-226 to arginine in the *E. coli* Na$^+$/H$^+$ antiporter NhaA resulted in a change in pH dependence of the antiporter.[26] On the other hand, mutation of conserved histidine residues located in the intracellular loop between M5a and M5b had no significant effect on the affinity for intracellular H$^+$.[27]

C. CELLULAR DISTRIBUTION OF NA$^+$/H$^+$ EXCHANGERS

NHE1 is ubiquitously expressed. It is found in nearly all mammalian cells including nonpolarized cells such as fibroblasts and A431 epidermoid cells. In polarized cells including intestinal and renal epithelial cells, NHE1 is localized to the basolateral membrane.[4,7]

NHE2, NHE3, and NHE4 are more restricted in message and protein distribution and are predominantly expressed in stomach, intestine, and kidney, respectively.[6,8] NHE2 message is present in large amounts in kidney, intestine, stomach, adrenal gland and much less in trachea and skeletal muscle. In rabbit kidney, expression in the medulla exceeds that in the cortex. In the gastrointestinal tract of rabbit the ascending colon has the most message followed by jejunum, ileum, duodenum, and descending colon. In rat, NHE2 mRNA is expressed at much lower levels in kidney than in intestine. By immunocytochemistry, NHE2 is localized to the brush border membrane of rat jejunum, ascending, and descending colon; human jejunum, ileum, ascending, and descending colon; and rabbit jejunum, ileum, ascending, and descending colon.[20] In kidney, NHE2 is found in the brush border membranes of proximal tubule cells (Yip and Tse, unpublished results). Interestingly, only NHE1 and NHE2 messages are found in the cultured renal inner medullary collecting duct cell lines which only express basolateral Na$^+$/H$^+$ exchange activity, implying that NHE1 and NHE2 are colocalized on the basolateral membranes in these cells.[28] Thus, it is possible that NHE2 can be expressed in either apical and basolateral membranes in certain types of epithelial cells. The function of NHE2 in intestine is not known. NHE2 is not involved in the postprandial increase in ileal water and Na$^+$ absorption which is due to brush border Na⁺/H⁺ exchange[29] (also see below). In chicken intestine, low Na⁺ diet led to an increase in colonic brush border NHE2 and Na⁺/H⁺ exchange. We therefore suggest a role of NHE2 as a back-up for Na⁺ absorption, especially in colon.

NHE3 message is found exclusively in kidney, intestine, and stomach.[5,7,12] By immunocytochemical analysis, NHE3 is found on the brush border membrane of jejunum, ileum, and ascending colon in human and rabbit, but is only present in descending colon of man, which is consistent with the absence of NHE3 message in rabbit descending colon.[30] NHE3 is also found in the brush border membrane of the proximal tubule cells and the medullar thick ascending limb cells (MTAL) (Yip and Tse, unpublished results). The location of NHE3 expression coincides with its potential function in transepithelial Na⁺ absorption where it functions as a part of neutral NaCl absorption-that is NHE3 colocalizes with neutral NaCl absorption in all GI segments.

NHE4 message is present in the order of magnitude of stomach (maximum gastric antrum) > proximal small intestine = cecum = proximal colon with much smaller amounts in the uterus, brain, kidney, and skeletal muscle.[7] Yet, there is no information on cellular immunocytochemistry of this protein, although it has been suggested to be on the basolateral membrane of renal inner medullary tubules.[31] The physiological function of NHE4 is not known, and it has been postulated to be involved in volume regulation.[32] This is based on evidence which showed that (a) when cloned NHE4 is expressed in PS120 cells, it is inactive unless stimulated by hyperosmolarity; and (b) by in situ hybridization, NHE4 is found in inner renal medullary collecting tubules where the tissue has high osmolarity.

D. REGULATION OF NA⁺/H⁺ EXCHANGERS

D.1 CLONED EXCHANGERS

D.1.1 Short-Term Regulation of Cloned Na⁺/H⁺ Exchangers by Growth Factors/Protein Kinases

NHE1, NHE2, and NHE3 have largely been characterized in PS120 and AP-1 fibroblasts, which are deficient in endogenous Na⁺/H⁺ exchange, and a human colonic carcinoma cell line, Caco-2. Despite the similarity in their structures, the Na⁺/H⁺ exchanger isoforms differ greatly in their kinetic properties and the growth factor/protein kinase elicited regulation of the Na⁺/H⁺ exchange activity. Table 4.1 summarizes the effects of serum, PMA, FGF, cAMP, cGMP, okadaic acid, a calmodulin inhibitor W13, and hyperosmolarity on Na⁺/H⁺ exchanger activities of the stably expressed NHEs. All growth factor/protein kinase regulation of NHE1 is via a change in the affinity for intracellular H⁺ ($K_i(H^+_i)$), whereas NHE2 and NHE3 are regulated by changes in V_{max}.[3] Serum, FGF, and okadaic acid stimulate NHE1, NHE2, and

Table 4.1. Effect of various agents on Na$^+$/H$^+$ exchange rate of the cloned rabbit NHEs expressed in PS120 fibroblasts

Agent	NHE1	NHE2	NHE3
FBS	↑ (K′)	↑ (V_{max})	↑ (V_{max})
FGF	↑ (K′)	↑ (V_{max})	↑ (V_{max})
α-thrombin	↑	↑ (V_{max})	↑ (V_{max})
PMA	↑(K′)	↑ (V_{max})	↓ (V_{max})
8-Br-cAMP	No effect	No effect	No effect
8-Br-cGMP	No effect	No effect	No effect
Okadaic acid	↑ (K′)	↑ (V_{max})	↑ (V_{max})
W13	No effect	↑ (V_{max})	↑ (V_{max})
Hyperosmolarity	↑	↓ (V_{max})	↓ (V_{max})

↑, up regulation; ↓, downregulation

NHE3. PMA stimulates NHE1 and NHE2, but inhibits NHE3. NHE1 is not affected by W13 whereas NHE2 and NHE3 are inhibited. On the other hand, cAMP has no effect on cloned exchangers expressed in PS120 fibroblasts and Caco-2 cells. However, cAMP stimulates NHE3 expressed in AP-1 cells.[34]

Na$^+$/H$^+$ exchange plays a role in the process of Na$^+$ uptake by some osmotically shrunken cells.[35-37] Hyperosmolarity stimulates NHE1.[37] Although the regulation by hyperosmolarity is ATP-dependent, the phosphorylation state of NHE1 was not affected during this activation.[38] It has recently been determined that hyperosmolarity inhibits NHE3 expressed in AP-1 and PS120 cells.[39,40] The inhibition of NHE3 by cell shrinkage has also been observed in cultured renal proximal tubule cells, LLC-PK1 and OK,[41] and the medullary thick ascending limb.[42] Hyperosmolarity appears to have opposite effects on NHE2 expressed in AP-1 cells[39] and PS120 fibroblasts;[40] NHE2 is stimulated in AP-1 cells (rat NHE$_2$) but is inhibited in PS120 cells (rabbit NHE). These results suggest that in addition to the isoform specific growth factor/protein kinase regulation, the host cells also play a role in regulating Na$^+$/H$^+$ exchangers possibly through different cellular signaling pathways. In this context, it has been reported that when cloned NHE1 is stably expressed in OK cells, which have an endogenous amiloride resistant Na+/H$^+$ exchanger on the apical membrane but lack any basolateral Na$^+$/H$^+$ exchanger, it is expressed partially in the basolateral membrane. This transfected NHE1 on the basolateral membrane of OK cells is inhibited by protein kinase C and is stimulated by protein kinase A, effects different from stably expressed NHE1 in PS120 and Caco-2 cells.

D.1.2 The Role of the C-Terminus of Na^+/H^+ Exchangers in Growth Factor/Protein Kinase Regulation

The role of the cytoplasmic C-terminal domain of NHE1 in growth factor/kinase regulation of NHE1 has been previously reviewed.[43] In our laboratory, we have generated a series of C-terminal truncation mutants of NHE3 to identify the role of C-terminus of NHE3 in regulation of NHE3 activity.[44] We have identified separate, nonoverlapping stimulatory and inhibitory regions within the cytoplasmic C-terminus of NHE3 (Fig. 4.2). The domain between amino acid 455 and 585 is responsible for the stimulation by FGF, OA, and serum, whereas the domain between a.a. 586 and 832 is necessary for inhibition by PMA and calmodulin. Moreover, within the stimulatory and inhibitory domains, there are discrete subdomains for each of these growth factors and protein kinases. The simulatory domain consists of separate regions for serum (a region N-terminal to aa 475), FGF (aa 475-509) and OA (aa 510-585) subdomains whereas the inhibitory domain contains separate PMA (aa 586-689) and calmodulin (aa 756-832) subdomains. Although direct binding of CaM to the region distal to a.a. 756 is yet to be demonstrated, the potential autoinhibitory role of this region is intriguing in parallel to the stimulation of NHE1 by CaM binding. All stimulatory and inhibitory effects of these growth factors/protein kinases are independent and additive and act via a V_{max} effect on the wild-type exchanger or truncation mutants. However, it is not known how the cytoplasmic tail interacts with the membrane-spanning transporter domain to cause a V_{max} change in activity.

D.1.3 Chimeric Na^+/H^+ Exchangers

The mechanisms of the differences in the kinetics and regulation of the exchanger isoforms are not known. Studies were done to determine (1) which domain, the transmembrane N-terminus or cytoplasmic C-terminus, plays the critical role in determining the V_{max} vs $K'(H^+_i)$ regulatory effect of the Na^+/H^+ isoforms, and (2) if the second messenger regulation of the exchangers can be modified by swapping the cytoplasmic C-termini, which are thought to be solely responsible for the protein kinase regulation.[45] Chimeric Na^+/H^+ exchangers were constructed by exchanging the N- and C-termini among three cloned rabbit Na^+/H^+ exchangers (NHE1-3). The chimeric exchangers were named MxCy such that M and C denote the Membrane bound N-terminus and the Cytoplasmic C-terminus, respectively, and x and y are 1, 2 or 3 designating the parent Na^+/H^+ exchanger isoforms. All the chimeric cDNAs resulted in functional Na^+/H^+ exchangers when expressed in PS120 fibroblasts with the activities comparable to the wild-type exchangers.

To determine any modification in protein kinase regulation of the chimeric exchangers as a result of swapping of the cytoplasmic

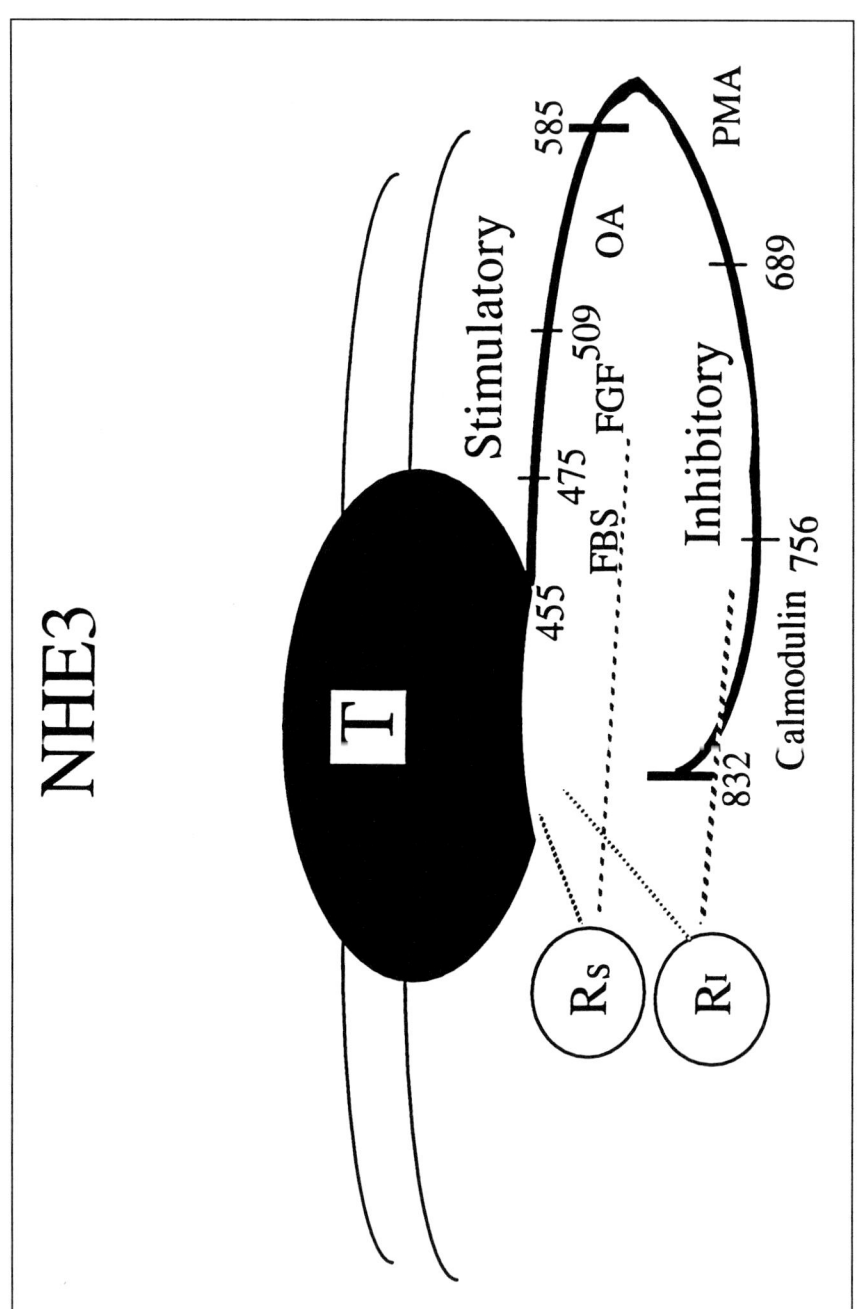

Fig. 4.2. The stimulatory and inhibitory regions of the cytoplasmic C-terminus of NHE3. The cytoplasmic tail of NHE3 contains separate sites for stimulation by FBS, FGF and OA and for inhibition by PMA and calmodulin. The effect of FBS may be mediated either through the most proximal part of the cytoplasmic tail or the membrane spanning portion of the exchanger. The stimulatory and inhibitory effects are independent and additive. The mechanism by which the cytoplasmic tail interacts with the membrane spanning N-terminal domain to cause the change in exchange activity is not known. However, the interaction may not be due to conformational changes within the NHE3 protein in response to phosphorylation of amino acid residues in the cytoplasmic tail. Alternatively, it may involve intermediates, shown here as postulated regulatory proteins R_s and R_I, which might mediate interaction between the N- and C-termini of NHE3. (From Levine et al, J Biol Chem 1995; 270:13736-13725 with permission.)

domains, effects of PMA, FGF, and fetal bovine serum (FBS) were studied. Kinetic analysis of the initial rates of the pH recovery revealed that FBS caused an increase in V_{max} for M3C1 without a significant effect on the apparent affinity for intracellular H^+ or the Hill coefficient (n_{app}) (Fig. 4.3A). FBS also stimulated M1C2 (Fig. 4.3B). However, FBS increased the affinity of intracellular H^+ without a significant effect on V_{max} or Hill coefficient. It appears that the increase in V_{max} by FBS on M3C1 originated from M3 since the parent form of M3, NHE3, displays a V_{max} increase by FBS. Similarly, the $K'(H^+_i)$ effect by FBS on M1C2 probably results from M1 since NHE1 is the only isoform which shows a $K'(H^+_i)$ effect of serum. Thus, the previously reported V_{max} vs $K'(H^+_i)$ effect of serum originates from the membrane-bound N-termini of the exchangers.

Phorbol myristate acetate is known to stimulate NHE1 by increasing the affinity for H^+, and NHE2 by increasing V_{max} (Table 4.1). In contrast, NHE3 is inhibited by PMA through a change in V_{max}, a feature also seen with the ileal brush border Na^+/H^+ exchanger. PMA did not affect M1C3, M1C2, M2C1, and M3C1, all of which consist of one domain from the housekeeping isoform NHE1 and the other domain from an epithelial isoform. On the other hand, PMA resulted in changes in the transport activities of M3C2 and M2C3. V_{max} of M3C2 was increased by PMA without any changes in $K'(H^+_i)$ and n_{app} (Fig. 4.4A). M2C3, on the other hand, was inhibited with a decrease in V_{max} (Fig. 4.4B). Of note, M3C2 and M2C3 consist of a membrane-bound N-terminus of an epithelial exchanger and a cytoplasmic C-terminus of another epithelial exchanger. The stimulation of M3C2 and inhibition of M2C3 by PMA followed the effect of PMA on NHE2 and NHE3, respectively.

The response of the chimeric exchangers to FGF was similar in that only M3C2 and M2C3 which are made up of epithelial isoforms are stimulated by FGF with an increase in V_{max}, whereas the kinetic properties of M2C1, M1C2, M1C3, and M3C1 were not affected by FGF.

The above observation suggests that the regulation of the chimeras by PMA and FGF is dictated by the C-terminus, but this requires a specific interaction between the membrane domain and the cytoplasmic domain of a Na^+/H^+ exchanger. The lack of regulation on M1C3, M1C2, M2C1, and M3C1 by PMA and FGF is due to the inability of the N- and C-termini of the chimeric proteins to interact. On the other hand, M2C3 and M3C2 are regulated by PMA and FGF because there is a common structural element between C2 and C3, and also between M2 and M3, which is involved in V_{max} regulation of these epithelial Na^+/H^+ exchangers. Similarly, the previously reported cAMP-dependent stimulation of a chimera with the N-terminus of NHE1 and the cytoplasmic C-terminal domain of βNHE is due to a specific interaction between the membrane domain of NHE1 and the cytoplasmic domain of βNHE, which allows a $K'(H^+_i)$ regulation of Na^+/H^+

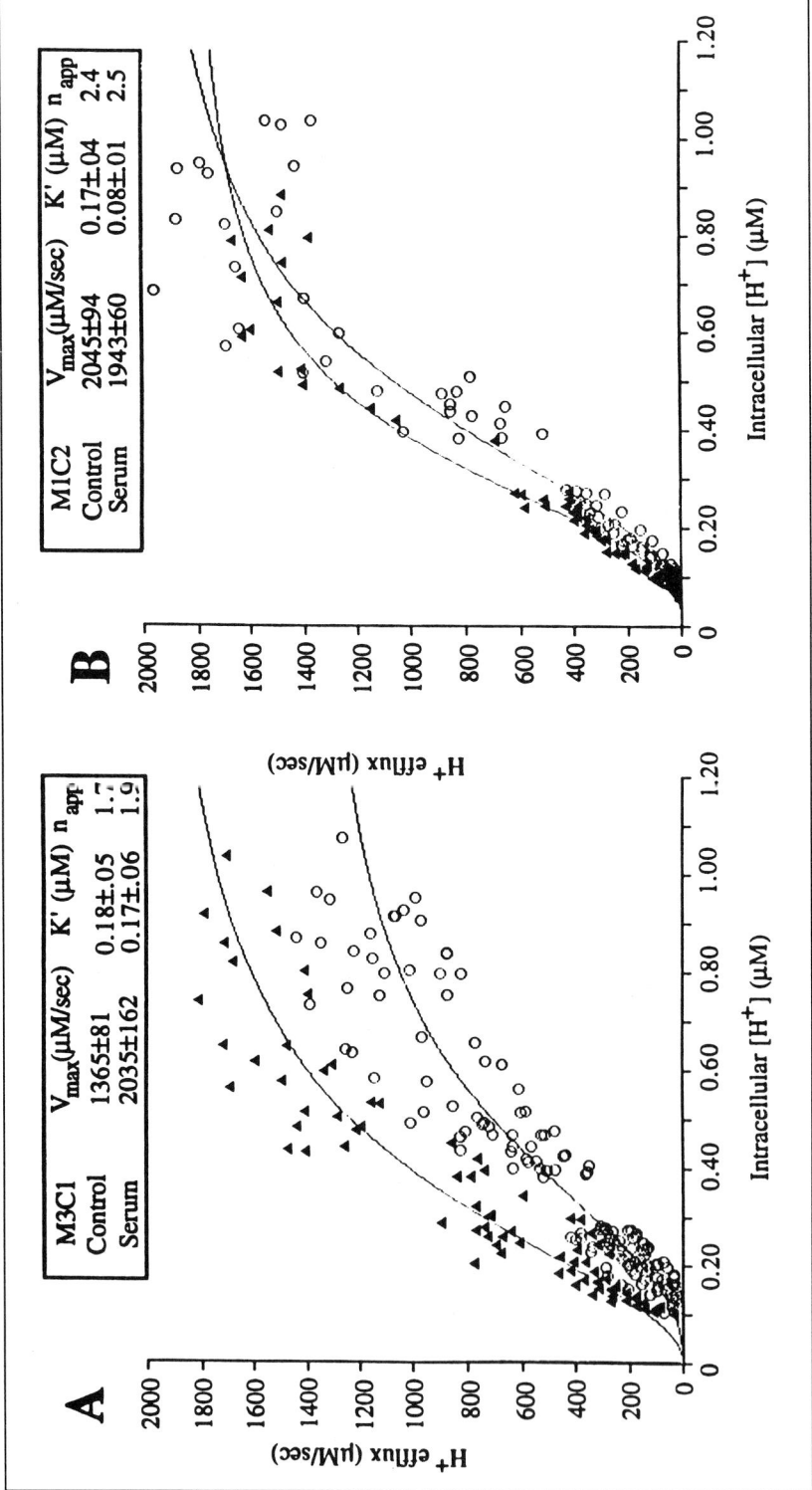

Fig. 4.3. Effect of fetal bovine serum (FBS) on Na^+/H^+ exchange activity of PS120 cells transfected with chimeric Na^+/H^+ exchangers. Control (○) cells were acidified with NH_4Cl and allowed to recover in Na^+ medium, while treated cells (▲) were similarly acidified, then perfused with Na^+ medium containing dialyzed 10% FBS. Na^+/H^+ efflux rates were calculated at various pH_i, and lines were fit to the data using an allosteric model,[33] and kinetic parameters (V_{max}, $K'([H^+])$ and n_{app}) were estimated. (A) PS120/M3C1 treated with serum showed an increase in V_{max} (49%) without any significant change in K' or Hill coefficient (n_{app}). (B) Stimulation of PS120/M1C2 cells by serum was characterized by a 47% decrease in K' without an effect on V_{max} or n_{app}. (From Yun et al, Proc Natl Acad Sci USA 1995; 92:10723-10727 with permission.)

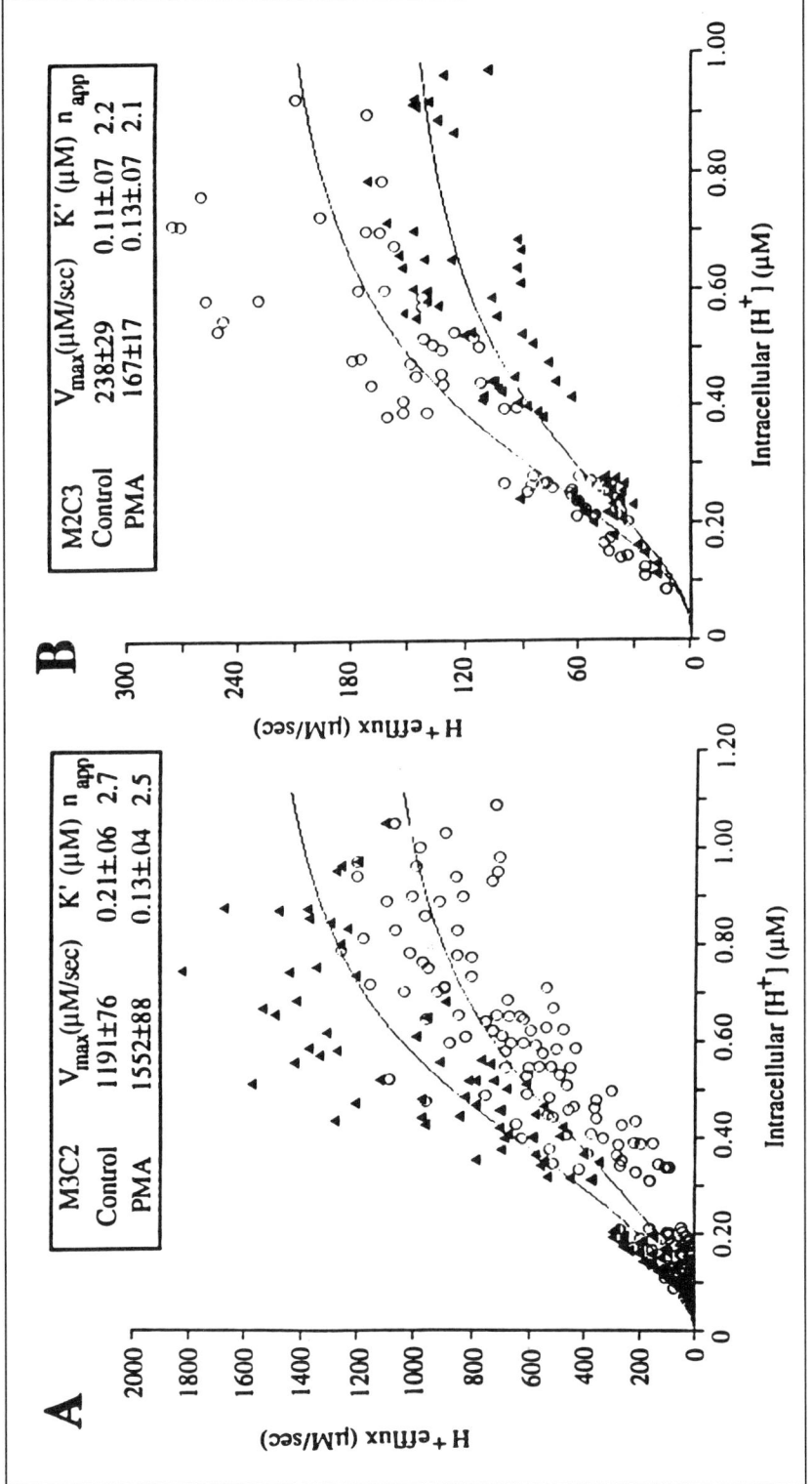

Fig. 4.4. Phorbol ester (PMA) stimulates Na^+/H^+ exchange activity in PS120/M3C2 cells. (A) Treatment of PS120/M3C2 with 1 μM PMA (s) 5 minutes before addition of Na^+ stimulated with an increase in V_{max} (30%) without significant change in K' or n_{app}. (B) In contrast, PS120/M2C3 was inhibited by PMA (s) with about 30% decrease in V_{max}. PMA did not have any effect on PS120/M1C2, M2C1, M1C3 and M3C1 (data not shown). (From Yun et al, Proc Natl Acad Sci USA 1995; 92:10723-10727 with permission.)

exchange.[46] βNHE has a high homology to the C-terminus with NHE1 and is probably an analogous form of mammalian NHE1.

We propose to functionally classify the cloned NHEs into two types based on ability of the N- and C-termini of NHEs to interact to allow protein kinase regulation (Table 4.2); NHE1 and βNHE which have housekeeper functions are called type I. Type IIs are epithelial and consist of NHE2 and NHE3 which also show limited tissue distribution. In which class NHE4 falls is not clear since it has no basal activity and little hyperosmolar stimulated activity.

This requirement for presence of specific parts of members of a transport protein gene family to allow regulation is similar to a member of the chloride channel gene family, the volume activated ClC-2 chloride channel. In this chloride channel, a specific interaction between the N- and C-termini is necessary for response by the channel to voltage and volume.[47] Determining what parts of the Na^+/H^+ exchanger N- and C-termini are involved in this interaction should provide further insights into the ways these proteins carry out basal and regulated Na^+/H^+ exchange.

It has been shown that the stimulation of NHE1 by growth factors/proteins kinases (α-thrombin, epidermal growth factor (EGF), phorbol myristate acetate (PMA), okadaic acid (OA), and fetal bovine serum (FBS)) is partially through an increase in phosphorylation.[21,48] Phosphorylation studies of NHE1 by α-thrombin, EGF, and OA have shown that NHE1 is phosphorylated only on serine residues. However, the deletion of the major phosphorylation sites (aa 636-815) reduced the growth factor-induced regulation of NHE1 only by 50% and preserved the high pH sensitivity. On the other hand, deletion of a region between aa 567-635 abolished the regulation and the high intracellular pH sensitivity of NHE1. Taken together, these data suggest the existence of a mechanism of NHE1 regulation by growth factors not involving phosphorylation of the exchanger. This led to a hypothesis that the stimulation of Na^+/H^+ exchange of NHE1 by growth factors requires "regulatory or accessory proteins" that interact with NHE1 at the region between aa 567-635.

We have recently determined by immunoprecipitation and two-dimensional phosphopeptide mapping analysis of NHE3 stably expressed

Table 4.2. Classes of cloned Na^+/H^+ exchangers based on ability to cross regulate in response to growth factor/protein kinase

	Isoforms
Type 1	NHE1, βNHE
Type 2	NHE2, NHE3

in PS120 fibroblasts that the growth factor/protein kinase-induced changes in NHE3 activity are not associated with changes in phosphorylation of NHE3 (Yip, Donowitz and Tse, submitted). The absence of change in the phosphorylation of NHE3 by protein kinases suggests that protein kinases regulate NHE3 through an indirect mechanism such as involvement of regulatory proteins as implied for NHE1. It is worth noting that a regulatory protein essential for the protein kinase A-mediated inhibition of the rabbit renal brush border membrane Na^+/H^+ exchanger has recently been identified.[49] In light of these observations, the lack of regulation in M1C2, M1C3, M2C1, and M3C1 chimeras by PMA and FGF may be due to the requirement of different sets of regulatory proteins by the epithelial (NHE2 and NHE3) and housekeeping (NHE1) isoforms. The similarity in regulatory proteins recruited by both NHE2 and NHE3 could explain the regulation of M3C2 and M2C3. Identification of specificity of the regulatory proteins may further determine the differential nature of the growth factor/protein kinase-induced regulation for the different Na^+/H^+ exchanger isoforms.

How the interaction with a regulatory protein affects the Na^+/H^+ exchange activity of NHE3 is not known. It seems plausible that the phosphorylation of the regulatory protein by protein kinase results in an interaction with a domain within the cytoplasmic C-terminus of the exchanger, which in turn interacts with either the transport site or "$H^{\pm}$sensor" located within the membrane bound N-terminus of NHE3. A potential drawback of this model is that it seems physically strenuous for the several regulatory regions identified along the full-length of the NHE3 C-terminus to converge onto a common active site upon the interaction with regulatory proteins. An alternate mode of interaction requires the C-terminus to simply act as an anchor for multiple proteins. We predict that there is more than one regulatory protein for NHE3. We hypothesize that binding of a regulatory protein to a part of the C-terminus of NHE3 brings the active site of the regulatory protein into the vicinity of the transport site, which results in modulation of the Na^+/H^+ exchange activity.

D.2 INTACT INTESTINAL TISSUE STUDIES

Intestinal Na^+/H^+ exchangers have been characterized in the most detail via vesicle Na^+ uptake experiments. Two characteristics have been emphasized: relative apical membrane amiloride resistance compared to the basolateral membrane and presence of intracellular H modifier site. Knickelbein et al[50,51] pointed out that rabbit ileal brush border Na^+/H^+ exchanger differed from basolateral Na^+/H^+ exchanger by being more resistant to inhibition by amiloride. Intestinal apical membrane K_i's for amiloride range from 30 μM to 300 μM while reported basolateral Na^+/H^+ exchanger has a K_i for amiloride of 29 μM (4-fold less than the brush border exchanger studied at the same time) (Table 4.3).

Table 4.3. Amiloride resistance of intestinal apical membrane Na+/H+ exchangers

SPECIES	Brush Border Membrane	
	ORGAN	K_i AMILORIDE
chicken	small intestine	44 μM
human	small intestine	30 μM, 99 μM, 140μM
rabbit	jejunum	230 μM, 300 μM
	Basolateral Membrane	
human	small intestine	29 μM

How do these aspects of vesicle characterization of intestinal brush border Na+/H+ exchange correlate with characteristics of NHE2 and NHE3? The amiloride sensitivity of intestinal brush border Na+/H+ exchange is characteristic of NHE3 and not NHE2. As shown earlier NHE3 and NHE2 expressed in PS120 fibroblasts have K_i's of 39 μM and 1 μM, respectively. Moreover, the amiloride inhibitory kinetics were not complex as would be predicted if both NHE2 and NHE3 were contributing to brush border Na+/H+ exchange.

An allosteric relationship between intracellular H+ ion concentration and rate of Na+/H+ exchange has been demonstrated for the human small intestinal basolateral membrane Na+/H+ exchanger and brush border exchangers, establishing the presence of an intracellular modifier site. In the small intestinal study (published only in abstract form), the effect of pH_i on Na+ efflux and Na+/H+ exchanger appeared more complex than the presence of a modifier site would predict. In other studies, the modifier site was not found for rat colon apical membranes and in rabbit ileal apical and basolateral membrane Na+/H+ exchanger.[50,52] It is likely these results are technical with only a limited number of intracellular pH's studied. A consistent finding, however, is that the brush border Na+/H+ exchanger is not totally off when the intravesical pH is 7.5 (extracellular pH either 7.5 or 5.5). This is what would be predicted for an intestinal transport protein involved in Na+ absorption, since Na+ absorption continues at pH 7.5.

Brush border Na+/H+ exchange along with apical membrane Cl−/HCO_3^- exchange are thought to make up neutral NaCl absorption. Neutral NaCl absorption, which is the major pathway for basal absorption and the major transport process under neurohumoral/paracrine/inflammatory regulation in the intestine, occurs in human ileum and colon, but apparently not in human jejunum; in rabbit jejunum, ileum, and proximal but not distal colon; and in rat small intestine and proximal and distal colon. NHE3 colocalizes with neutral NaCl absorption consistent with a role for NHE3 in this process. In contrast, NHE2 is present in rabbit descending colon brush border in spite of the absence of neutral NaCl absorption in this segment.

The role of the brush border Na^+/H^+ exchanger in neutral NaCl absorption has been evaluated in intact tissue. In vivo studies of Thiry-Vella loops in dogs have demonstrated that basal water and Na^+ absorption in jejunum and ileum is at least partially due to a brush border Na^+/H^+ exchanger (Na^+/H^+ exchanger inhibited by luminal amiloride) while the meal stimulated increase in absorption in jejunum is not mediated by Na^+/H^+ exchange.[29] In contrast, in ileum the postprandial increase in water/Na^+ absorption is at least partially dependent on apical Na^+/H^+ exchange. The percent of the postprandial increase in ileal Na^+ absorption attributable to Na^+/H^+ exchange appears to depend on the concentration of luminal substrate to stimulate apical linked Na^+-substrate uptake mechanisms, providing a mechanism in addition to Na^+/H^+ exchange. However, these substrates are probably not present normally in ileum in significant concentrations. Based on sensitivity to amiloride and dimethylamiloride, NHE3 but not NHE2 or NHE1 contributes to basal and the postprandial increase in canine ileal water/Na^+ absorption (Maher, Donowitz and Yeo, unpublished results). A role for NHE2 in normal basal or postprandial ileal water/Na^+ absorption has not been established, while NHE3 contributes to both basal and the postprandial increase in water/Na^+ absorption.

Neurohumoral/paracrine extracellular regulators acutely regulate the rate of neutral NaCl absorption while over a longer period of time glucocorticoids increase neutral NaCl absorption and brush border Na^+/H^+ exchange in small intestine and colon while hyperaldosteronism inhibits rat distal colon functional apical membrane Na^+/H^+ exchange and induces apical membrane Na^+ channel activity.[53-55] Details of agents which up- and downregulate neutral NaCl absorption are provided in a recent review.[56] We have studied the effect of glucocorticoids on ileal brush border Na^+/H^+ exchange. Treatment of rabbits with methylprednisolone for 24 and 72 hrs increased ileal brush border Na^+/H^+ exchange ~100%, whereas aminogluthetimide treatment, which blocks glucocorticoids led to a ~50% decrease in Na^+/H^+ exchange.[57] Quantitation of message of NHE1, NHE2, and NHE3 showed that methylprednisolone stimulated the NHE3 mRNA level by 4- to 6-fold without any changes in the mRNA level of NHE1 and NHE2. These studies, in addition to the amiloride sensitivity, also suggest that NHE3 is the Na^+/H^+ exchanger involved in ileal NaCl absorption and in glucocorticoid-stimulated brush border Na^+/H^+ exchange.

Brush border Na^+/H^+ exchange in intact systems is acutely regulated by protein kinases and growth factors.[58-64] As shown in Table 4.4, the regulation of brush border Na^+/H^+ exchange by growth factors/protein kinases using either brush border vesicles or intact tissues has been duplicated by regulation of cloned NHEs in PS120 fibroblasts and, while less well characterized, in apical membrane of Caco-2 cells. In

Table 4.4. Regulation of NaCl absorption and Na⁺/H⁺ exchange

	Rabbit ileum intact tissue (NaCl absorption)	Rabbit ileal brush border vesicles (Na⁺/H⁺ exchange)	Caco-2/NHE3 brush border (Na⁺/H⁺ exchange)	PS120/NHE3 (Na⁺/H⁺ exchange)	AP-1/NHE3 (Na⁺/H⁺ exchange)
cAMP	↓	O	O	O	↓
Ca²⁺ basal	↓	↓	↓	↓	ND
C kinase	↓	↓	↓	↓	↓
EGF	↑	↑	↑	↑	ND

↑, upregulation; ↓, downregulation; O, no effect; ND, no data.

both PS120 cells and Caco-2/NHE3 cells, elevating Ca^{2+} alone by Ca^{2+} ionophores or thapsigargin does not affect NHE3 and NHE2. However, activating protein kinase C inhibits NHE3 and stimulates NHE2 by V_{max} mechanisms. The inhibition of NHE3 by protein kinase C is similar to intact tissue and vesicles where activating C kinase inhibits brush border Na⁺/H⁺ exchange. Therefore, NHE3 and not NHE2 is consistent with the characteristic protein kinase C inhibition of brush border Na⁺/H⁺ exchange. Of note, ionophore A23187 does inhibit intact tissue brush border Na⁺/H⁺ exchange, but this is associated with activation of brush border C kinase in rabbit ileum. Furthermore, at basal Ca^{2+} levels calmodulin inhibits ileal apical membrane Na⁺/H⁺ exchange and similarly regulates NHE3 and NHE2 in PS120 cells and NHE3 at the apical membrane of Caco-2 cells. While this regulation by calmodulin in rabbit ileal Na⁺ absorbing cells was initially thought to be due to their high basal free Ca^{2+} level (approximately 450 nM), the similar regulation of NHE3 in PS120 cells in which basal Ca^{2+} is approximately 50 nM, indicates that the basal calmodulin inhibition of NHE3 is a property of the Na⁺/H⁺ exchanger and not due to the free Ca^{2+} level.

The most striking discrepancy between the various studies is cAMP regulation. In intact intestine tissue, including rabbit ileum, elevation of cAMP inhibits neutral NaCl absorption. It has not been established whether the effect is exerted primarily on the Na⁺/H⁺ or the Cl^-/HCO_3^- exchanger. In brush border vesicle studies with rabbit ileum we have been unable to show inhibition of Na⁺/H⁺ exchange by cAMP under experimental conditions that we could show inhibition by calmodulin and protein kinase C and stimulation by EGF. Elevation in cAMP also failed to affect NHE2 or NHE3 in PS120 cells and on the apical membrane of Caco-2 cells, seemingly supporting lack of direct regulation of brush border Na⁺/H⁺ exchange by cAMP.[33] This conflicting observation is striking but must be interpreted with the realization that not only the exchanger isoform but the cell in which the exchanger is expressed contributes to the regulation. In the AP-1 fibroblasts, elevating cAMP inhibited NHE3 and stimulated NHE1 and NHE2.[66]

Concerning stimulation of brush border Na+/H+ exchange, EGF acting by basolateral membrane receptors stimulates rabbit ileal brush border Na^+/H^+ exchange as well as NHE3 expressed on the apical membrane of Caco-2 cells by a mechanism that involves a brush border tyrosine kinase based on sidedness of effects of the tyrosine kinase inhibitor, genistein.[63] Genistein inhibits the effect of basolateral EGF and carbachol when added to the apical surface of Caco-2/NHE3 cells and rabbit ileum, but only blocks the effect of EGF and not that of carbachol from the basolateral surface. That is, genistein is effective from the basolateral surface when a basolateral tyrosine kinase is activated in signal transduction (EGF and not carbachol).

REFERENCES

1. Clark JD, Limbird LE. Na⁺-H⁺ exchanger subtypes: a predictive review. Am J Physiol 1991; 261:C953-C963.
2. Tse C-M, Levine S, Yun C, et al. Structure/function studies of the epithelial isoforms of the mammalian Na/H exchanger gene family. J Membr Biol 1993; 135:93-108.
3. Sardet C, Franchi A, Pouysségur J. Molecular cloning, primary structure and expression of the human growth factor-activatable Na/H antiporter. Cell 1989; 56:271-280.
4. Tse C-M, Ma AI, Yang VW, et al. Molecular cloning of cDNA encoding the rabbit ileal villus cell basolateral membrane Na/H exchanger. EMBO J 1991; 10:1957-1967.
5. Tse C-M, Brant SR, Walker SM, et al. Cloning and sequencing of a rabbit cDNA encoding an intestinal and kidney-specific Na/H exchanger isoform NHE-3). J Biol Chem 1992; 267:9340-9346.
6. Tse C-M, Levine SA, Yun CHC, et al. Cloning and expression of a rabbit cDNA encoding a serum-activated ethylisopropylamiloride-resistant epithelial Na⁺/H⁺ exchanger isoform (NHE-2). J Biol Chem 1993; 268:11917-11924.
7. Orlowski J, Kandasamy RA, Shull GE. Molecular cloning of putative members of the Na/H exchanger gene family. J Biol Chem 1992; 267:9331-9339.
8. Wang Z, Orlowski J, Shull G. Primary structure and functional expresseion of a novel gastrointestinal isoform of the rat Na/H exchanger. J Biol Chem 1993; 268:11925-11928.
9. Reilly RF, Hildebradt F, Biemesderfer D, et al. cDNA cloning and immunolocalization of a Na/H exchanger in LLC-PK₁ renal epithelial cells. Am J Physiol 1991; 261:F1088-F1094.
10. Counillon L, Pouysségur J. Nucleotide sequence of the Chinese hamster Na/H exchanger NHE1. Biochim Biophys Acta 1993; 1172:343-345.

11. Rao DD, Dahdal RY, Layden TJ, et al. cDNA cloning of Na/H exchanger isoforms (NHE2 and NHE3) from human stomach and intestine. Gastroenterology 1994; 106.
12. Brant SR, Yun CHC, Donowitz M, et al. Cloning, tissue distribution and functional analysis of a human Na/H exchanger isoform, NHE3. Am J Physiol 1995; 269:C198-C206.
13. Benos DJ. Amiloride: chemistry, kinetics, and structure-activity relationships. In: Grinstein S, ed. Na/H exchange. CRC, 1988;121-136.
14. Ives HE, Yee VJ, Warnock DG. Mixed type inhibition of the renal Na/H antiporter by Li^+ and amiloride. J Biol Chem 1983; 258:9710-9716.
15. Astarie C, David-Dufilho M, Devynck MA. Direct characterization of the Na/H exchanger in human platelets. FEBS Lett 1990; 277:235-238.
16. Scholz W, Albus U, Lang HJ, et al. Hoe 694, a new Na/H exchange inhibitor and its effects in cardiac ischaemia. Br J Pharmacol 1993; 109:562-568.
17. Yu FH, Shull G, Orlowski J. Functional properties of the rat Na/H exchanger NHE2 isoform expressed in Na/H exchanger-deficient Chinese hamster ovary cells. J Biol Chem 1993; 268:25536-25541.
18. Kyte J, Doolittle RF. A simple method for displaying the hydrophobic character of a protein. J Mol Biol 1982; 157:105-132.
19. Engelman DM, Goldman A, Steitz TA. The identification of helical segments in the polypeptide chain of Bacterorhodopsin. Methods Enzymol 1982; 88:81-89.
20. Yun CHC, Tse C-M, Nath SK, et al. Mammalian Na^+/H^+ exchanger gene family:structure and function studies. Am J Physiol 1995; 269:G1-G11.
21. Sardet C, Counillion L, Franchi A, et al. Growth factors induce phosphorylation of the Na/H antiporter, a glycoprotein of 110 kD. Science 1990; 218:1219-1221.
22. Counillon L, Franchi A, Pouysségur J. A point mutation of the Na/H exchanger gene (NHE1) and amplification of the mutated allele confer amiloride resistance upon chronic acidosis. Proc Natl Acad Sci USA 1993; 90:4508-4512.
23. Yun CHC, Little PL, Nath SK, et al. Leu143 in the putative fourth membrane spanning domain is critical for amiloride inhibition of an epithelial Na^+/H^+ exchanger isoform (NHE-2). Biochem Biophys Res Comm 1993; 193:532-539.
24. Counillon L, Pouysségur J, Reithmeier RAF. The Na/H exchanger NHE1 possesses N- and O-linked glycosylation restricted to the first N-terminal extracellular domain. Biochem 1994; 33:10463-10469.
25. Fafournoux P, Noel J, Pouysségur J. Evidence that Na/H exchanger isoform NHE1 and NHE3 exist as stable dimers in membranes with a high degree of specificity for monomers. J Biol Chem 1994; 269:2589-2596.
26. Gerchman Y, Olami Y, Rimon A, et al. Histidine-226 is part of the pH sensor of NhaA, a Na/H antiporter in *Escherichia coli*. Proc Natl Acad Sci USA 1993; 90:1212-1216.
27. Noel J, Pouysségur J. Hormonal regulation, pharmacology, and membrane sorting of vertebrate Na/H exchanger isoforms. Am J Physiol 1995; 268:C283-C296.

28. Soleimani M, Singh G, Bizal GL, et al. Na/H exchanger isoforms NHE2 and NHE1 in inner medulary collecting duct cells. J Biol Chem 1994; 269:27973-27978.
29. Yeo CJ, Barry M, Gontarek JD, et al. Na/H exchange mediates meal-stimulated ileal absorption. Surgery 1994; 116:388-395.
30. Hoogerwerf S, Levine S, Montgomery JM, et al. NHE2 and NHE3 are intestinal brush border proteins: immunolocalization of NHE2 and NHE3 in human and rabbit intestine. Am J Physiol 1996; In press
31. Depaloi AM, Chang EB, Bookstein C, et al. In situ hybridization of mRNA and immunolocalization of Na^+/H^+ exchanger isoform NHE-3 and NHE-4 in rat kidney. J Am Soc Nephrol 1993; 4:836 (abstr.).
32. Bookstein C, Depaoli M, Xie Y, et al. A unique sodium-hydrogen exchange isoform (NHE4) of the inner medullary of the rat kidney in induced by hyperosmolarity. J Biol Chem 1994; 269:29704-29709.
33. Levine SA, Montrose MH, Tse C-M, et al. Kinetics and regulation of three cloned mammalian Na/H exchangers stably expressed in a fibroblast cell line. J Biol Chem 1993; 268:25527-25535.
34. Moe OW, Amemiya M, Yamaji Y. Activation of protein kinase A acutely inhibits and phosphonylate Na1H exchanger NHE-3. J Clin Invest 1995; 96:2187-2194
35. Grinstein S, Cohen S, Goetz JD, et al. Osmotic and phorbol ester-induced activation of Na/H exchange: possible role of protein phosphorylation in lymphocyte volume regulation. J Cell Biol 1985; 101:269-276.
36. Grinstein S, Cohen S, Goetz JD, et al. Activation of the Na/H antiport by changes in cell volume amd by phorbol ester; possible role of protein kinase. Curr Topics Membr Transport 1986; 26:115-136.
37. Grinstein S, Woodside M, Sardet C, et al. Activation of the Na/H antiporter during cell regulation. J Biol Chem 1992; 267:23823-23828.
38. Goss GG, Woodside M, Wakabayashi S, et al. ATP dependence of NHE1, the ubiquitous isoform of the Na/H antiporter. J Biol Chem 1994; 269:8741-8748.
39. Kapus A, Grinstein S, Wasan S, et al. Functional charaterization of three isoforms of the Na/H exchanger stably expressed in Chinese hamster ovary cells. J Biol Chem 1994; 269:23544-23552.
40. Nath SK, Hang CYH, Levine SA, et al. Hyperosmolarity inhibits the cloned epithelial specific Na/H exchange isoform, NHE2 and NHE3: an effect opposite to that of the house keeping isoform NHE1. Am J Physiol 1996; in press.
41. Soleimani M, Bookstein C, McAteer JA, et al. Effect of high osmolarity on Na/H exchange in renal proximal tubule cells. J Biol Chem 1994; 269:15613-15618.
42. Watts BA III, Good DW. Apical membrane Na/H exchange in rat medullary thick ascending limb. J Biol Chem 1994; 269:20250-20255.
43. Wakabayashi S, Sardet C, Fafournoux P, et al. Structure-function of the growth factor-activatable Na/H exchange (NHE-1). Rev Physiol Biochem Pharmacol 1992; 119:157-186.

44. Levine SA, Nath S, Yun CHC, et al. Separate C-terminal domains of the epithelial specific brush border Na$^+$/H$^+$ exchanger isoform NHE3 are involved in stimulation and inhibition by protein kinases/growth factors. J Biol Chem 1995; 270:13716-13725.
45. Yun CHC, Tse C-M, Donowitz M. Chimeric Na/H exchanger: an epithelial membrane bound N-terminal domain requires an epithelial cytoplasmic C-terminal domain for regulation by protein kinases. Proc Natl Acad Sci USA 1995; 92:10723-10727.
46. Borgese F, Malapert M, Fievet B, et al. The cytoplasmic domain of the Na/H exchangers (NHEs) dictates the nature of the hormonal response: behavior of a chimeric human NHE1/trout BNHE antiporter. Proc Natl Acad Sci USA 1994; 91:5431-5435.
47. Grunder S, Thiemann A, Pusch M, et al. Regions involved in the opening of ClC-2 chloride channel by voltage and cell volume. Nature 1992; 360:759-762.
48. Sardet C, Fafournoux P, Pouysségur J. α-thrombin, epidermal growth factor, and okadaic acid activate the Na/H exchanger, NHE-1, by phosphorylating a set of common sites. J Biol Chem 1991; 266:19166-19171.
49. Weinman EJ, Steplock D, Shenolikar S. cAMP-mediated inhibition of the renal brush border membrane Na/H exchanger requires a dissociable phosphoprotein cofactor. J Clin Invest 1993; 92:1781-1786.
50. Knickelbein RG, Aronson PS, Dobbins JW. Properties of distinct luminal and basolateral Na/H exchangers in rabbit ileum. FASEB J 1988; 2:A941.
51. Knickelbein RG, Aronson PS, Dobbins JW. Membrane distribution of sodium-hydrogen and chloride-bicarbonate exchangers in crypt and villus cell membrane from rabbit ileum. J Clin Invest 1988; 82:2158-2163.
52. Rajendra VM, Binder HJ. Characterization of Na-H exchange in apical membrane vesicles of rat colon. J Biol Chem 1990; 265:8408-8414.
53. Donowitz M, Walsh MJ. Regulation of mammalian small intestinal electrolyte transport. In: Johnson LR, ed. Physiology of the Gastrointestinal Tract. Raven, 1987;1351-1388.
54. Powell DW. Immunophysiology of intestinal electrolyte transport. In: Handbook of Physiology. The Gastrointestinal System, Intestinal Absorption and Secretion. Sect. 6. Am Physiol Soc 1991:591-641.
55. Sellin JH, Field M. Physiologic and pharmacologic effects of glucocorticoids on ion transport across rabbit ileal mucosa in vitro. J Clin Invest 1981; 67:770-778.
56. Chang EB, Rao M. Intestinal water and electrolyte transport: Mechanisms of physiological and adaptive responses. In: Johnson LR, ed. Physiology of the Gastrointestinal Tract. 3. Raven, 1994:2027-2081.
57. Yun CHC, Gurubhagavatula S, Levine SA, et al. Glucocorticoid stimulation of ileal Na$^+$ absorptive cell brush border Na/H exchange and association with an increase in message for NHE-3, an epithelial Na/H exchanger isoform. J Biol Chem 1993; 268:206-211.

58. Rood RP, Emmer E, Wesolek, et al. Regulation of the rabbit ileal brush-border Na/H exchange by an ATP-requiring Ca^{2+}/calmodulin-mediated process. J Clin Invest 1988; 82:1091-1097.
59. Cohen ME, Reinlib L, Watson AJM, et al. Rabbit ileal villus cell brush border Na/H exchange is regulated by Ca^{2+}/calmodulin-dependent protein kinase II, a brush border membrane protein. Proc Natl Acad Sci USA 1990; 87:8990-8994.
60. Emmer E, Rood RP, Wesolek KH, et al. Role of calcium and calmodulin in the regulation of the rabbit ileal brush border membrane Na/H antiporter. J Membr Biol 1989; 108:207-215.
61. Rood RR, Donowitz M. Regulation of small intestinal Na^+ absorption by protein kinase: implications for therapy of diarrheal diseases. Viewpoints on Digestive Diseases 1990; 22:1-5.
62. Donowitz M, Cohen ME, Gould M, et al. Elevated intracellular Ca^{2+} acts through protein kinase C to regulate ileal NaCl absorption. Evidence for sequential control by Ca^{2+}/calmodulin and protein kinase C. J Clin Invest 1989; 83:1953-1962.
63. Donowitz M, Montgomery JLM, Walker MS, et al. Brush-border tyrosine phosphorylation stimulated ileal neutral NaCl absorption and brush-border Na/H exchange. Am J Physiol 1994; 266:G647-G656.
64. Opleta-Madsen K, Hardin J, Gall DG. Epidermal growth factor up regulates intestinal electrolyte and nutrient transport. Am J Physiol 1991; 260:G807-G814.
65. Cuunillon L, Scholz W, LAng HS, et al. Pharmacological characterization of stably transfected Na^+/H^+ antiporter isoforms using amiloride analog and a new inhibitor exhibitions anti-ischemic properties. Mol Pharmacol 1993; 44:1041-1045.
66. Kandasamy RA, Yu FH, Harris R, et al. Plasma membrane Na^+/H^+ exchanger isoforms (NHE-1,-2,and -3) are differentially responsive to second messenger agonists of the protein kinase A and C pathogens. J Biol Chem 1995; 270:29029-29216.

CHAPTER 5

REGULATION OF EXPRESSION OF THE Na⁺/H⁺ EXCHANGER (NHE1) PROMOTER

Larry Fliegel and Jason R.B. Dyck

A. INTRODUCTION

The Na⁺/H⁺ exchanger (NHE1) is a growth factor-activatable antiporter which is ubiquitously expressed in mammalian cells. This transmembrane glycoprotein functions mainly as a pH regulatory protein by exchanging extracellular sodium for intracellular protons. Because the exchanger is a Na⁺ dependent transporter, it also plays a significant role in the control of cell volume.[1] The level of expression of the Na⁺/H⁺ exchanger varies with a number of environmental stimuli. Studies have shown that NHE1 mRNA levels are increased due to a number of treatments including chronic acid loading and treatments causing cellular differentiation.[2-4] For example, during retinoic acid induced differentiation of human leukemic cells (HL-60), there is an 8.3-fold increase in NHE1 transcription.[2] In addition, a number of studies have shown that mRNA levels and activity of the Na⁺/H⁺ exchanger are increased in response to chronic acid treatment. In tissues such as the kidney, there are adaptive mechanisms by which cells can upregulate NHE1 in response to chronic acid load. Both the V_{max} and mRNA levels of NHE1 have been shown to increase due to chronic metabolic acidosis in intact animals' kidneys and in renal cell lines.[5-7] In primary cultures of isolated myocytes chronically low external pH affects either the regulation or the amount of NHE1 allowing for a more rapid recovery from an acute acid load.[8]

To attempt to understand the mechanisms involved in regulation of the NHE1 gene, studies have recently begun on the regulation of the human and mouse promoters. The recent cloning of these two

The Na⁺/H⁺ Exchanger, edited by Larry Fliegel. © 1996 R.G. Landes Company.

genes has contributed a great deal to the understanding of how the NHE1 gene is regulated.[9,10] Miller and coworkers[9] were the first group to isolate the upstream region of the human NHE1 gene. Their work has contributed greatly to the characterization of the human Na+/H+ exchanger gene. They have identified the intron-exon boundaries, the start sites of transcription and have provided the sequence of the 5' untranslated region along with approximately 1.3 kb of the promoter/enhancer region.[9] Following this work, another group has identified regions of the promoter that can bind nuclear proteins.[11] These proximal regulatory elements of the NHE1 promoter have provided evidence towards the trans-acting factors involved in regulating the human NHE1 gene.

More recently, we examined the mouse NHE1 gene.[10] A 1.1 kb fragment upstream of the 5'-untranslated region was isolated and characterized. This study identified specific DNA motifs characteristic of promoter and enhancer elements lying within the 1.1 kb promoter region. In addition, a transcription factor has been described which is involved in the regulation of the mouse NHE1 gene.[10,12] This transcription factor is involved in regulation of the gene during cellular differentiation of at least one cell type.

The purpose of this review is to summarize current information on the isolation and characterization of the NHE1 promoter. In addition, we compare the human, mouse and rabbit NHE1 gene. This analysis will help provide an understanding of the regions of DNA that are responsible for the control of NHE1 expression. Furthermore, predictions will be made as to the importance of specific cis-regions in the regulation of the NHE1 genes. In doing so, it is hoped that a better understanding of the NHE1 promoter can be obtained, and future directions of study will emerge. The regulation of the NHE1 promoter is a field in its infancy. It is hoped that reviews such as this one will stimulate interest in this important area.

B. CLONING AND ANALYSIS OF THE NHE1 PROMOTER

The first Na+/H+ exchanger genomic clone was isolated by Miller and associates.[9] Library screening of genomic clones was used to isolate the 5' flanking regions of the human gene.[9] The human Na+/H+ exchanger clone contained 1377 bases upstream of the start site of transcription. Several putative proximal regulatory elements were identified in the human NHE1 gene. These include two CAT boxes, three GC boxes, a cyclic AMP response element and three AP-1 sites[9] (Fig. 5.1a). Of these sites, there have been only two studies that have examined some of the trans-acting factors involved in regulation of the human gene.[11,13] The first study examined the role of the transcription factor AP-1 in acid-induced increases in Na+/H+ exchanger expression in MCT cells (mouse proximal tubule cells).[13] It was shown that acid incubation

Regulation of Expression of the Na+/H+ Exchanger (NHE1) Promoter

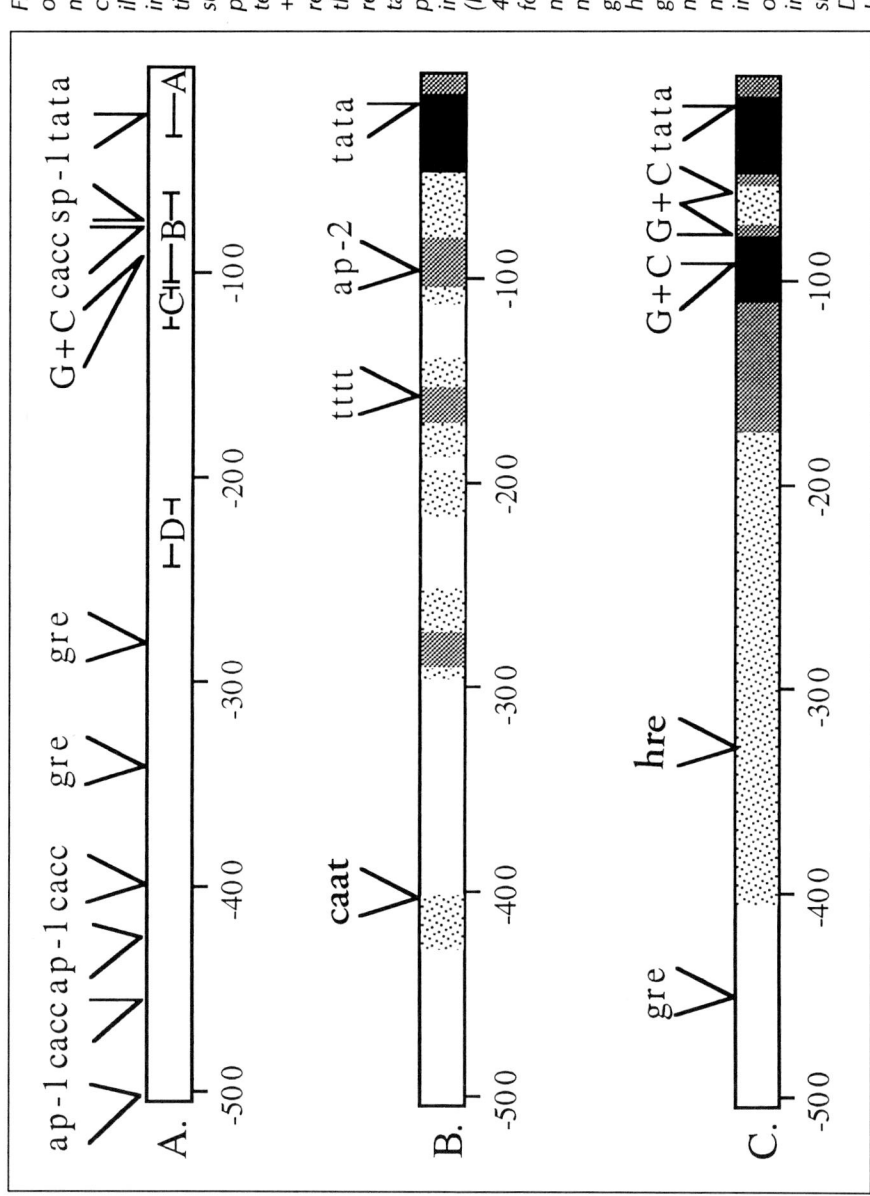

Fig. 5.1 A,B,C. Schematic diagrams of the initial 510 bp of the human,[9] mouse[10] and rabbit[11] Na+/H+ exchanger promoters. Figure 5.1 a, b, c illustrates the putative sites of binding of the transcription factors identified earlier.[10,15,18] Potential consensus sequences, ap-1, AP-1 binding protein; cacc, CACCC-binding protein; caat, CAAT binding protein; G+C (G+C box); gre, glucocorticoid response element; hre, half site for thyroid hormone, vitamin D or retinoids; sp-1, SP-1 binding protein, tata, TATA box; tttt, a conserved polyT segment with unknown binding proteins. The elements A, B, C, D (Fig. 5.1A) illustrate the positions of 4 protected regions identified using footprint experiments with the human gene.[11] Shaded elements of the mouse gene (Fig. 5.1B) indicate regions with high, medium or low homology to the corresponding region of the human gene (high, ■; medium, ▨; low, ▦). Shaded elements of the rabbit gene (Fig. 5.1C) indicate regions with high, medium or low homology to the corresponding region of the mouse gene. Analysis of homology was made using the DNA analysis program MacVector™. High, medium and low homology were defined as regions with over 95%, 85% and 65% identity respectively over 13 residues.

causes an increase in antiporter activity and in transcription factor AP-1 activity. The role of the three NHE1 AP-1 sites was examined by using a reporter plasmid constructed with six tandem AP-1 binding sites. Upon treatment with acid, there was an increase in transcription from the reporter gene. This result indicates that acid can increase transcription from genes containing tandem AP-1 sites. However, it was not demonstrated that the NHE1 promoter itself is activated by acid treatment.

Kolyada et al[11] also studied the human NHE1 gene. Footprint analysis of the gene indicated that there were four principal protected regions in the most proximal regions of the promoter. These were identified as A, -31 to -9; B, -108 to -65; C, -124 to -111; and D, -239 to -215 (Fig. 5.1a). Deletion of the promoter up to area D did not result in major reductions in activity of the promoter in HepG2, NIH 3T3 or VSM A7r5 cells. The effect was cell specific indicating that the NHE1 gene is regulated in a different manner in different cell types. In HepG2 and VSM A7r5 cells, deletion up to the D region caused stimulation of the activity of the promoter, however in NIH 3T3 cells it caused a modest reduction in activity of the promoter. Deletion or substitution of nucleotides within area D caused large decreases in activity of the promoter.[11] This region of DNA was shown to bind nuclear factor(s), and this binding was responsible for approximately 70% of transcriptional activity.[11] Bandshift analysis implicated C/EBP as the family of transcription factors that may be responsible for this regulation.[11] However further experiments with cotransfection of transcription factors are necessary to confirm the involvement of these proteins.

The mouse NHE1 gene varies from the human gene. We initially isolated the mouse NHE1 promoter by screening libraries with cDNA probes.[10] Sequencing and analysis of a 1.1 kb fragment of the promoter showed some differences and some similarities to the human gene. Similar to the human NHE1 gene, the mouse NHE1 promoter-enhancer region also possesses a number of putative regulatory elements. These include two CAT boxes, an SP-1 site, a cyclic AMP response element binding site (CREB) and an AP-2 site (Fig. 5.1b).[10] Deletion of all these sites except the AP-2 consensus sequence reduces the level of transcription activity of a reporter gene by between 30 and 70% depending on cell type.[10,12]

The AP-2 site of the mouse gene directly affects the transcriptional regulation of the mouse NHE1 gene.[10] Deletion or mutation of the site resulted in dramatic losses of activity of the mouse NHE1 promoter. Gel mobility shift analysis showed that the transcription factor AP-2 can bind to this region, and DNase I footprinting analysis showed that this region is protected by nuclear extracts from NIH 3T3 cells. In addition, transfection of an AP-2 expression plasmid into AP-2 deficient cells resulted in increased activity of the NHE1 promoter.[10]

It is of note that the AP-2 site of the mouse promoter is involved in the regulation of the Na+/H+ exchanger during differentiation of embryonal carcinoma cells.[12] On retinoic acid induced differentiation we found an early and rapid 10-fold increase in NHE1 transcription. The increases in expression preceded major changes in cell morphology. This differentiation is caused by retinoic acid similar to HL60 cellular differentiation.[2] Because it has been shown that retinoic acid induced differentiation in P19 cells also produces increased AP-2 levels, this transcription factor may be an important mechanism of regulation of the NHE1 promoter.[14]

Recently the NHE1 promoter of the rabbit gene was isolated and characterized.[15] The transcription initiation start-sites were identified at a similar location to the human gene. The sequence of the proximal 5' flanking region showed similarity to the human sequence and contained a TATA box, (G + C) boxes and a putative AP-1 site. It was shown that the rabbit promoter was active in LLC-PK1 cells. It was interesting that a 4.7 kb fragment of the promoter showed only 40% of the activity of a 1.1-kb fragment of the promoter. This suggested the presence of negative regulatory elements in the upstream region of the clone. In addition, the 1.1-kb fragment showed orientation specific activity. In the antisense orientation there was only basal luciferase activity of a reporter construct.[15]

C. COMPARISON OF THE MOUSE, HUMAN AND RABBIT NHE1 PROMOTERS

A comparison of the mouse, human and NHE1 promoters is shown in Figure 5.1. Highly conserved regions may be important in regulation of expression in these species. The more proximal regions of the genes show a stronger conservation between three species. The consensus sites for the TATA box vary by only one base pair (i.e., TATAAG human, and TATAAA in the mouse and rabbit). In both the human and the mouse NHE1 genes, there are two start sites for transcription.[9,10] In the rabbit only one principle start site was identified, but a second start site was not ruled out.[15] The exact role of the two start sites is not known. It is possible that they increase the likelihood of transcription by the provision of a back-up initiation site. The region of DNA lying between the start sites of transcription and the TATA box is also quite conserved among species. This segment of DNA contains exactly 25 base pairs in the human and the mouse NHE1 genes, and 19 of the 25 base pairs are identical. Furthermore, the sequence of DNA from the TATA box to just 5' of the AP-2 site of the mouse gene[10] is conserved in the human gene.[9] A similar high degree of conservation is shown in the rabbit gene.[15] The region of the mouse gene that binds the transcription factor AP-2 is highly conserved in location and sequence in all three species. This region was identified as a G + C region in the rabbit promoter[15] (Fig. 5.1c). Generally speaking the more distal from the start site, the less conservation between species. It appears that the proximal 200 bps of the promoter NHE1 gene are quite conserved. The human and mouse genes only begin to vary at regions 5' to the AP-2 site. The rabbit gene also begins to vary to a greater extent past this region. The conservation of the AP-2 site suggests that AP–2 may be important in regulating not only the mouse gene, but also the human and rabbit NHE1 gene. Future experiments may examine whether the transcription factor AP-2 has an important role in the human and rabbit promoter.

Regions lying more distal to the promoter are, however, less conserved. They may confer different regulatory effects on the two NHE1 genes. It is of note that many of the putative transcription factor binding sequences first identified in the human promoter are not conserved in the mouse gene. It is also of interest that both region D and region C of the human gene[11] are not well conserved in the mouse gene. We have recently noted one region of potential importance. A pyrimidine rich stretch of T's is present and is highly conserved in the human, mouse and rabbit gene (Fig. 5.1b). Deletion of this conserved region can result in partial loss of NHE1 promoter activity in at least some cell types (L. Fliegel; submitted for publication). The role of this poly T rich region still has to be elucidated, although Blaurock et al[15] have suggested that such homopyrimidine repeats could be involved in triple helical DNA formation. Future experiments may explore this possibility.

Other more distal regions of the NHE1 gene are not well conserved. There are few regions of identity with the exception of bp -980 to -880 of the mouse gene which contains regions of identity with bp -1277 to -1151 of the human gene (not shown). Experimental results have shown that the more distal regions may contain negative regulatory elements.[15,16] Future experiments are necessary to determine which part of these regions serves this function in the NHE1 promoter.

D. SUMMARY AND FUTURE DIRECTIONS

Clearly there is still a great deal to be learned in the area of regulation of NHE1 expression. Figure 5.2 illustrates a summary of some of the above mentioned studies. Many different pathways of regulation are as yet unclear. One area still open to investigation is the precise regulation of the gene during acidosis induced increases in NHE1 mRNA. Although the transcription factor AP-1 may have a role to play there,[13] it has not yet been shown that it can directly activate the NHE1 gene during stimulation by acidosis. In fact, in recent studies,[16] we have shown that acidosis does not activate a reporter construct containing the proximal 1.1 kb of the mouse promoter in either MCT cells or isolated myocytes. Though protein kinase C clearly plays a role in this mechanism,[13] future studies are necessary to elucidate the exact mechanism of action of acidosis on the NHE1 promoter.

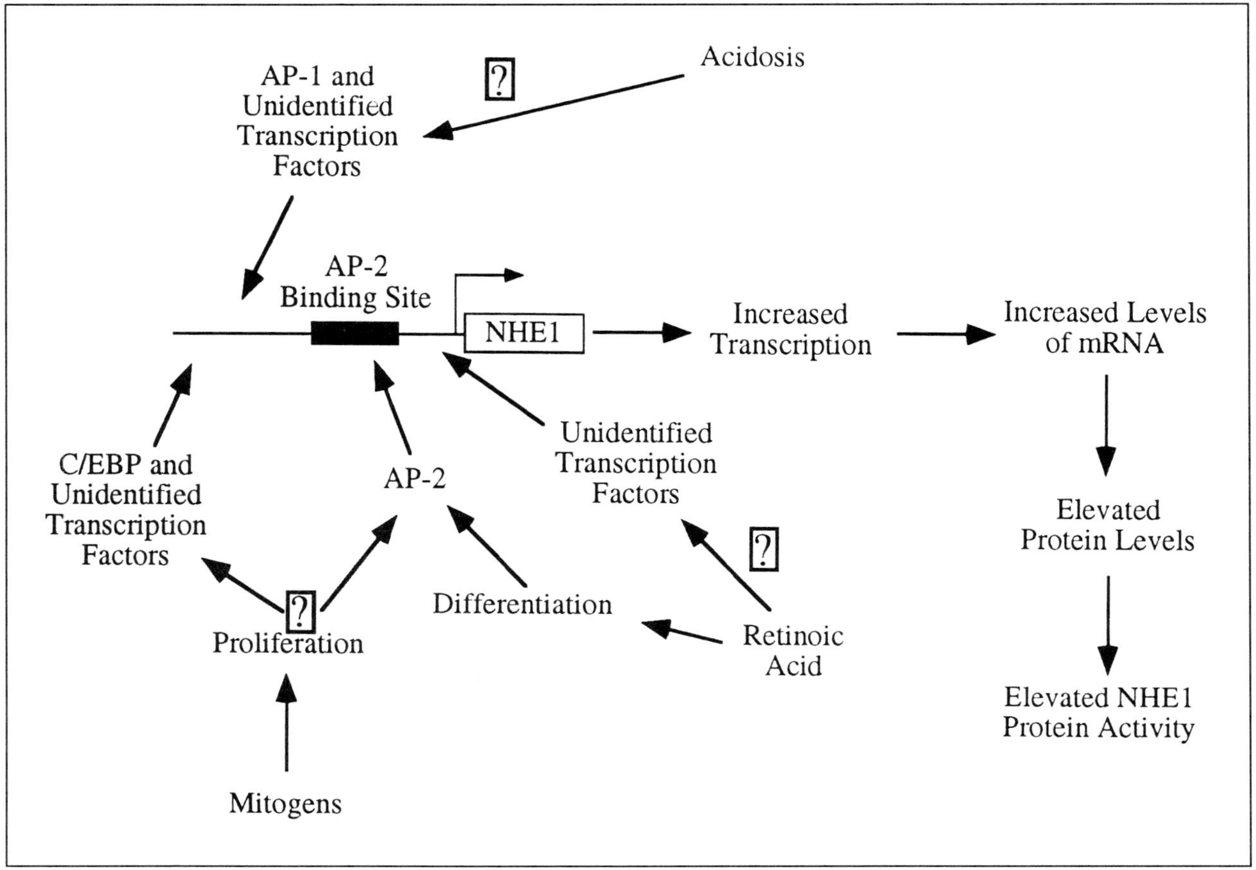

Fig. 5.2. Model for the regulation of the Na⁺/H⁺ exchanger. The possible mechanisms of induction of expression of the NHE1 gene are indicated. The NHE1 promoter/enhancer region is indicated by the horizontal line. The black box represents the region of DNA which binds the transcription factor AP-2. The start site of transcription is indicated by the right angled arrow. Straight arrows indicate the sequence of events involved in regulation of NHE1 expression. Question marks represent possible events involved in induction of expression and possible areas for future investigation.

Further studies involving cellular differentiation and the exchanger should involve other models of cellular differentiation. This would indicate if the role of NHE1 in differentiation is universal or if it is specific to neuroectodermal cells. In addition, the detailed mechanisms of activation of the NHE1 gene during retinoic acid induced differentiation of neuroectodermal cells still require further elaboration. We have shown that 5 μM of retinoic acid can directly induce the exchanger gene 4-fold with the presence of a functional retinoic acid receptor.[12] The mechanism responsible for this regulation is not known. The mouse NHE1 gene does not have a typical retinoic acid receptor binding site present in the 1.1 kb promoter/enhancer region, and other proteins may mediate this effect. It should be noted that the rabbit gene has a half-site for retinoid binding (Fig. 5.1c). However, this was outside the AP-2 region which was necessary for activation by retinoic acid.[12]

The Na^+/H^+ exchanger has long been suggested to be involved in cell growth and proliferation. During cellular proliferation, the exchanger is responsible for an elevation of intracellular pH. In some cell types such as vascular smooth muscle cells, hyperplastic agonists have been reported to result in a 25-fold increase in exchanger mRNA.[17] This increase may play a permissive role in the growth of certain cell types. The mechanisms involved in this increase have not been established. Preliminary studies have suggested that mitogenic stimulation can activate the NHE1 promoter in some cell types (unpublished observations), however the exact region and elements involved are unknown.

Thus there is still a great deal unknown on the mechanisms of regulation of expression of NHE1. Future studies can elaborate on these many facets of regulation of the promoter. In addition, it will be of great interest to compare the regulation of the different Na^+/H^+ exchanger isoforms and the mechanisms involved in their tissue specificity.

Acknowledgments

This research was supported by grants from the Heart and Stroke Foundation of Canada and from the MRC program grant in the molecular biology of membranes #PG-11440. Special thanks to those in the Dr. Fliegel's laboratory who helped with these projects especially Robert Haworth and Weidong Yang.

References

1. Grinstein S, Clarke CA, Rothstein A. Activation of Na^+/H^+ exchange in lymphocytes by osmotically induced volume changes and by cytoplasmic acidification. J Gen Physiol 1983; 82:619-638.
2. Rao GN, Sardet C, Pouysségur J, et al. Na^+/H^+ antiporter gene expression increases during retinoic acid-induced granulocytic differentiation of HL60 cells. J Cell Physiol 1992; 151:361-366.
3. Horie S, Moe O, Miller RT, et al. Long term activation of protein kinase C causes chronic Na/H antiporter stimulation in cultured proximal tubule cells. J Clin Invest 1992; 89:365-372.

4. Krapf R, Pearce D, Lynch C, et al. Expression of rat renal Na/H antiporter mRNA levels in response to respiratory and metabolic acidosis. J Clin Invest 1991; 87:747-51.
5. Igarashi P, Freed MI, Ganz B, et al. Effects of chronic metabolic acidosis on na-H exchangers in LLC-PK1 renal epithelial cells. Am J Physiol 1992; 263:F83-F88.
6. Moe OW, Tejedor ALevi, M, Seldin DW, et al. Dietary NaCl modulates Na($^+$)-H$^+$ antiporter activity in renal cortical apical membrane vesicles. Am J Physiol 1991; 260:F130-F137.
7. Akiba T, Rocco VK, Warnock DG. Parallel adaptation of the rabbit renal cortical sodium/proton antiporter and sodium/bicarbonate cotransporter in metabolic acidosis and alkalosis. J Clin Invest 1987; 80:308-315.
8. Dyck JRB, Maddaford T, Pierce GN, et al. Induction of expression of the sodium-hydrogen exchanger in rat myocardium. Cardiovascular Res 1995; 29:203-208.
9. Miller RT, Counillon L, Pages G, et al. Structure of the 5'-flanking regulatory region and gene for the human growth factor-activatable Na/H exchanger NHE-1. J Biol Chem 1991; 266:10813-10819.
10. Dyck JRB, Silva NLC L, Fliegel L. Activation of the Na$^+$/H$^+$ exchanger gene by the transcription factor AP-2. J Biol Chem 1995; 270:1375-1381.
11. Kolyada AY, Lebedeva TV, Johns CA, et al. Proximal regulatory elements and nuclear activities required for transcription of the human Na/H exchanger (NHE-1) gene. Biochim Biophys Acta 1994; 1217:54-64.
12. Dyck JRB, Fliegel L. Specific activation of the Na$^+$/H$^+$ exchanger during neuronal differentiation of embryonal carcinoma cells. J Biol Chem 1995; 270:10420-10427.
13. Horie S, Moe O, Yamaji Y, et al. Role of protein kinase C and transcription factor AP-1 in the acid-induced increase in Na/H antiporter activity. Proc. Natl Acad Sci USA 1992; 89:5236-40.
14. Philipp J, Mitchell PJ, Malipiero U, et al. Cell type-specific regulation of expression of transcription factor AP-2 in neuroectodermal cells. Dev Biology 1994; 165:602-614.
15. Blaurock NC, Reboucas NA, Kusnezov JL, et al. Phylogenetically conserved sequences in the promoter of the rabbit sodium-hydrogen exchanger isoform 1 gene (NHE1 SLC9A1). Biochim Biophys Acta 1995; 1262:159-163.
16. Yang W, Dyck JRB, Wang H, et al. Regulation of the NHE-1 promoter in the mammalian myocardium. Am J Physiol 1995; in Press.
17. Rao GN, Sardet C, Pouyssegur J, et al. Differential regulation of Na$^+$/H$^+$ antiporter gene expression in vascular smooth muscle cells by hypertrophic and hyperplastic stimuli. J Biol Chem 1990; 265:19393-19396.
18. Fafournoux P, Ghysdael J, Sardet C, et al. Functional expression of the human growth factor activatable Na$^+$/H$^+$ antiporter (NHE-1) in baculovirus-infected cells. Biochemistry 1991; 30:9510-95155.

CHAPTER 6

ROLE OF THE Na$^+$/H$^+$ ANTIPORTER ISOFORMS IN CELL VOLUME REGULATION

Lamara D. Shrode, Ana G. Cabado,
Greg G. Goss and Sergio Grinstein

A. INTRODUCTION

The regulation of cellular volume is one of the fundamental properties for all cells. Most mammalian cells are freely permeable to water and for this reason, the alteration of cellular osmolyte content is expected to lead to changes in water content and cell volume. To preclude deleterious changes, the cells defend their volume by manipulating their content of osmolytes' principally inorganic ions. Under normal conditions, there is a tendency for animal cells to swell due to the passive gain of inorganic ions down their electrochemical gradient. To counter this ongoing net gain, there are constitutive ion extrusion pathways which have evolved to extrude monovalent ions. In addition, cells also possess the ability to regulate their cell volume acutely when subjected to anisosmolar conditions. Exposure to hyposmolar solutions results in cell swelling due to the net influx of water. To reverse this effect, there is a regulated loss of electrolytes and a concomitant loss of osmotically obliged water in a process termed regulatory volume decrease (RVD). In general, the mechanisms responsible for RVD involve activation of Cl$^-$ and K$^+$ conductive pathways, coupled K$^+$/Cl$^-$ cotransport and/or loss of organic osmolytes such as taurine, inositol, betaine or glutamine by activation of organic solute transporters.[1-3] Conversely, cells are able to re-swell under conditions where exposure to hypertonic conditions results in cell shrinkage due to the loss of water from the cell. The cell gains water from the surrounding media by first increasing its cellular osmolyte levels via activation of specific cellular

The Na$^+$/H$^+$ Exchanger, edited by Larry Fliegel. © 1996 R.G. Landes Company.

ion transport systems in a process termed regulatory volume increase (RVI). The systems principally activated during RVI are either the furosemide-sensitive $Na^+/K^+/2Cl^-$ cotransporter, or a combination of the amiloride-sensitive Na^+/H^+ exchanger (NHE) followed by a secondary alkalinization-dependent activation of the Cl^-/HCO_3^- exchanger.[4] In this chapter, we outline briefly the importance of cellular volume control and focus primarily on some of the recent advances in understanding the role that the various NHE isoforms play in RVI. Furthermore, we will speculate as to the mechanisms involved in the activation and regulation of NHE during RVI. This chapter is not intended as an exhaustive review of the field of cellular volume control. More comprehensive analyses of the topic can be found elsewhere.[2,3,5-7]

B. CLINICAL SIGNIFICANCE AND PATHOPHYSIOLOGY OF VOLUME REGULATION

Most mammalian cells are bathed in an exquisitely controlled medium of approximately 300 mosmol/l, which is thought to be isosmolar with the cell interior. However, in many areas of the body, the cells are regularly exposed to both hyper- and hypo-osmotic challenges. For example, cells of the kidney inner medulla must be able to withstand variable hyperosmotic conditions ranging from 300 up to 1400 mosmol/l. In addition, cells in the circulatory system (e.g., lymphocytes, monocytes, neutrophils) are regularly exposed to hyperosmotic shock while passing through the inner medulla of the kidney via the vasa recta. Therefore, it is vitally important for these systems to possess mechanisms for cellular volume regulation.

Disturbances in cellular volume control are also commonly encountered in clinical medicine. Anisosmolar conditions commonly occur in patients with unregulated diabetes mellitus, during acute renal failure or after excessive ethanol ingestion, while cytotoxic cell swelling occurs after bouts of ischemia or hypoxia and following severe injury.[2,8] In addition, treatment-induced alterations in plasma osmolarity are common during intravenous administration of water, NaCl, glycerol or mannitol.[8] These treatments can have devastating effects if carried out improperly, as the formation of rebound edema is a potential occurrence. The organ principally affected by rapid osmoregulatory disturbances is the brain. Due to the confined nature of the brain, changes in cellular volume by acute exposure to anisosmolar conditions can result in severe neurologic impairment including seizures, coma and neurologic death.[2] These pathophysiological situations stress the importance of maintaining cell volume under anisosmolar conditions.

C. ROLE OF NA^+/H^+ EXCHANGER IN RVI

RVI is a complex phenomenon which involves the simultaneous reduction in permeability to K^+ and Cl^- via conductive channels and an increase in the uptake of Na^+ and Cl^- from the environment. There are two basic categories of RVI, the so-called primary RVI in which cells shrink in response to hyperosmotic challenge and subsequently re-swell towards their original volume. Secondary RVI occurs in cells which have recovered from exposure to hyposmotic media. In these cells, there has already been a net loss of osmolytes, principally K⁺ and Cl⁻, resulting in a return to near-normal cell volume. If the cells are then rapidly returned to a medium of normal osmolarity (≈300 mOsmol), they will shrink due to the reduced intracellular ion content relative to the external osmolarity. Such shrunken cells tend to swell once again back to normal size. This secondary RVI is also commonly referred to as RVI after RVD. Mechanistically, primary and secondary RVI are similar, although it has been demonstrated that in some cell types there is no noticeable primary RVI in response to hyperosmotic shock, yet secondary RVI after RVD occurs quite rapidly.[9,10]

One of the most perplexing aspects of RVI is the variability in the underlying mechanisms reported in different cells types. In avian rat and human red cells,[7,11] and in Erhlich ascites tumor cells,[12] Na⁺/Cl⁻ uptake during RVI is mediated by the loop diuretic (e.g., furosemide) sensitive Na⁺/K⁺/2Cl⁻ cotransporter first described by Geck et al.[13] In contrast, in many other cell types including lymphocytes,[14] *Amphiuma* red cells[15,16] and dog red cells,[17] RVI occurs by a mechanism involving NHE coupled to anion exchange. The reasons for these differential responses under comparable conditions is not readily apparent. Moreover, the mechanisms responsible for the activation for each of these pathways remain unknown.

Involvement of NHE in the RVI response was initially proposed on the basis of a number of experimental observations. The first indication of involvement of NHE was that, in a number of cells including *Amphiuma* and dog red cells, RVI is blocked by the addition of amiloride and its 5' substituted derivatives, a pharmacological profile that is the hallmark of the NHE.[15] Secondly, in dog red cells undergoing RVI the net gain of Na⁺ and Cl⁻ was found to occur via an electroneutral transport pathway, thereby eliminating the possibility that Na⁺ uptake involved conductive pathways.[18] Another supporting piece of evidence involved observations that in cells undergoing RVI in HCO_3^--deficient media, there was a concomitant cytoplasmic intracellular alkalization which was also blocked by amiloride and its derivatives.[17] Indirect evidence for NHE induced alteration of intracellular pH was first obtained by Parker[17] who demonstrated that when induced in nominally HCO_3^--free, lightly buffered media, RVI was associated with an extracellular acidification. Furthermore, it was also demonstrated that extracellular Cl⁻ must be present in order for RVI to proceed.[17] Based on these observations, a general model for the NHE dependent activation of RVI was proposed by Cala.[15] In this model exposure of cells to hyperosmotic solutions results in the immediate activation of the NHE, which in turn elevates the intracellular pH (pH_i) due to a net loss of intracellular protons. The increase in pH_i (and the consequent increase in intracellular HCO_3^-) then promotes activation of the Cl^-/HCO_3^- exchanger. The overall effect is a net gain of Na^+ and Cl^- by the cell, which is followed by osmotic water uptake and regulatory cell swelling.

Perhaps the most compelling evidence for the involvement of the NHE in mediating RVI is from the work of Rotin and Grinstein.[10] In their report, a NHE-deficient CHO cell line was isolated by the proton-suicide technique first described by Pouysségur and colleagues.[19] Using this NHE-deficient cell line and taking advantage of the pH sensitive dye BCECF, the former authors directly tested the hypothesis that the NHE was involved in RVI. Figure 6.1A compares the properties of normal (WT5) and antiport deficient (AP-1) cells. A prepulse with 30 mM NH_4Cl results in a rapid, reproducible acidification of the intracellular compartment. In wild-type (WT5) cells the acidification persists in a Na^+-free medium, but is rapidly corrected upon the re-addition of 20 mM NaCl. The involvement of NHE in this alkalinization response is demonstrated by its sensitivity to the amiloride analog EMA [N-ethyl-N-(1-methyl-ethyl)amino amiloride], a potent inhibitor of the NHE (WT5 + EMA). Importantly, antiport-deficient cells (AP-1) are unable to alkalinize upon re-addition of Na^+ to the medium. Figure 6.1B documents the involvement of the NHE in mediating the response to hyperosmotic shock. Cells were resuspended in nominally HCO_3^--free medium and exposed to a RVI after RVD regime. Under these conditions, WT5 cells were able to spontaneously alkalinize while both the absence of NHE (in AP-1 cells) and the presence of NHE inhibitors (WT5 + EMA) eliminated this response. The subsequent increase in cell volume (RVI) after shrinkage could then be quantified by measurement of the cellular volume at various times after induction of the osmotic shock. As shown in Figure 6.1C, while WT5 cells were able to increase their cell volume towards normal, cells which are devoid of the NHE were unable to re-swell. Jointly, the data presented point to the involvement of the NHE in the mediation of RVI. In the following sections we will discuss the mechanisms involved in the regulation of the NHE and its isoforms, and will speculate about the potential pathways involved.

D. MECHANISMS OF ACTIVATION OF THE NHE

A variety of growth factors are known to activate NHE[20] by increasing the exchanger's affinity for protons.[21] Because growth factor activation of NHE requires ATP, it was hypothesized early on that a phosphorylation-dependent step is involved.[20] Sardet et al[22] subsequently showed that epidermal growth factor, thrombin, phorbol esters and serum all induced phosphorylation of NHE. Later, Wakabayashi et al[23] demonstrated that the cytoplasmic domain of the exchanger was responsible for mediating growth factor signals, presumably in a

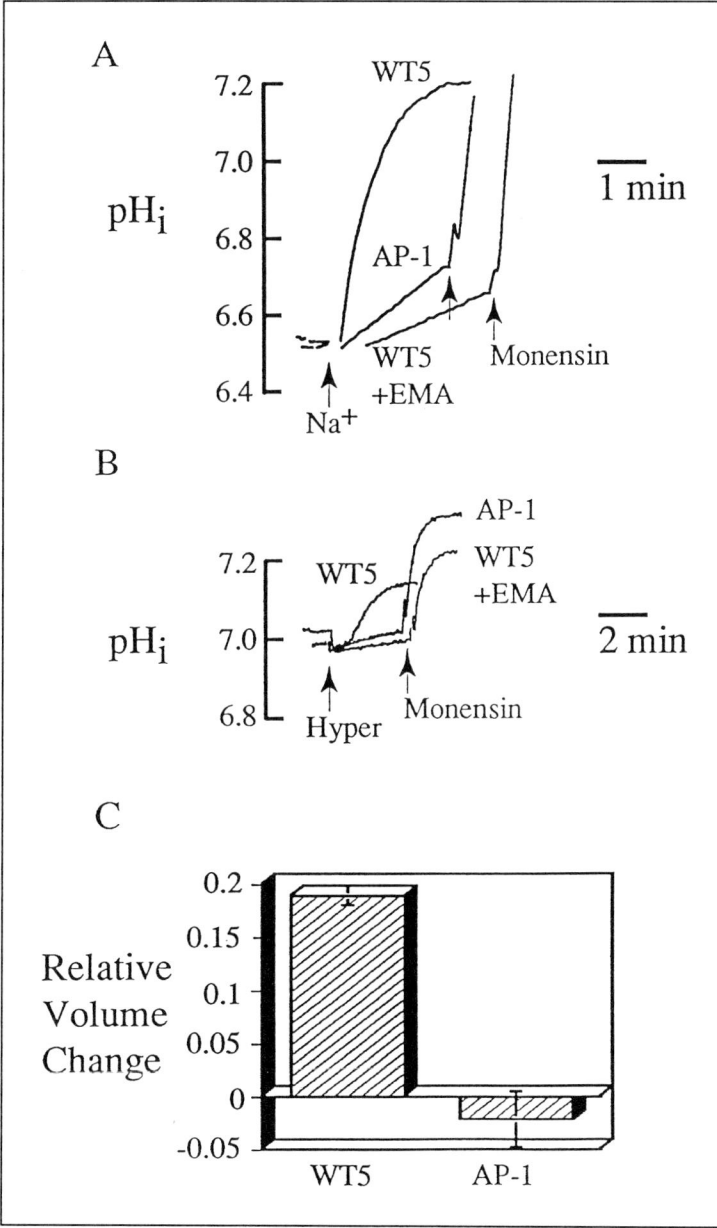

Fig. 6.1. Requirement for Na+/H+ exchange during RVI in CHO cells.
(A) Isolation of antiport deficient mutants was demonstrated by loading approx. 1 x 10⁶ wild-type CHO (WT5) or antiport deficient (AP-1) with BCECF to measure intracellular pH (pH_i). Cells were then acidified and suspended in either Na+-free N-methyl-glucamine (NMG) media or NMG plus ethyl (methylethyl) amino amiloride (EMA; 20 μM). Addition of Na+ (20 mM) and monensin (5 μM) is indicated by arrows.
(B) Approx. 1 x 10⁶ wild-type CHO (WT5) or antiport deficient (AP-1) were loaded with BCECF and subsequently resuspended in isotonic medium in the presence or absence of 20 μM EMA. Where indicated, the cells were made hypertonic (490 mOsmol) by the addition of 100 mM NaCl. Addition of monensin (5 μM) is indicated by arrows.
(C) Relative volume change measured using a Coulter counter expressed as change from control in both WT5 and AP-1 cells 5 minutes after initiation of RVI. All figures are representative of at least three independent experiments. All experiments were carried out at 37 °C. See text for definitions.

phosphorylation-dependent manner. However, recent data suggest that direct phosphorylation of NHE cannot completely account for growth factor activation, since deletion of the phosphorylated residues is not fully inhibitory.[24]

As discussed above, the NHE is also activated during RVI in a variety of cell types. Like growth factor activation, shrinkage-induced activation of NHE results in an increase in the exchanger's affinity for internal H^+.[25] In addition, hyperosmotic activation has a comparable time course and results in the same degree of alkalinization as that observed in growth factor activation.[26] Since the characteristics of the activation of the Na^+/H^+ exchanger by growth factors and by shrinkage are so similar, the latter was postulated to occur via a phosphorylation-dependent step as well. Accordingly, shrinkage-induced activation of NHE in lymphocytes was found to be ATP-dependent,[26] suggesting the involvement of a kinase. Shrode et al also found that shrinkage-induced activation of NHE was ATP-dependent in primary rat astrocytes[27] and C6 glioma cells.[28] Contrary to these predictions, however, hyperosmotic treatment did not increase the phosphorylation level of NHE1 in fibroblasts.[29]

While direct phosphorylation of NHE does not occur upon cell shrinkage, the ATP-dependence of shrinkage-induced activation nevertheless suggests the involvement of a phosphorylation-dependent step. Exposure to phorbol esters, which directly activate protein kinase C (PKC), and hyperosmotic stress both result in similar phosphorylation patterns.[30] Thus it was proposed that PKC may be involved in shrinkage-induced activation of NHE. However, exposing C6 glioma cells to phorbol ester prior to hyperosmotic exposure had no effect on the rate of shrinkage-induced activation of NHE.[31] Additional alkalinization was observed when phorbol ester treatment followed hyperosmotic activation. Moreover, hyperosmotic stress further stimulated NHE after it had previously been activated by a phorbol ester in an osteoblast-like rat cell line.[32] These findings argue against a role for PKC.

While acute exposure to phorbol esters activates PKC, prolonged exposures can downregulate this enzyme. This approach was used to determine if PKC is required for activation of NHE upon hyperosmotic stress. Shrinkage-induced activation of NHE still occurred in PKC-depleted lymphocytes.[30] Likewise, hyperosmotic stress also activated NHE in rat astrocytes after PKC was depleted.[27] Together, the above data indicate that hyperosmotic stress activates NHE via a PKC-independent pathway.

Another kinase that might be involved in the shrinkage-induced activation pathway is cAMP-dependent protein kinase (PKA). In principle, cell shrinkage could increase adenylate cyclase activity resulting in an increase in cAMP necessary to activate PKA. However, except in trout red blood cells,[33] cAMP has little effect or inhibits NHE activity. Indeed, an increase in cAMP failed to mimic shrinkage-induced activation of NHE in osteoblast-like cells[32] and in dialyzed barnacle muscle fibers.[34] In addition, hyperosmotic stress did not increase levels of cAMP in rat astrocytes.[27] Thus, hyperosmotic activation of NHE seems to occur independently of cAMP and PKA.

Many signal transduction pathways are Ca^{2+}-dependent. Indeed, an increase in intracellular Ca^{2+}, but not PKC, mediates activation of NHE by thrombin in WS-1 fibroblasts.[35] Therefore, shrinkage-induced activation of NHE may involve a similar Ca^{2+}-dependent pathway. Nonetheless, several studies suggest that Ca^{2+} concentration increases are not required for hyperosmotic activation. Grinstein et al[36] showed that a cytosolic Ca^{2+} increase upon hyperosmotic stress did occur in rat thymic lymphocytes. However, this increase was inhibited by amiloride, suggesting that it is a result of NHE activation, rather than its cause. Also, exposing osteoblast-like cells to hyperosmotic stress had no effect on intracellular Ca^{2+} levels. In fact, when Dascalu et al[32] simulated a Ca^{2+} increase by applying the Ca^{2+} ionophores ionomycin or A23187, no effect on shrinkage-induced activation of NHE was observed. Finally, hyperosmotic stress had no effect on Ca^{2+} levels in rat astrocytes.[27]

While overall cytoplasmic Ca^{2+} levels did not increase upon cell shrinkage, local changes near the plasma membrane could play a role in shrinkage-induced activation of NHE. In an attempt to test this theory, astrocytes were loaded with BAPTA, a calcium chelator, and then exposed to a hyperosmotic stress. Shrinkage-induced activation of NHE persisted in BAPTA-loaded cells,[27] and similar results were obtained in C6 glioma cells.[28] Thus, increasing cytosolic Ca^{2+} is neither sufficient nor necessary for activation of NHE during RVI.

Calmodulin-dependent (CaM) kinases are another potential class of kinases that might be responsible for hyperosmotic activation of NHE. Calmodulin-dependent kinases are most often activated when forming a tripartite complex with Ca^{2+} and calmodulin. Because the experiments summarized above seem to discount a role for Ca^{2+}, CaM kinase would not appear to be important. However, calmodulin has been shown, at least in yeast, to perform some functions in the nominal absence of Ca^{2+}.[37] Furthermore, cell shrinkage may reduce the pKa of calmodulin for Ca^{2+}. Therefore, it is possible that calmodulin could activate a kinase in the absence of, or at reduced levels of, calcium. Several studies suggest that a calmodulin-dependent step is involved in shrinkage-induced activation of NHE. In lymphocytes, trifluoperazine, a calmodulin antagonist, inhibited shrinkage-induced activation of NHE.[26] Trifluoperazine also inhibits shrinkage-induced activation of NHE in rat astrocytes.[27] Two other calmodulin inhibitors, chlorpromazine and W-7, inhibit hyperosmotic activation of NHE in osteoblast-like rat cells,[32] rat astrocytes[27] and C6 glioma cells.[28] Therefore, a calmodulin-dependent step may be implicated in the hyperosmotic activation pathway.

One of the first calmodulin-dependent kinases discovered was myosin light chain kinase (MLCK). MLCK phosphorylates, and thus activates, myosin light chain (MLC) during both smooth muscle and nonmuscle contraction. Recently, Takeda et al[38] presented data showing increased phosphorylation of MLC in glomerular mesangial cells upon cell shrinkage. In addition, an increase in phosphorylation of MLC was also observed in shrunken endothelial cells, which was rapidly reversed upon the return to isosmotic solutions.[39] MLCK has also been implicated in the shrinkage-induced activation pathway in astrocytes. A reversible, time-dependent increase in MLC phosphorylation occurred when astrocytes were exposed to a hyperosmotic solution.[27] Moreover, ML-7 an inhibitor of MLCK, was found to inhibit shrinkage-induced activation of NHE in astrocytes, thus providing additional support of the involvement of MLCK. ML-7 also inhibits shrinkage-induced activation of $Na^+/K^+/2Cl^-$ cotransport in endothelial cells.[39] It is therefore possible that a common pathway activates different transporters involved in restoring cell volume following shrinkage.

The involvement of MLCK in the activation pathway provides support for the idea that the cytoskeleton plays a role in regulating NHE. In mesangial cells, formation of stress fibers, which terminate at focal adhesions, is associated with an increase in MLC phosphorylation.[40] Recently, Grinstein et al[41] showed that NHE1 is localized near focal adhesions in PS127A cells. In addition, Goss et al[42] showed that ATP-depletion reduces the number of stress fibers and redistributed NHE from the focal accumulation sites. Thus, the ATP-dependence of shrinkage-induced activation of NHE could be explained by the ATP requirement for MLC phosphorylation. MLC phosphorylation upon cell shrinkage might lead to cytoskeletal rearrangement, which could in turn result in the activation of NHE. Alternatively, cytoskeletal rearrangement may be the cause of both NHE activation and the observed increase in MLC phosphorylation.

In addition to the cytoskeleton, other ancillary proteins might be associated with NHE and thus may be involved in shrinkage-induced activation of the antiporter. Goss et al[43] recently detected a 24 kD protein that co-immunoprecipitates with the ubiquitous isoform of NHE. This protein, and possibly others yet unidentified, may be a convergence point for the various activators of NHE, including growth factors and hyperosmotic exposure.

Mitogen-activated protein kinases (MAPK), also known as extracellular signal-regulated kinases (ERK), are implicated in activating various effectors upon exposure to growth factors and vasoactive peptides.[44] MAPK require phosphorylation of both threonine and tyrosine residues by MAPK-kinase in order to become activated. Current evidence suggests that MAPK or MAPK-like kinases are activated upon hyperosmotic stress. Itoh and his colleagues[45] showed that two MAPK, of molecular weights of 49 and 42 kD, and MAPK-kinase were activated by hyperosmotic stress in a time- and dose-dependent manner in MDCK cells. Recently, a family of genes responsible for yeast survival in high osmolarity glycerol (HOG) was identified by Brewster et al.[46] The product of the HOG1 gene had sequence similarity to members of the MAPK family. Moreover, HOG4 is identical to PBS2, a gene that encodes a member of the MAPK-kinase family. Interestingly, hyperosmotic stress leads to the PBS2-dependent tyrosine phosphorylation of HOG1.

Han et al[47] have demonstrated in murine macrophage-like cells that hyperosmotic stress leads to tyrosine phosphorylation of p38, a protein that has 52% homology with HOG1. p38 and HOG1 both possess a TGY sequence at the site of phosphorylation, while MAPK family members have a conserved TEY sequence in this position.[48] In addition, p38 and HOG1 are six amino acids shorter than MAPK in a loop between two potentially critical subdomains (see ref. 47 for discussion). Thus, it is likely that p38 and HOG1 are members of a distinct group of MAPK-related proteins. Further support for this argument comes from the finding that HOG1 is phosphorylated specifically by PBS2, but not by other MAPK-kinases.[46]

Jnk1, the c-Jun N-terminal kinase, though distinct from p38, is also similar to the yeast HOG1 gene product. In fact, when Jnk1 (also called SAPK) is introduced into HOG1-deficient yeast, which cannot survive under hyperosmotic conditions, the transformants continued to grow. By contrast, transfection with ERK2 (MAPK) did not provide protection against hyperosmotic stress.[49]

For these reasons, HOG1, p38 and Jnk1 have been assembled into a family of stress-activated protein kinases. They are all seemingly activated by cell shrinkage, yet are members of distinct, parallel cascading pathways (Fig. 6.2). MAPK (ERK) are activated by MAPK-kinase (also called MEK), which is in turn activated by Raf. By analogy, Jnk (SAPK) isoforms are activated by SAPK-kinase or SEK, which are themselves activated by MEK-kinase (MEKK).[50] The steps upstream of p38 are analogous, as detailed in Figure 6.2. Cell shrinkage has been found to activate several steps and multiple pathways in single cell types. Thus, Matsude et al[51] reported that shrinkage increases the activity of MAPK (ERK1 and 2) and of two Jnk forms (p54 and p46) in rat 3Y1 cells. Moreover, upstream members of the pathways, including MEK, Raf and MEKK are reportedly also stimulated by hypertonicity in fibroblasts and PC12 cells.[51] It remains to be defined whether the MAPK, p38 and/or SAPK pathways lead to activation of NHE upon cell shrinkage. To date, no evidence has directly linked shrinkage-induced activation of the antiporter and activation of any of these pathways. Activation of the two MAPK in MDCK cells[45] is not inhibited by the MLCK inhibitor ML-9. This suggests, if these kinases are members of a common hyperosmotic stress activation pathway, that MAPK and MAPK kinase may be upstream of MLCK. Alternatively, the lack of inhibition by ML-9 may suggest that MAPK and MLCK are members of two independent pathways.

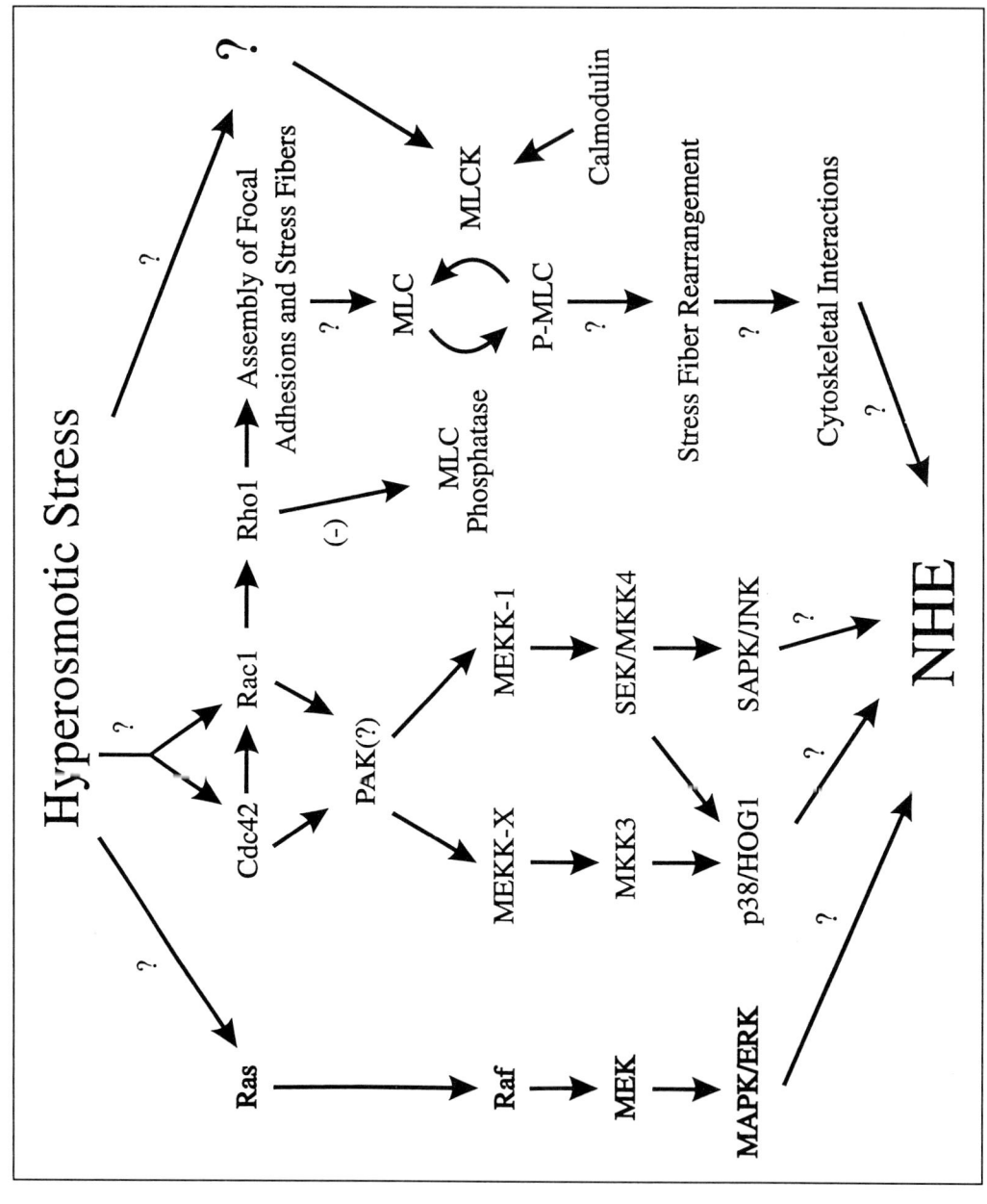

Fig. 6.2. Schematic drawing of the signaling pathways potentially involved in shrinkage-induced activation of NHE. Three of the pathways involve proteins of the MAPK superfamily, while the fourth involves myosin light chain (MLC), MLC kinase and presumably the cytoskeleton.

In summary, the signal transduction pathway(s) that link changes in cell volume with activation of NHE are still unknown. Current evidence suggests that various kinases, including MLCK and one or more stress-activated kinases may be involved. Further studies are obviously necessary to fully characterize and understand the signaling cascade leading to stimulation of NHE during volume regulation.

E. ISOFORMS OF THE Na⁺/H⁺ EXCHANGER

E.1 CLONING, SEQUENCING AND HOMOLOGY

As detailed elsewhere in this volume, the first mammalian NHE was cloned by Sardet et al.[52] Using genetic complementation with exchanger-deficient cell lines, these authors were able to clone a cDNA encoding the human ubiquitous isoform of the exchanger, NHE1. Three related isoforms of NHE and homologue from different species were subsequently cloned by cross-hybridization using various libraries.[53-60]

NHE1 has been cloned from human, rabbit, rat, pig, mouse and Chinese hamster and contains 815-820 amino acid residues.[52-54,61-63] NHE2 and NHE3 have been identified in human, rabbit and rat.[54,55,57,58,64,65] They contain 809-813 and 831-834 amino acids, respectively. NHE4 has been only cloned from the rat and encompasses 717 amino acids.[54] Recently a novel member of the NHE family, named NHE5, was partially sequenced.[66]

The primary and secondary structures of all cloned mammalian NHE isoforms bear similarities. NHE2 and NHE4 are the most closely related isoforms with ≈57% identity at the amino acid level.[55,59] NHE3 is the least related isoform, sharing ≈40% identity with NHE1, NHE2 and NHE4.[54,55,65] NHE5 exhibits 59-73% identity with the other members of the family, being more closely related to NHE3.[66]

E.2 TISSUE AND CELLULAR DISTRIBUTION

Based on Northern analysis and ribonuclease protection assays, mRNA encoding NHE1 was found in nearly all mammalian cells,[53,54] except in two renal cell types.[67] In polarized epithelial cells, NHE1 is expressed predominantly in the basolateral membrane.[68] NHE2, NHE3 and NHE4 exhibit a more restricted distribution. NHE2 message is expressed in the gastrointestinal tract, although it was also found at low levels in liver, kidney, brain, heart, testes, uterus, skeletal muscle and lung.[54,55,57,59] NHE3 mRNA is expressed in small intestine, colon, kidney, stomach and also in testes, ovary and brain.[54,58,65] In renal and intestinal epithelial cells, NHE3 is found in apical membranes.[60,69,70] NHE4 mRNA message is most abundant in stomach, small intestine and colon. Lesser amounts are also found in kidney, brain, uterus and skeletal muscle.[54,71] mRNA encoding NHE5 was identified in brain, testes, spleen and skeletal muscle. No message was found in epithelial tissues.[66]

E.3 FUNCTIONAL AND PHARMACOLOGICAL PROPERTIES

Aside from the existence of several isoforms, the analysis of NHE physiology is further complicated by the coexistence of more than one isoform in the same cell (e.g., NHE3 in apical vs. NHE1 in basolateral membranes of epithelial cells). Moreover, the same isoform can vary its behavior depending on the cellular context. Heterologous expression in antiport deficient cells has revealed functional differences when the same isoform is expressed in two different lines.

Although the isoforms are all inhibited by amiloride and its analogs, they exhibit different sensitivities to these drugs. In addition, a new NHE inhibitor, HOE694 [(3-methylsulphonyl-4-piperidinobenzoyl) guanidine methanesulphonate], was found to discriminate between isoforms much more effectively than amiloride and its derivates.[63] NHE1 and NHE2 isoforms are nearly equally sensitive to amiloride, whereas NHE2 is ≈30-fold less sensitive to HOE694 than NHE1. NHE3 exhibits the lowest affinity for both amiloride and HOE694.[56,58,63,72]

When heterologously expressed in NHE-deficient cells, the individual isoforms are also distinguished by their kinetic properties. They all display similar kinetic characteristics towards external Na$^+$, obeying classical Michaelis-Menten behavior with a Hill coefficient ≈1. However, NHE3 exhibited higher affinity for external Na$^+$ than NHE1,[72] consistent with its presumed role as the apical exchanger responsible for Na$^+$ reabsorption in the kidney and gastrointestinal tract. The kinetics with respect to internal H$^+$ are also similar for the three isoforms, all having a Hill coefficient of ≥ 2, reflecting allosteric H$^+$ activation. Orlowski et al[72] found that NHE1 is more sensitive to changes in the intracellular H$^+$ concentration than NHE3, a finding consistent with the physiological role of NHE1 in maintaining pH$_i$ homeostasis.

The isoforms studied thus far are differentially regulated by phosphorylation, in accordance with the presence of distinct consensus sequences for unique kinases (reviewed in ref. 73). This may explain the variability of the results obtained in tissues that likely express different isoforms. Varying effects of cAMP on NHE activity have been reported in mammalian tissues, including osteoblasts where NHE activity has been studied extensively and the isoform identified as NHE1 (see Table 6.1). Both stimulation and inhibition of endogenously expressed NHEs were found. Yet, when expressed heterologously in a fibroblast line NHE1, NHE2 and NHE3 failed to respond to cAMP.[74,75] By contrast, when β–NHE, an exchanger cloned from trout red blood cells, was expressed in the same cell type it was clearly stimulated by cAMP.[74] In a more recent study NHE1 and NHE2 stably expressed in CHO cells were stimulated by agonists that increase cAMP, while NHE3 was found to be inhibited.[76] This effect is similar to that seen in apical membranes of epithelia,[77] where cAMP inhibits this isoform through an accessory protein.[78,79] Structural analyses using a series of mutants of NHE3 identified a cytosolic region between amino acids 579 and

Table 6.1. Effect of hyperosmolarity, PKC, PKA and Ca^{+2} on the isoforms of the Na^+/H^+ exchanger. (+) stimulation, (–) inhibition, (T) transfected, (E) endogenously expressed

ISOFORM	PKC	PKA	Ca+2	Hypertonic shock	Ref.
NHE1					
Fibroblasts (T)	(+)				22
Fibroblasts (T)	(+)	No effect			53, 75
Fibroblasts (T)		No effect			74
Fibroblasts (T)				(+)	29
Fibroblasts (T)			(+)	(+)	84, 24
CHO (T)	(+)			(+)	89
CHO (T)				(+)	10
CHO (T)				(+)	92
CHO (T)	(+)	(+)			76
Fibroblasts and CHO (T)				(+)	90
OK (T)	(–)	(–)			83
Osteoblasts (E)		(+)			97
Smooth muscle cells (E)				(+)	98
LLC-PK$_{20}$ (E)	(–)	(+)			99
MCT (E)	(+)	(–)			100
RKPC-2 (E)	(+)	(–)			82
NHE2					
Fibroblasts (T)	(+)	No effect			75
Fibroblasts (T)				(–)	93
CHO (T)				(+)	92
CHO (T)	(+)	(+)			76
MCT (E)	(–)	(+)			100
RKPC-2 (E)	(+)	(–)			82
NHE3					
Fibroblasts (T)	(–)	No effect			80
Fibroblasts (T)	(–)				58
Fibroblasts (T)			(–)		80
CHO (T)				(–)	92
Fibroblasts (T)			No effect		102
CHO (T)	(–)	(–)			76
Renal epithelial cells (E)				(–)	96
OK (E)	(–)	(–)			77
Apical Na^+/H^+ in kidney				(–)	94
Apical Na^+/H^+ in kidney				(–)	95
NHE4					
Fibroblasts (T)				(+)	71

684 that is essential for regulation by PKA.[101] Interestingly, this same region is also required for regulation of NHE3 by PKC (see next).[80]

It is generally accepted that PKC modulates NHE activity, stimulating in most cases (see Table 6.1; reviewed in ref. 81). For instance, phorbol esters stimulated NHE1 and NHE2 endogenously expressed in kidney cells.[82] In agreement, fibroblasts transfected with NHE1 and NHE2 are also stimulated by phorbol esters. Deletion studies on NHE1 showed that a region between amino acids 567-635 is central to the PKC response, even though the major phosphorylation sites are located elsewhere (636-815).[24] This finding suggests phosphorylation-independent activation of NHE1, resembling the activation by osmotic shrinkage.[29] Unexpectedly, when NHE1 was transfected into an epithelial line, namely OK cells, it was inhibited by PKC.[83] These findings imply that both the isoform and the cellular context determine NHE responsiveness to kinases.

In contrast to the above results, PKC agonists were shown to inhibit NHE activity in apical membranes of native epithelia, which likely express NHE3. Accordingly, when NHE3 is heterologously expressed in fibroblasts, its activity is inhibited by PKA.[76] Studies using NHE3 mutants indicated that the regulatory site(s) for PMA response are located between residues 585-689 and that this region is different from those implicated in regulation by volume and by other stimuli.[80]

As mentioned briefly earlier, the effect of Ca^{2+} on NHE is controversial. Elevation of intracellular Ca^{2+} has been reported to activate the exchanger in some cases and to inhibit in others (see Table 6.1). The heterogeneity may again be due to differential expression of isoforms or to the cellular context. Using fibroblasts transfected with NHE1, Bertrand et al show that this isoform is activated by CaM, due to displacement of an "autoinhibitory domain".[84,85] In contrast apical NHE (likely NHE3) is inhibited by CaM kinase II.[86,87] In concordance with this, it was recently demonstrated that NHE3 is also a CaM-binding protein although in the transfectants NHE3 does not respond to Ca^{2+}.[102]

F. RESPONSIVENESS OF THE INDIVIDUAL ISOFORMS TO OSMOLARITY

Osmotic shrinkage activates NHE in a variety of cell types[10,29,84,88-90] (reviewed in ref. 91). However, osmotic activation of NHE is not universal and is not common to all isoforms. The responsiveness of the individual isoforms to changes in cell volume has been studied using heterologous expression in stable transfectants. In this system NHE1 was reported to be stimulated by hyperosmotic shock and inhibited by hypotonicity.[92] Several studies have sought the critical region(s) of NHE1 molecule implicated in the osmotic response. The intracellular application of an antibody against the carboxy-terminal 157 amino acids did not affect the osmotic activation of the exchanger.[89] Consistently, using mutagenesis analysis in cells stably transfected with truncated forms of human and rat NHE1, Bianchini et al[90] were able to show that the terminal 117 residues are not essential for osmotic regulation. The volume-sensitive site must reside in a region proximal to the N-terminus, as stimulation was also seen in mutants truncated at positions 582 and 566. Consistent results were obtained by Orlowski et al (Orlowski et al unpublished observations). On the contrary, Bertrand et al showed that deletion of the autoinhibitory domain of NHE1 (636-656) reduced by 80% the cytoplasmic alkalinization in response to hyperosmotic stress.[84] The source of this apparent discrepancy may be the constitutive stimulation that occurs in these mutants.

The hypertonic response of NHE2 varies in different systems. When transfected into CHO cells, NHE2 is stimulated,[92] yet it is inhibited in PS120 fibroblasts.[93] Once again, these findings stress the importance of the cellular context to the biological response of NHE.

NHE3 was found to be consistently inhibited by hypertonic cell shrinkage and unaffected by hypotonic challenge in both heterologous expression models[92,93] and in kidney epithelial cells that most likely express this isoform.[94-96] This inhibitory effect of hyperosmolarity on NHE3 appears not to be mediated by PKA, PKC or CaM[95,96] but may instead depend on a tyrosine kinase.[95]

Studies using truncated mutants of NHE3 have yielded contradictory results regarding the location of the volume-sensitive domain(s). Some authors have suggested that the elements implicated in hyperosmotic-induced inhibition are located within the amino-terminal membrane domain.[93] Contrasting results were obtained by Orlowski et al who found no inhibition in the truncated mutant 638 or in shorter mutants (Orlowski et al; unpublished observations).

Since NHE3 is localized in apical membranes of several epithelial cell types, inhibition of NHE3 when cells are shrunk might be of functional and clinical importance. On one hand, inhibition of NHE3 might contribute to impaired proximal tubule acidification in conditions of increased plasma osmolarity, such as when blood sugar is elevated, during renal failure or in cases of methanol and ethylene glycol toxicity.[96] Inhibition of NHE3 might also contribute to natriuresis during glucosuria and during mannitol infusion preventing sodium and cell volume increase. On the other hand, it is known that the thick ascending limb of the Loop of Henle reabsorbs part of the bicarbonate filtered at the glomerulus. The H^+ secretion required for this bicarbonate absorption is mediated by apical membrane NHE. Hyperosmolarity inhibits bicarbonate absorption by inhibiting apical exchanger, facilitating excretion of acid in the hyperosmotic environment of the renal medullary cells.[94] Finally, hyperosmotic-induced inhibition of apical NHE3 in the intestine may have a relevant role in electrolyte or sorbitol-induced diarrhea, favoring loss of sodium and water.

Curiously, NHE4 is generally refractory to activation in heterologous expression, however it can be stimulated under hyperosmotic conditions.[71] The authors of this report point out that due to its renal localization in regions of high osmolarity, this isoform may play a specialized role in the kidney contributing to volume regulation in extreme conditions.[71]

G. CONCLUDING REMARKS

In conclusion, NHE plays a central role in cell volume regulation. While the precise mechanisms of activation remain to be elucidated, important inroads have been made in this field, and several candidate pathways are currently under investigation. It is also clear that different isoforms respond differently to osmotic stress, and there is some evidence that the responsiveness of any single isoform can in addition be altered by the cellular context in which it is expressed. Lastly, volume-induced regulation of NHE is important not only in the control of cell size, but potentially also in the physiology and pathophysiology of acid-base regulation and in salt and water homeostasis at the systemic level.

REFERENCES

1. Hoffmann EK, Simonsen LO. Membrane mechanisms in volume and pH regulation in vertebrate cells. Physiol Rev 1989; 69:315-382.
2. Strange K. Maintenance of cell volume in the central nervous system. Pediatr Nephrol 1993; 7:689-697.
3. Grinstein S, Foskett JK. Ionic mechanism of cell volume regulation in leukocytes. Ann Rev Physiol 1990; 52:399-414.
4. Al-Habori M. Cell volume and ion transport regulation. Int J Biochem 1994; 26:319-334.
5. Sardaki B, Parker JC. Activation of ion transport pathways by changes in cell volume. Biochim Biophys Acta 1991; 1071:407-427.
6. McCarty NA, O'Neil RG. Calcium signaling in cell volume regulation. Physiol Rev 1992; 72:1037-1061.
7. O'Neil WC. Volume-sensitive Cl-dependent K transport in human erythrocytes. Am J Physiol 1987; 253:C833-C888.
8. Pollock AS, Arieff AI. Abnormalites of cell volume regulation and their functional consequences. Am J Physiol 1980; 239:F195-F205.
9. Eveloff JLG, Warnock D. Activation of ion transport systems during cell volume regulation. Am J Physiol 1987; 252:F1-F10.
10. Rotin D, Grinstein S. Impaired cell volume regulation in Na^+-H^+ exchange-deficient mutants. Am J Physiol 1989; 257:C1158-C1165.
11. Duhm J, Gobel BO. Na^+-K^+ transport and volume of rat erythrocytes under dietary K^+ deficiency. Am J Physiol 1984; 246:C20-29.
12. Hoffmann EK, Lambert IH, Simonsen LO. Mechanisms of volume regulation in Ehrlich ascites tumor cells; role of internal Ca^{2+}. J Memb Biol 1988; 78:211-222.
13. Geck P, Pietryzyk C, Burckhardt BC, et al. Electrically silent cotransport of Na^+, K^+ and Cl^- in Ehrlich cells. Biochim Biophys Acta 1980; 600:432-447.

14. Grinstein S, Rothstein A, Sardaki B, et al. Responses of lymphocytes to anisotonic media: volume regulating behaviour. Am J Physiol 1984; 246:C204-C215.
15. Cala PM. Cell volume regulation by *Amphiuma* red blood cells: the role of Ca^{2+} as a modulator of alkali metal/H^+ exchange. 1983; 82:761-784.
16. Cala PM. Volume regulation by *Amphiuma* red blood cells: strategies for identifying alkali metal/H^+ transport. Federation Proc 1985; 44:2500-2507.
17. Parker JC. Volume-responsive sodium movements in dog red blood cells. Am J Physiol 1983; 244:C324-C330.
18. Parker JC, Castranova V. Volume-responsive sodium and proton movements in dog red blood cells. J Gen Physiol 1984; 84:379-401.
19. Pouysségur J, Sardet C, Franchi A, et al. A specific mutation abolishing Na^+/H^+ antiport activity in hamster fibroblasts precludes growth at neutral and acidic pH. Proc Natl Acad Sci 1984; 81:4833-4837.
20. Grinstein S, Rothstein A. Mechanisms of regulation of the Na^+/H^+ exchanger. J Membr Biol 1986; 90:1-12.
21. Paris S, Pouyssegur J. Growth factors activate the Na^+/H^+ antiporter in quiescent fibroblasts by increasing its affinity for intracellular H^+. J Biol Chem 1984; 259:10989-10994.
22. Sardet C, Counillon L, Franchi A, et al. Growth factors induce phosphorylation of the Na^+/H^+ antiporter, a glycoprotein of 110 kDa. Science 1990; 247:723-726.
23. Wakabayashi S, Fafournoux P, Sardet C, et al. The Na^+/H^+ antiporter cytoplasmic domain mediates growth factor signals and controls "H-sensing." Proc Natl Acad Sci USA 1992; 89:2424-2428.
24. Wakabayashi S, Bertrand B, Shigekawa M, et al. Growth factor activation and "H^+-sensing" of the Na^+/H^+ exchanger isoform 1 (NHE1). Evidence for an additional mechanism not requiring direct phosphorylation. J Biol Chem 1994; 269:5583-5588.
25. Green J, Yamaguchi DT, Kleeman CR, et al. Selective modification of the kinetic properties of Na^+/H^+ exchanger by cell shrinkage and swelling. J Biol Chem 1988; 263:5012-5015.
26. Grinstein S, Cohen S, Goetz JD, et al. Osmotic and phorbol ester-induced activation of Na^+/H^+ exchange. Possible role of protein phosphorylation in lymphocyte volume regulation. J Cell Biol 1985; 269-276.
27. Shrode LD, Klein JD, O'Neill WC, et al. Shrinkage-induced activation of Na^+/H^+ exchange in primary rat astrocytes: role of myosin light chain kinase. Am J Physiol 1995; 269:C257-C266.
28. Shrode LD, Klein JD, O'Neill WC, et al. Involvement of myosin light chain kinase in shrinkage-induced activation of Na^+/H^+ exchange in C6 glioma cells. In preparation.
29. Grinstein S, Woodside M, Sardet C, et al. Activation of the Na^+/H^+ antiporter during cell volume regulation. Evidence for a phosphorylation-independent mechanism. J Biol Chem 1992; 267:23823-23828.
30. Grinstein S, Mack E, Mills GB. Osmotic activation of the Na^+/H^+ antiport in protein kinase C-depleted lymphocytes. Biochem Biophys Res Comm 1986; 134:8-13.

31. Jean T, Frelin C, Vigne P, et al. The Na^+/H^+ exchange system in glial cell lines: Properties and activation by an hyperosmotic shock. Eur J Biochem 1986; 160:211-219.
32. Dascalu A, Nevo Z, Korenstein R. Hyperosmotic activation of the Na^+/H^+ exchanger in a rat bone cell line: temperature dependence and activation pathways. J Physiol 1992; 456:503-518.
33. Baroin A, Garcia-Romeu F, Lamarre T, et al. A transient sodium-hydrogen exchange system induced by catecholamines in erythrocytes of rainbow trout, *Salmo gairdneri*. J Physiol 1984; 356:21-31.
34. Davis BA, Hogan EM, Boron WF. Role of G proteins in stimulation of Na-H exchange by cell shrinkage. Am J Physiol 1992; 262:C533-C536.
35. Hendey B, Mamrack MD, Putnam RW. Thrombin induces a calcium transient that mediates an activation of the Na^+/H^+ exchanger in human fibroblasts. J Biol Chem 1989; 264:19540-19547.
36. Grinstein S, Rothstein A, Cohen S. Mechanism of osmotic activation of Na^+/H^+ exchange in rat thymic lymphocytes. J Gen Physiol 1985; 85:765-788.
37. Geiser JR, van-Tuinen D, Brockerhoff SE, et al. Can calmodulin function without binding calcium? Cell 1991; 65:949-959.
38. Takeda M, Homma T, Breyer MD, et al. Volume and agonist-induced regulation of myosin light-chain phosporylation in glomerular mesangial cells. Am J Physiol 1993; 264:F421-F426.
39. Klein JD, O'Neill WC. Myosin light chain phosphorylation in endothelial cells is regulated by cell volume and correlates with volume-regulatory Na-K-2Cl cotransport. J Gen Physiol 1993; 102:18a.
40. Kreisberg JI, Venkatachalam MA, Radnik RA, et al. Role of myosin light-chain phosphorylation and microtubules in stress fiber morphology in cultured mesangial cells. Am J Physiol 1985; 249:F227-F235.
41. Grinstein S, Woodside M, Waddell TK, et al. Focal localization of the NHE-1 isoform of the Na^+/H^+ antiport: assessment of effects on intracellular pH. Embo J 1993; 12:5209-5218.
42. Goss GG, Woodside M, Wakabayashi S, et al. ATP dependence of NHE-1, the ubiquitous isoform of the Na^+/H^+ antiporter. Analysis of phosphorylation and subcellular localization. J Biol Chem 1994; 269:8741-8748.
43. Goss G, Orlowski J, Grinstein S. Co-immunoprecipitation of a 24 kDa protein with NHE-1, the ubiquitous isoform of the Na^+/H^+ exchanger. Am J Physiol 1995; in press.
44. Davis RJ. The mitogen-activated protein kinase signal transduction pathway. J Biol Chem 1993; 268:14553-14556.
45. Itoh T, Yamauchi A, Miyai A, et al. Mitogen-activated protein kinase and its activator are regulated by hypertonic stress in Madin-Darby canine kidney cells. J Clin Invest 1994; 93:2387-2392.
46. Brewster JL, Valoir Td, Dwyer ND, et al. An osmosensing signal transduction pathway in yeast. Science 1993; 259:1760-1763.
47. Han J, Lee J-D, Bibbs L, et al. A MAP kinase targeted by endotoxin and hyperosmolarity in mammalian cells. Science 1994; 265:808-811.
48. Nishida E, Gotoh Y. The MAP kinase cascade is essential for diverse signal transduction pathways. Trends Biochem Sci 1993; 18:128-131.

49. Galcheva-Gargova Z, Derijard B, Wu I-H, et al. An osmosensing signal transduction pathway in mammalian cells. Science 1994; 265:806-808.
50. Yan M, Dai T, Deak JC, et al. Activation of stress-activated protein kinase by MEKK1 phosphorylation of its activator SEK1. Nature 1994; 372:798-800.
51. Matsuda S, Kawasaki H, Moriguchi T, et al. Activation of protein kinase cascades by osmotic shock. J Biol Chem 1995; 270:12781-12786.
52. Sardet C, Franchi A, Pouyssegur J. Molecular cloning, primary structure, and expression of the human growth factor-activable Na^+/H^+ antiporter. Cell 1989; 56:271-280.
53. Tse C-M, Ma AI, Yang VW, et al. Molecular cloning and expression of a cDNA enconding the rabbit ileal villus cell basolateral Na^+/H^+ exchanger. EMBO J 1991; 10:1957-1967.
54. Orlowski J, Kandansamy RA, Shull GE. Molecular cloning of putative members of the NHE gene family. cDNA cloning, deduced amino acid sequence, and mRNA tissue expression of the rat NHE exchanger NHE-1 and two structurally related proteins. J Biol Chem 1992; 267:9331-9339.
55. Wang Z, Orlowski J, Shull GE. Primary structure and functional expression of a novel gastrointestinal isoform of the rat Na/H exchanger. J Biol Chem 1993; 268:11925-11928.
56. Tse C-M, Levine S, Yun C, et al. Structure/function studies of the epithelial isoforms of the mammalian Na^+/H^+ exchanger gene family. J Membr Biol 1993; 135:93-108.
57. Tse C-M, Levine SA, Yun CHC, et al. Cloning and expression of a rabbit cDNA encoding a serum-activated ethylisopropylamiloride-resistant epithelial Na^+/H^+ exchanger isoform (NHE-2). J Biol Chem 1993; 268: 11917-11924.
58. Brant SR, Yun CHC, Donowitz M, et al. Cloning, tissue distribution, and functional analysis of the human Na^+/H^+ exchanger isoform, NHE3. Am J Physiol 1995; 269:C198-C206.
59. Collins JF, Honda T, Knobel S, et al. Molecular cloning, sequencing, tissue distribution, and functional expression of a Na^+/H^+ exchanger (NHE-2). Proc Natl Acad Sci USA 1993; 90:3938-3942.
60. Amemiya M, Yamaji Y, Cano A, et al. Acid incubation increases NHE3-mRNA abundance in OKP cells. Am J Physiol 1995; 269:C126-133.
61. Reilly RF, Hildebradt F, Biemesderfer D, et al. cDNA cloning and immunolocalization of a Na/H exchanger in LLC-PK1 renal epithelial cells. Am J Physiol 1991; 261:F1088-F1094.
62. Takaichi K, Balkovetz DF, Van Meir E, et al. Cloning, sequencing and expression of Na/H antiporter cDNAs from human tissues. Am J Physiol 1992; 262:C1069-C1076.
63. Counillon L, Scholz W, Lang HJ, et al. Pharmacological characterization of stably transfected Na^+/H^+ antiporter isoforms using amiloride analogs and a new inhibitor exhibiting anti-ischemic properties. Mol Pharmacol 1993; 44:1041-1045.

64. Rao DD, Dahdal RY, Layden TJ, et al. cDNA cloning of Na$^+$/H$^+$ exchanger isoforms (NHE2 and NHE3) from human stomach and intestine. Gastroenterology 1994; 106:A264.
65. Tse CM, Brant SR, Walker MS, et al. Cloning and sequencing of a rabbit cDNA encoding an intestinal and kidney-specific Na$^+$/H$^+$ exchanger (NHE3). J Biol Chem 1992; 267:9340-9346.
66. Klanke CA, Su YR, Callen DF, et al. Molecular cloning and physical and genetic mapping of a novel human Na$^+$/H$^+$ exchanger (NHE5/SLC9A5) to chromosome 16q22.1. Genomics 1995; 25:615-622.
67. Krapf G, Solioz M. Na$^+$/H$^+$ antiporter in RNA expression in single nephron segments of rat kidney cortex. J Clin Invest 1991; 88:783-788.
68. Biemesderfer D, Reilly RF, Exner M, et al. Immunocytochemical characterization of Na-H exchanger isoform NHE1 in rabbit kidney. Am J Physiol 1992; 263:F833-F840.
69. Biemesderfer D, Pizzonia J, Exner M, et al. NHE3: a Na$^+$/H$^+$ exchanger isoform of the renal brush border. Am J Physiol 1993; 265:F833-F840.
70. Bookstein C, DePaoli AM, Xie Y, et al. Na$^+$/H$^+$ exchangers, NHE-1 and NHE-3, of rat intestine. J Clin Invest 1994; 93:106-113.
71. Bookstein C, Musch MW, DePaoli A, et al. A unique sodium-hydrogen exchange isoform (NHE-4) of the inner medulla of the rat kidney is induced by hyperosmolarity. J Biol Chem 1994; 269:29704-29709.
72. Orlowski J. Heterologous expression and functional properties of amiloride high affinity (NHE-1) and low affinity (NHE-3) isoforms of the rat Na/H exchanger. J Biol Chem 1993; 268:16369-16377.
73. Fliegel L, Frohlich O. The Na$^+$/H$^+$ exchanger: an update on structure, regulation and cardiac physiology. Biochem J 1993; 296:273-285.
74. Borgese F, Sardet C, Cappadoro M, et al. Cloning and expression of a cAMP-activated Na$^+$/H$^+$ exchanger: evidence that the cytoplasmic domain mediates hormonal regulation. Proc Natl Acad Sci 1992; 89:6765-6769.
75. Levine SA, Montrose MH, Tse CM,, et al. Kinetics and regulation of three cloned mammalian Na$^+$/H$^+$ exchangers stably expressed in a fibroblast cell line. J Biol Chem 1993; 268:25527-25535.
76. Kandasamy RA, Yu FH, Harris R, et al. Plasma membrane Na/H exchanger isoforms (NHE-1,-2,-3) are differentially responsive to second messenger agonists of the protein kinase A and C pathways. J Biol Chem 1995; 270:29209-29216.
77. Azarani A, Goltzman D, Orlowski J. Parathyroid hormone and parathyroid hormone- related peptide inhibit the apical Na$^+$/H$^+$ exchanger NHE-3 isoform in renal cells (OK) via a dual signaling cascade involving protein kinase A and C. J Biol Chem 1995; 270:20004-20010.
78. Weinman EJ, Shenolikar S, Khan AM. cAMP-associated inhibition of Na$^+$/H$^+$ exchanger in rabbit kidney brush-border membranes. Am J Physiol 1987; 252:F19-F25.
79. Weinman EJ, Steplock D, Wang Y, et al. Characterization of a protein cofactor that mediates protein kinase A regulation of the renal brush border membrante Na$^+$-H$^+$ exchanger. J Clin Invest 1995; 95:2143-2149.

80. Levine SA, Nath SK, Yun CH, et al. Separate C-terminal domains of the epithelial specifici brush border Na^+/H^+ exchanger isoform NHE3 are involved in stimulation and inhibition by protein kinases/growth factors. J Biol Chem 1995; 270: 13716-13725.
81. Nöel J, Pouysségur J. Hormonal regulation, pharmacology, and membrane sorting of vertebrate Na^+/H^+ exchanger isoforms. Am J Physiol 1994; 268:C283-C296.
82. Mrkic B, Tse C-M, Forgo J, et al. Identification of PTH-responsive Na/H exchanger isoforms in a rabbit proximal tubule cell line (RKPC-2). Pflugers Arch 1993; 424:377-384.
83. Helmle-Kolb C, Counillon L, Roux D, et al. Na^+/H^+ exchange activities in NHE1- transfected OK cells: cell polarity and regulation. Pflugers Arch 1993; 425:34-40.
84. Bertrand B, Wakabayashi S, Ikeda T, et al. The Na^+/H^+ exchanger isoform 1 (NHE1) is a novel member of calmodulin binding proteins. Identification and characterization of calmodulin-binding sites. J Biol Chem 1994; 269:13703-13709.
85. Wakabayashi S, Bertrand B, Ikeda T, et al. Mutation of calmoldulin-binding site renders the Na^+/H^+ exchanger (NHE1) highly H^+ sensitive and Ca^{2+} regulation-defective. J Biol Chem 1994; 269:13710-13715.
86. Cohen ME, Reinlib L, Watson AJM, et al. Rabbit ileal villus cell brush border Na^+/H^+ exchange is regulated by Ca^{+2}/calmodulin-dependent protein kinase II, a brush border membrane protein. Proc Natl Acad Sci USA 1990; 87:8990-8994.
87. Chakraborty M, Chatterjee D, Gorelick F, et al. Cell cycle-dependent and kinase-specific regulation of the apical Na/H exchanger and the Na,K-ATPase in the kidney cell line LLC- PK_1 by calcitonin. Proc Natl Acad Sci USA 1994; 91:2115-2119.
88. Grinstein S, Woodside M, Goss GG, et al. Osmotic activation of the $Na^+/H+$ antiporter during volume regulation. Biochem Soc Trans 1994; 22:512-516.
89. Winkel GK, Sardet C, Pouysségur J, et al. Role of cytoplasmic domain of the Na^+/H^+ exchanger in hormonal activation. J Biol Chem 1993; 268:3396-3400.
90. Bianchini L, Kapus A, Lukacs G, et al. Responsiveness of mutants of the NHE-1 isoform of the Na^+/H^+antiport to osmotic stress. Am J Physiol 1995; 269:C998-C1002.
91. Demaurex N, Grinstein S. Na^+/H^+ antiport: Modulation by ATP and role in cell volume regulation. J Exp Biol 1994; 196:389-404.
92. Kapus A, Grinstein S, Wasan S, et al. Functional characterization of three isoforms of the Na^+/H^+ exchanger stably expressed in Chinese hamster ovary cells. ATP dependence, osmotic sensitivity, and role in cell proliferation. J Biol Chem 1994; 269:23544-23552.
93. Yun CHC, Tse CM, Nath SK, et al. Mammalian Na^+/H^+ exchanger gene family: structure and function studies. Am J Physiol 1995; 269:G1-G11.

94. Watts BA, Good DW. Apical membrane Na$^+$/H$^+$ exchange in rat medullary thick ascending limb. pH$_i$-dependence and inhibition by hyperosmolality. J Biol Chem 1994; 269:20250-20255.
95. Good DW. Hyperosmolarity inhibits bicarbonate absorption in rat medullary thick ascending limb via a protein-tyrosine kinase-dependent pathway. J Biol Chem 1995; 270:9883-9889.
96. Soleimani M, Bookstein C, McAteer JA, et al. Effect of high osmolality on Na$^+$/H$^+$ exchange in renal proximal tubule cells. J Biol Chem 1994; 269:15613-15618.
97. Azarani A, Orlowski J, Goltzman D. Parathyroid hormone and parathyroid hormone-related peptide activate the Na$^+$/H$^+$ exchanger NHE-1 isoform in osteoblastic cells (UMR-106) via a cAMP-dependent pathway. Evidence for crosstalk between the protein kinase A and C pathway. J Biol Chem 1995; in press.
98. Soleimani M, Singh G, Bizal GL, et al. Na$^+$/H$^+$ exchanger isoforms NHE-2 and NHE-1 in inner medullary collecting duct cells. Expression, functional localization, and differential regulation. J Biol Chem 1994; 269:27973-27978.
99. Casavola. Polarized expression of Na$^+$/H$^+$ exchange activity in LLC-PK1/PK20 cells. II. Hormonal regulation. Pfluegers Arch 1992; 420:282-289.
100. Mrkic B, Forgo J, Murer H, et al. Apical and basolateral Na/H exchange in ultured murine proximal tubule cells (MCT): effect of parathyroid hormone (PTH). J Membr Biol 1992; 130:205-217.
101. Cabado, AG, Yu FH, Kapus A, et al. Distinct structural domains confer cAMP sensitivity and ATP dependence to the Na$^+$/H$^+$ exchanger NHE3 isoform. J Biol Chem 1996; in press.
102. Wakabayashi S, Ikeda T, Noël J, Schmitt B, et al. Cytoplasmic domain of the ubiquitous Na$^+$/H$^+$ exchanger NHE1 can confer Ca^{2+} responsiveness to the apical isoform NHE3. J Biol Chem 1995; 270:2640-26465.

CHAPTER 7

CHARACTERISTICS OF THE PLASMA MEMBRANE Na^+/H^+ EXCHANGER GENE FAMILY

John Orlowski and Gary Shull

In cells of higher organisms, Na^+/H^+ exchanger (NHE) activity is localized to both the plasma membrane[1,2] and the mitochondrial inner membrane.[3] During the last several years, significant advances have been made in our understanding of the structure-function relationships of the ubiquitous plasma membrane Na^+/H^+ exchanger following determination of its primary structure by Sardet et al[4] in 1989. Much less is known about the mitochondrial Na^+/H^+ exchanger and a more detailed characterization of its molecular features awaits its isolation by molecular cloning.

The physiological importance of the plasma membrane NHE to cellular homeostasis is now widely acknowledged and well documented.[1,2,5-7] It participates in a variety of cellular processes, including the control of intracellular pH (pH_i), maintenance of cellular volume, transepithelial Na^+ reabsorption, and facilitation of cell proliferation in response to growth factor stimulation. Na^+/H^+ exchanger activity is also acutely and chronically regulated by a diverse range of molecular stimuli, including various growth factors, peptide hormones, neurotransmitters, cytokines, steroids, thyroxine, lectins, osmolarity and cellular adhesion.

It is now clear that there is more than one species of NHE molecule. These transporters are derived from multiple genes, with no clear evidence to date of alternative splicing of a single mRNA transcript. They possess a primary structure containing conserved structural

---

The Na^+/H^+ Exchanger, edited by Larry Fliegel. © 1996 R.G. Landes Company.

motifs believed to represent the transmembrane spanning segments as well as cytoplasmic regions that are quite divergent and most likely involved in regulation. The isoforms examined to date can be readily distinguished on the basis of their sensitivity to various pharmacological agents and, to a lesser extent, by their affinities for Na^+, H^+ and other monovalent cations. Molecular biological and immunological studies suggest that the NHE isoforms are located in particular tissues and cell types as well as targeted to distinct membrane locations. Many cell types examined so far express more than one isoform in variable ratios, and their membrane distribution may depend on the growth state of the cells. Their response to extracellular signals is also variable and is partly dependent on the cell type and membrane location. Although there is some evidence pertaining to their specific physiological functions, further studies are needed to define their distinct roles in a variety of physiological and pathological processes.

This review highlights some of the more recent studies that have defined some of the molecular characteristics of this particular family of cation transporters. The reader is also directed to other recent published reviews on this topic.[8-10]

A. MOLECULAR HETEROGENEITY

A.1 ISOLATION OF MULTIPLE ISOFORMS

The Na^+/H^+ exchanger cDNA that encodes the NHE1 isoform was initially isolated by Sardet et al[4] using an elegant genetic complementation protocol. The procedure involved the transfection of human genomic DNA into chemically-mutagenized hamster fibroblasts that were devoid of endogenous NHE activity, followed by selection for their ability to survive an NH_4Cl-induced intracellular acidification. Subsequently, the same isoform was cloned from other human tissue cDNA libraries,[11,12] consistent with its wide-spread tissue distribution. Its homolog has since been cloned from a variety of other eukaryotic species, including rat,[13] rabbit,[14,15] pig,[16] hamster,[17] trout,[18] crab (genbank accession No. U09274), nematode[19] and yeast.[20] Further screening of cDNA and genomic libraries using NHE1 cDNA fragments as probes under low stringency hybridization conditions resulted in the identification and isolation of related clones that encode distinct isoforms of the exchanger (consecutively termed NHE2, NHE3, NHE4 and NHE5).

To date, homologous gene products corresponding to NHE2 have been isolated from rat[21] and rabbit[22] gastrointestinal cDNA libraries. In addition, Collins et al[23] reported the cloning of a variant rat NHE2 cDNA from an intestinal library that lacked coding sequences for the N-terminal 116 amino acids. This region encompasses the first two putative transmembrane α-helical segments, and the variant cDNA is postulated to be derived from an alternatively-spliced transcript. Instead of the *N*-terminal coding sequences, the cDNA contains a

546-nucleotide untranslated region. The first 490 nucleotides show no close similarity to the other NHE isoforms while the last 56 nucleotides precisely match a coding segment of the full-length NHE2 cDNA. This putative NHE2 variant is capable of mediating Na^+-H^+ exchange in a heterologous expression system,[23,24] although the activity is minimal compared to the full-length rat NHE2 reported by Wang and colleagues.[21] Although it is possible that the variant NHE2 cDNA encodes a functional and physiologically relevant form of NHE2, an alternative possibility is that it is derived from a processing intermediate RNA. Such a possibility, which remains to be determined, is indicated by the fact that the point at which sequence divergence occurs closely resembles a consensus intron-exon boundary acceptor sequence (i.e., $(^T/_C)_n N(^C/_T)AG/G$ where $n > 11$).[25] Moreover, the position of this potential splice-site junction is analogous to the intron-exon boundary between the first intron and the second exon of the human NHE1 gene.[26] In this regard, it has been generally observed that members of gene families retain similar genomic organization. In total, these observations suggest that the variant rat NHE2 cDNA may have been synthesized from a partially processed transcript, and that it contains an intron sequence at its 5' end.

A second isoform has also been cloned from the nematode *Caenorhabditis elegans* (genbank accession No. U21317). It shares 41% amino acid identity to the nematode *C. elegans* NHE1[19] but does not appear to align itself with any particular mammalian isoform.

NHE3 has been cloned from rat,[13] rabbit,[27] opossum[28] and human[29] renal and/or intestinal cDNA libraries. At present, a cDNA clone corresponding to NHE4 has been found only in a rat stomach library.[13] Most recently, a fifth isoform was identified in humans by partial genomic cloning and polymerase chain reaction (PCR) analysis.[30] The primary structures of these isoforms range in length from 717 to 834 amino acids, with calculated relative molecular masses (M_r) of ~81 k to ~93 k (Table 7.1).

Evolutionarily, the NHE1 isoform shows considerable divergence among multicellular eukaryotic organisms (summarized in Fig. 7.1). For example, the nematode *C. elegans* NHE1 exhibits only 37% amino acid identity to human NHE1. In single cell species such as the yeast *Schizosaccharomyces pombe*, the primary structure of the Na^+/H^+ exchanger *Sod2* protein shares minimal similarity to its homolog in higher organisms. Pairwise comparisons among mammalian isoforms also reveals variable degrees of relatedness. NHE1 shares 48% amino acid identity with NHE2 and NHE4, while the latter two isoforms exhibit approximately 60% identity to each other. NHE3 is more divergent, sharing ~40% identity with NHE1, NHE2 and NHE4. Sequence analysis of a partial PCR product of human NHE5 reveals that it has highest sequence similarity (~73%) to NHE3.

Table 7.1. Characteristics of mammalian Na^+/H^+ exchanger isoform genes

Characteristics	NHE1	NHE2	NHE3	NHE4	NHE5	NHE3P
Amino acid residues	815-822	809-813	831-834	717	?	–
Calculated M_r	~91 K	~91 K	~93 K	~81 K	?	–
Tissue expression	Ubiquitous	Many tissues	Kidney, stomach and intestine	Stomach	Brain, spleen, testis, and skeletal muscle	–
Chromosomal Location:						
Human	1p35-p36.1	2	5p15.3	2	16q22.1	10
Rat	5	9	1	9	?	–
Mouse	4	?	?	?	?	?
Comments					partial sequence	pseudogene

These data are summarized from references 4, 13, 14, 21, 22, 27, 29-33.

A.2 CHROMOSOMAL MAPPING

These isoforms represent unique gene products that are dispersed throughout the mammalian genome (Table 7.1). In humans, NHE1 maps to chromosome 1p35-p36.1.[31,32] The human NHE3 cDNA has been used as a probe to map the NHE3 gene to chromosome 5p15.3 in human-rodent somatic cell hybrids. In an independent study, Szpirer et al[33] used a rat NHE3 cDNA to screen a different panel of human-rodent somatic cell hybrids and obtained two hybridization signals in the human genome, one on chromosome 5 and the other on chromosome 10. However, recent unpublished data suggest that this latter NHE3-like gene (termed NHE3P) on chromosome 10 most likely represents a pseudogene (see genbank accession No. U16020). Rat cDNA probes have also been used on somatic cell hybrids to map both the NHE2 and the NHE4 loci to human chromosome 2,[33] although more refined localization remains to be undertaken. The presence of NHE2 and NHE4 on the same chromosome and their higher amino acid similarity to each other suggests that they may have arisen from a recent gene duplication event. Fluorescence in situ hybridization analysis of human metaphase chromosomes and analysis of high-resolution human-mouse somatic cell hybrids have localized the NHE5 gene to the cytogenetic subband 16q22.1.[30] In rats, analyses of rat-mouse somatic cell hybrids assign the general locations of NHE1, NHE2, NHE3 and NHE4 to chromosomes 5, 9, 1 and 9, respectively.[33] The identification of these loci provide the foundation for further investigations into possible genetic disorders involving defects in the Na^+/H^+ exchanger.

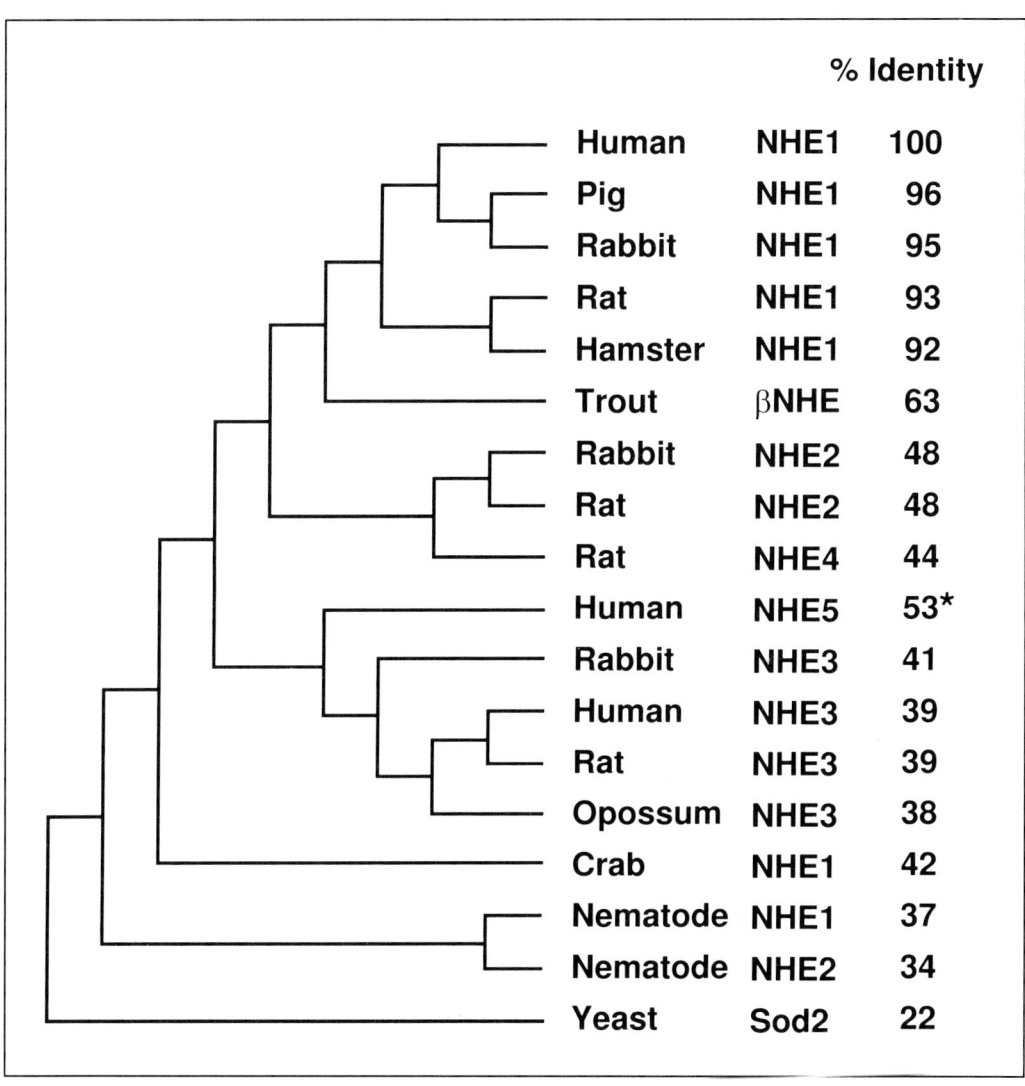

Fig. 7.1. Evolutionary relationship of the members of the eukaryotic Na+/H+ exchanger gene family. The amino acid sequences were aligned using the CLUSTAL W alignment algorithm[126] and the phylogenetic tree was constructed using the Neighbor-Joining method in the Molecular Evolutionary Genetics Analysis (MEGA) program.[127] The percentage of amino acid identity was calculated by pairwise comparisons of the different exchangers to human NHE1. The asterisk indicates that the value was calculated using a partial amino acid sequence.

B. GENOMIC ORGANIZATION

At present, information on the genomic organization of the NHE genes has been derived primarily from studies of NHE1. The human NHE1 gene spans approximately 70 kilobases and contains 12 exons that range in length from 71 to 1132 nucleotides.[26] The 5'-flanking promoter/enhancer region contains a classical consensus TATA box and potential consensus sequences for a variety of regulatory transcription factors (e.g., CCAAT-box binding protein (C/EBP), AP-1, SP-1, cAMP responsive element-binding protein (CREB) and glucocorticoid receptor). Regions capable of binding some of these *trans*-acting factors have been defined,[34] although their functional importance has yet to be rigorously demonstrated.

The 5'-flanking promoter/enhancer regions of the mouse[35] and rabbit[36] *Nhe1* gene have also been characterized and contain similar potential *cis*-acting elements. In addition, the transcription factor AP-2 directly binds and regulates transcription of the mouse *Nhe1* gene[35] and appears to play an important role in the activation of *Nhe1* transcription during retinoic acid-induced neuronal differentiation of mouse P19 embryonal carcinoma cells.[37]

Much less information is available concerning the genomic organization of the other NHE genes. Partial genomic characterization of mouse *Nhe2* and *Nhe4* shows that their intron/exon organization in the 5' region containing exon 2 is similar to human NHE1.[30] Although limited in scope, these data are consistent with these genes diverging from a closely related ancestral gene (see Fig. 7.1). In contrast, this exon in rat *Nhe3* and human NHE5 is split at an identical position by an apparent intron that has no counterpart in the other NHE genes. This suggests that the NHE3 and NHE5 genes are more closely related evolutionarily, consistent with their higher percentage of amino acid identity in pairwise comparisons. As such, these latter isoforms can be classified as a distinct sub-branch of the NHE gene family.

C. TISSUE EXPRESSION

The NHE isoforms are expressed in isoform-specific patterns and at varying levels in different tissues and cells (summarized in Table 7.1). Analysis of mRNA abundance reveals that NHE1 is present in virtually all mammalian tissues and cells examined,[13-15] although some exceptions have been reported. Using microdissected nephron segments, Krapf and Solioz detected NHE1 mRNA in rat glomeruli, proximal tubule S_1 and S_2 segments from juxtamedullary nephrons, cortical thick ascending limb, distal convoluted tubule and cortical collecting duct segments, but not in proximal tubule S_1 and S_2 segments from superficial and midcortical nephrons.[38] As well, NHE1 was not expressed in an opossum kidney cell line which exhibits many of the functional characteristics of proximal tubule cells.[39] Whether these exceptions truly indicate the absence of NHE1 mRNA in some proximal tubule epithelial cells or reflect the detection limitations of the assays employed is unclear. Biochemical and immunological analyses have localized NHE1 to the basolateral membrane of various polarized epithelial cells,[14,40,41] although some exceptions have been reported.[42] The generally ubiquitous nature of NHE1 expression is consistent with this isoform fulfilling essential "housekeeping" functions such as the maintenance of intracellular pH and cellular volume.

The NHE2 isoform also has a wide tissue distribution, although it is somewhat more restricted. In adult rat tissues, its mRNA is expressed most abundantly in the gastrointestinal tract, with substantially lower amounts present in skeletal muscle, brain, kidney, testis, uterus, heart and lung, whereas it is absent in liver and spleen.[21] In comparison to rat, rabbit NHE2 mRNA has a similar broad pattern of tissue expression, although there are some differences in the relative tissue abundance between the two species. Unlike rat, rabbit NHE2 mRNA was not detected in brain and heart.[22] Thus, the transcriptional control of this isoform is dependent not only on the tissue but also the species. Aside from its extensive tissue distribution, NHE2 also exhibits diverse membrane targeting in polarized renal epithelial cells, with studies reporting localization to the basolateral membrane of mouse inner medullary collecting duct (IMCD-3) cells[43] and to the apical membrane of SV-40 transformed rabbit S_2 proximal tubule (RKPC-2) cells.[44] The physiological roles of NHE2 in these various tissues are unknown. However, when heterologously expressed in Chinese hamster ovary cells, it is capable of regulating intracellular pH, cellular volume and cell proliferation in a manner resembling that of NHE1,[45] suggesting that it may also fulfill these roles in native tissues.

The tissue distribution of NHE3 mRNA generally shows a more restricted pattern of expression. In adult rats, the mRNA is expressed predominantly in the kidney and gastrointestinal tract.[13] Following longer autoradiographic exposures of the Northern blot, barely detectable levels of NHE3 mRNA are also observed in heart and brain. Similarly in rabbits, NHE3 mRNA is present at high levels only in kidney and intestine.[27] Human NHE3 is also highly abundant in kidney and small intestine.[29] However, unlike rat or rabbit, it is also expressed in lower abundance in numerous other tissues, including testis, ovary, prostate, thymus, leukocytes, brain, spleen and placenta. In polarized renal proximal tubule[46,47] and intestinal[41] cells, immunological analyses show that NHE3 protein is localized exclusively to the apical or luminal surface. As such, this isoform most likely plays a significant role in facilitating Na^+ reabsorption in these tissues. Furthermore, in renal proximal tubule cells, the corresponding luminal secretion of H^+ is necessary for HCO_3^- reabsorption.[48,49]

Expression of the NHE4 gene occurs predominantly in stomach.[13] Substantially lesser amounts were also initially detected in small and large intestine, but it has since been determined that this was due to cross-hybridization to NHE2 mRNA which is highly abundant in these tissues (J. Orlowski and G. Shull, unpublished observations). Barely detectable levels of NHE4 are also found in uterus, brain, kidney and skeletal muscle. In kidney, NHE4 mRNA is specifically expressed in the collecting tubule of the inner medulla.[50] Because this region is exposed to high osmolarity, it is suggested that this isoform may fulfill a specialized role in controlling the volume homeostasis of these cells.[50]

Northern blot analysis reveals a distinct pattern of expression for human NHE5.[30] Unlike the other isoforms, human NHE5 is not present to any significant extent in kidney or intestine but appears to reside primarily in a subset of nonepithelial tissues. Highest expression is found in brain and spleen, followed by lower levels in testis and skeletal muscle. The physiological function of this isoform has yet to be elucidated.

D. MEMBRANE TOPOLOGY

As mentioned above, the primary structures of these isoforms share ~40-70% amino acid identity (M_r ranging from ~81-93 K). Based on these primary sequences, computerized analyses have been used to calculate the hydrophobic and hydrophilic nature of the protein and predict potential α-helical regions (~20 amino acids in length sufficient to traverse a 40-Å membrane) and β-sheet structures. Comparison of their predicted secondary structures reveals similar plasma membrane topologies with 10-12 predicted N-terminal hydrophobic membrane-spanning (M) segments and a large C-terminal hydrophilic cytoplasmic region.[4,13,14] A tentative model showing 12 transmembrane segments is illustrated in Figure 7.2, although the precise topological organization remains to be defined empirically.

The most conserved regions of the NHE isoforms are the predicted membrane-spanning segments (M3 to M12) which range from 55-95% amino acid identity. The most highly conserved (95% identity) of these are M6 and M7, thereby implicating this region in the cation selective transport of Na^+ and H^+ across the membrane.

In contrast, the C-terminal regions are highly hydrophilic and exhibit the lowest degree of similarity (~24-31% identity), with the exception of NHE2 and -4 which exhibit ~56% identity. These regions contain several potential sites for phosphorylation by different protein serine/threonine kinases known to modulate NHE activity, including the cAMP-dependent protein kinase (PKA), Ca^{2+}/phospholipid-dependent protein kinase (PKC), and Ca^{2+}/calmodulin-dependent protein kinase II (CaM-PKII) (for review, see refs. 6 and 7), further suggesting that this region is oriented towards the cytoplasmic side of the membrane. Indeed, deletion of significant portions of these regions in NHE1[51,52] and NHE3[53] yields functional transporters that can no longer be regulated by growth factors and other mitogens. Likewise, antibodies generated against the putative C-terminal cytoplasmic region of NHE1 are capable of blocking growth factor regulation when injected into cells.[54]

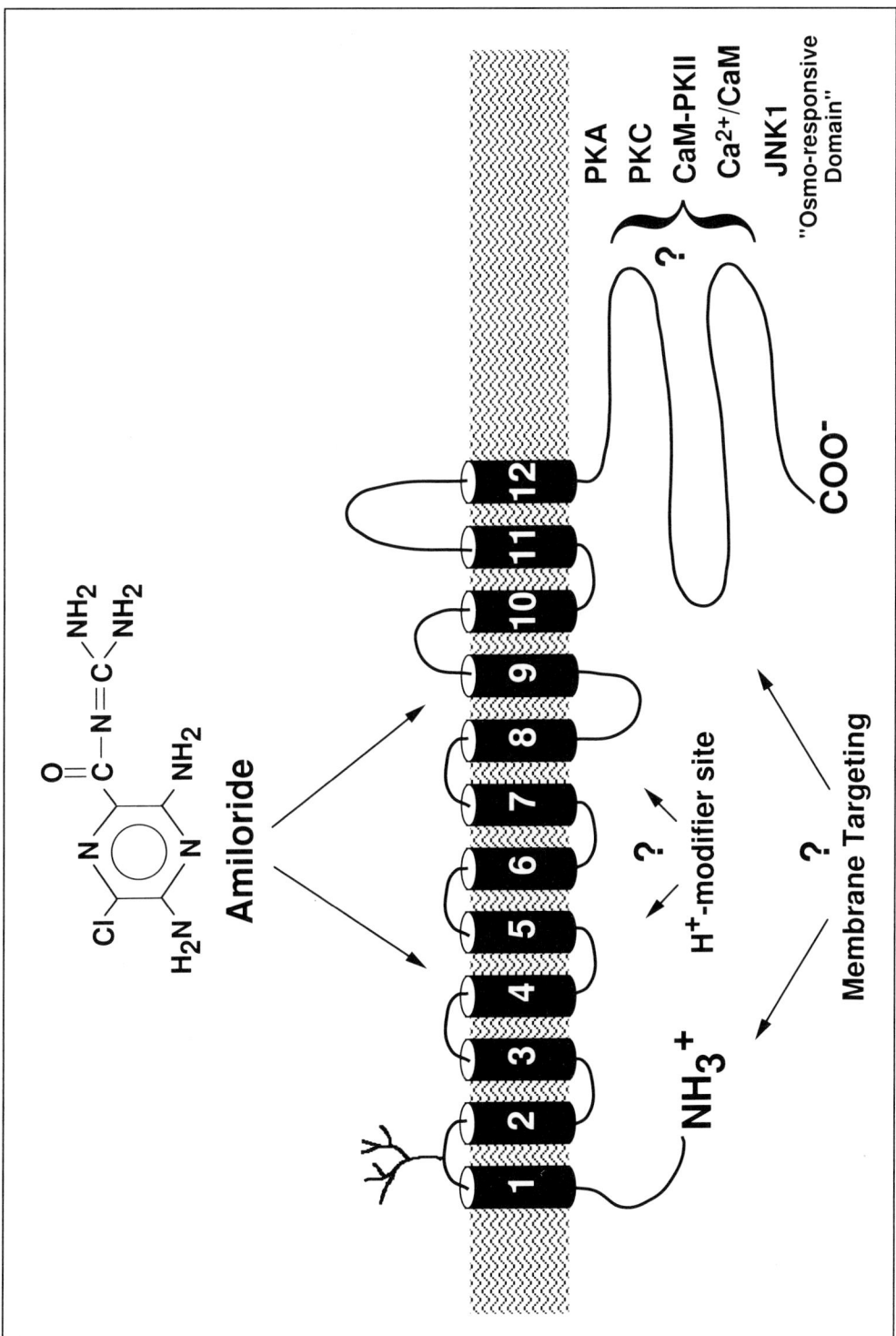

Fig. 7.2. Putative two-dimensional structural organization of the plasma membrane Na^+/H^+ exchanger. General locations of various structural features are illustrated based on molecular studies.

E. FUNCTIONAL CHARACTERISTICS

E.1 BIOCHEMICAL PROPERTIES

The rate of Na^+/H^+ exchange has generally been found to have a first order dependence on the external Na^+ concentration (Na^+_o) and exhibits simple, saturating Michaelis-Menten kinetics.[55] This kinetic relationship is observed for rat NHE1, NHE2 and NHE3 expressed in Chinese hamster ovary AP-1 cells, with apparent K_m values for Na^+_o of 10, 50 and 5 mM, respectively.[56,57] The higher Na^+ affinity of NHE3 compared to NHE1 and NHE2 suggests that it is better suited to mediating Na^+ uptake at low extracellular Na^+ concentrations, consistent with its presumed role in reabsorbing Na^+ from the lumen of renal proximal tubules and intestine. These results vary somewhat with those obtained for the rabbit NHE isoforms (NHE1, NHE2 and NHE3) which show similar affinities for Na^+_o ($K_m\sim$15-18 mM) when expressed in Chinese hamster lung PS120 fibroblasts.[58] These kinetic differences may result from the use of different measurement techniques ($^{22}Na^+$ influx versus H^+ efflux using a pH-sensitive fluorescent dye) or may reflect species and/or cell-type differences.

In contrast to Na^+_o, the H^+_i dependence of the Na^+/H^+ exchanger does not appear to follow simple Michaelis-Menten kinetics. Intracellular H^+ activates the Na^+/H^+ exchanger with a greater than first order dependence, implying the existence of more than a single binding site for intracellular protons.[59-61] As first suggested by Aronson et al,[59] this characteristic of the Na^+/H^+ exchanger could be accounted for by the existence of a H^+_i- sensitive modifier site in addition to the transport site; thereby allowing the exchanger to rapidly activate once intracellular pH drops below a certain threshold level. This feature is conserved in all NHE isoforms isolated from the different species examined to date, although the apparent H^+_i-sensitivity or set-point varies depending on the isoform.[51,56-58] At present, limited structural information (discussed in section G-3) is available concerning the domains or specific amino acids involved in either cation binding and transport or the internal H^+-modifier site that modulates the activity of the Na^+/H^+ exchanger.

E.2 PHARMACOLOGICAL PROPERTIES

The Na^+/H^+ exchanger has long been known to have different affinities for the diuretic compound amiloride and its analogs depending on the tissue or cell type examined.[7] Using a mammalian cell expression system, the mammalian NHE isoforms exhibit diverse affinities for amiloride and its analogs that range over two orders of magnitude and show the following order of sensitivity: NHE1>NHE2>>NHE3[56,57,62-64] (summarized in Table 7.2). The low affinity of NHE3 for amiloride compounds is consistent with this isoform being the low affinity form found on the apical surface of many polarized epithelial cells, most notably in renal

Table 7.2. Pharmacological properties of mammalian Na/H exchanger isoforms

Inhibitor	Inhibition Constants ($K_{0.5}$)								
	NHE1			NHE2			NHE3		
	Rat	Rabbit	Human	Rat	Rabbit	Human	Rat	Rabbit	Human
Amiloride compounds									
Amiloride	1.6 µM	1 µM	3 µM	1.4 µM	1 µM		100 µM	39 µM	49 µM
EIPA	15 nM	20 nM	22 nM	79 nM	1 µM		2.4 µM	8 µM	6.6 µM
DMA	23 nM		100 nM	250 nM			14 µM		
Benzamil	120 µM			320 µM			100 µM		
Other compounds									
HOE694			160 nM		5 µM		650 µM		
Cimetidine	26 µM		28 µM	330 µM			6.2 mM		
Clonidine	210 µM			42 µM			620 µM		
Harmaline	140 µM			330 µM			1.0 mM		

The apparent inhibition constants ($K_{0.5}$) are summarized from references 56, 57, 62-64, 120.

proximal tubule and intestinal epithelia. Recently, a new antagonist of NHE activity, HOE694, has also been found to inhibit the isoforms with a similar order of sensitivity as the amiloride compounds, but over a larger concentration range (three orders of magnitude).[64] The more selective binding properties of this compound have been exploited as a potential therapeutic agent in the treatment of cardiac ischemia and reperfusion injuries.[65]

In addition to amiloride analogs and HOE694, other pharmacological agents are known to inhibit the Na⁺/H⁺ exchanger. These include cimetidine, a histamine H_2-receptor antagonist,[66,67] clonidine, an α_2-adrenergic receptor agonist,[67,68] harmaline, a hallucinogenic drug known to inhibit amine oxidase and antagonize some Na⁺-dependent transport systems,[67,69,70] loperamide, an opiate receptor agonist,[71] and various guanidinium derivatives.[72] While these compounds are chemically unrelated to amiloride or HOE694, they generally possess either an imidazoline or guanidinium moiety and hence bear some structural similarity to amiloride. To date, the inhibitory effects of some of these compounds on the different NHE isoforms have been examined. Cimetidine and harmaline inhibit the rat NHE isoforms with an order of sensitivity that is similar to that of the amiloride compounds (i.e., NHE1>NHE2>NHE3).[56,57] In contrast, clonidine is more selective for NHE2.[56,57]

F. REGULATION OF Na⁺/H⁺ EXCHANGER ACTIVITY

A wide variety of molecular signals are known to modulate Na⁺/H⁺ exchanger activity either rapidly (seconds to minutes; including neurotransmitters, growth factors, peptide hormones, phorbol esters, chemotactic factors, lectins and osmotic shrinkage) or following a considerable latent period (hours to days; including acidosis, thyroid and steroid hormones) before the effects on the rate of transport are manifested.[2,6] The response of the Na^+/H^+ exchanger to these signals is complex and depends not only on the cell type but also on its location in the plasma membrane. This is reflected in part by the existence of multiple NHE isoforms as well as by the extensive molecular diversity of signal transduction pathways. Recent studies are now beginning to unravel this complexity and identify the stimuli which selectively modulate the individual NHE isoforms and the underlying mechanisms.

F.1 ACUTE REGULATION

Previous studies have convincingly demonstrated that the phospholipase C-diacylglycerol-PKC pathway constitutes a major signaling route for acute activation of the ubiquitous NHE1 isoform of the exchanger. Stable expression of human[73] and rabbit[58] NHE1 in PS120 fibroblastic cells or rat NHE1 in AP-1 cells[74] has shown that this isoform is rapidly activated following acute cell stimulation by phorbol esters as well as by growth factors and other mitogens. However, the precise molecular mechanism remains unclear.

Stimulation of NHE1 by phorbol esters and other growth-promoting agents increases phosphorylation of a common set of tryptic peptide fragments in the C-terminal region of the exchanger.[73,75] These data imply that the different agonists, which stimulate diverse signaling pathways, ultimately transmit their signals to a common protein kinase that phosphorylates and activates the exchanger. However, deletion of this region (amino acids 635-815) only partially impairs (50%) activation, whereas removal of another upstream region (amino acids 567-635), which does not contain any of the phosphorylation sites, completely abolishes activation by several growth-promoting agents.[52] The involvement of multiple regulatory regions to account for the stimulation of NHE1 by diverse agents is also supported by studies of Winkel et al[54] who demonstrated that microinjection of a polyclonal antibody raised against amino acids 658-815 of NHE1 ablates the stimulation mediated by endothelin-1 and α-thrombin, but is ineffective in preventing activation induced by phorbol ester and hyperosmotic medium. These data suggest that other mechanisms in addition to direct phosphorylation of NHE1 may play an important role in regulating its activity. One proposed mechanism is the participation of exchanger-associated regulatory factors, some of which may also be potential targets of protein kinases. In this regard, Ca^{2+}/calmodulin directly binds and activates the NHE-1 isoform.[76,77] In addition, the mammalian 70 kD heat-shock protein (hsp70)[78] and an unidentified 24 kD protein (G. Goss, J. Orlowski, S. Grinstein, Am J Physiol, in press) also associate in situ with NHE1, although their functional and potential regulatory significance has yet to be defined.

In contrast to PKC-mediated activation of NHE1, evidence supporting a role for cAMP in the regulation of NHE1 is rather sparse and contradictory and has led to the general perception that this isoform is not responsive to this second messenger. This view is supported by studies showing that human[18] and rabbit[58] NHE1 expressed in PS120 fibroblastic cells are unresponsive to cAMP analogs even though these cells have a functional PKA pathway.[18] However, when human NHE1 is stably transfected into opossum kidney cells, its activity is inhibited following activation of PKA or PKC, suggesting possible cell-specific regulatory effects.[79] In contrast, murine macrophages,[80] primary rat hepatocytes,[81] and rat osteoblastic UMR-106 cells[82] show significant cAMP-induced stimulation of NHE activity. Interestingly, these cells or tissues express only the NHE1 isoform.[13,21,83,84] Furthermore, rat NHE1 stably expressed in AP-1 cells is also stimulated by agonists that increase $cAMP_i$ accumulation.[74] Thus, NHE1 can be regulated by the cAMP/PKA pathway, but its response is greatly influenced by the cellular milieu.

In addition to rat NHE1, the trout red cell also expresses a Na^+/H^+ exchanger, called β–NHE, that has a primary structure with highest identity to that of mammalian NHE1 and is phorbol ester- and cAMP-activatable in PS120 fibroblasts.[18] The trout β–NHE contains two optimal consensus sites for phosphorylation by PKA (R-R/K-X-S*/T*) at Ser^{641} and Ser^{648} which, when simultaneously mutated to Gly, partially reduce (by ~72%) the ability of $cAMP_i$ to activate the exchanger.[85] The residual cAMP-activatable activity requires amino acids 559-661 which may contain cryptic PKA sites that have yet to be identified or, alternatively, may interact with cAMP/PKA-regulated accessory factors. Interestingly, mutation of the two serine residues does not alter the capacity of βNHE to be induced by phorbol ester, suggesting that the actions of PKC are not convergent with those of PKA and are mediated elsewhere in the exchanger. It is also worth noting that while rat NHE1 contains several putative PKC consensus sequences ($(R/K)_{1-3}$, X_{2-0})-S*/T*-(X_{2-0}, R/K_{1-3})) in its C-terminal region,[13] it does not contain a classical consensus site for PKA. However, since there is overlap in consensus sequence determinants among protein kinases,[86] one cannot exclude the potential for PKA phosphorylation of NHE1. Alternatively, as proposed for PKC regulation, the effects mediated by PKA may be transmitted through ancillary factors. Thus, while it is difficult at the present time to reconcile the variable regulation of NHE1 by increasing $cAMP_i$, several factors operating independently or in combination may account for these observations, such as cell-specific differences in the expression of signaling components, putative exchanger-associated regulatory factors, or perhaps species variation.

Unlike NHE1, much less is known about second messenger regulation of NHE2. In SV-40 transformed RKPC-2 cell lines, the native apically-localized NHE2 is inhibited by 8-bromo-cAMP, whereas it is stimulated by PMA.[44] However, when rabbit NHE2 is stably expressed in PS120 fibroblast cells, it is unresponsive to cell permeant cAMP analogs, although it is capable of being activated by phorbol esters as well as serum.[58] Rat NHE2 is also stimulated by phorbol ester when expressed in AP-1 cells.[74] However, its activity can also be enhanced by cAMP analogs.[74] Thus, the variable responsiveness of NHE2 to cAMP is probably dependent on the cell type. Unlike NHE1, the C-terminal cytoplasmic domain of rat and rabbit NHE2 contains several classical consensus sequences for phosphorylation by PKA as well as PKC. The question of whether kinase action is mediated through phosphorylation of these sites or possibly via cell-specific, exchanger-associated regulatory factors awaits future studies.

In contrast to NHE1 and NHE2, rat NHE3 in AP-1 cells shows decreased rates of transport in response to second messenger agonists of the PKA and PKC pathways.[74] The negative regulation by these pathways precisely mimics that observed for the endogenous, apically-located NHE3 isoform in renal proximal tubule OK cells.[39,87] Likewise, rabbit NHE3 in PS120 cells is also inhibited by acute exposure to PMA, although it does not respond to elevated cAMP$_i$ in these cells.[58] On the contrary, the rabbit renal Na$^+$/H$^+$ exchanger in isolated brush border membrane vesicles (presumably NHE3) is inhibited by PKA.[88] Thus, cAMP-mediated regulation of NHE3 is also influenced by the cellular environment.

More recent structural analyses by Levine and colleagues[53] show that a region between amino acids 585 and 689 of rabbit NHE3 mediates inhibition by PKC. Interestingly, a similar region in rat NHE3 is also essential for the cAMP response (A. Cabado, F. Yu, A. Kapus, G. Lukacs, S. Grinstein and J. Orlowski, J Biol Chem, in press). Examination of the cytoplasmic domain of rat and rabbit NHE3 reveals the presence of potential consensus sequences for PKA as well as for PKC within or in close proximity to this region. Thus, PKA and PKC may act at the same phosphorylation site or may phosphorylate discrete sites within this region which, nevertheless, similarly influence exchanger activity. Alternatively, these protein kinases may mediate their effects indirectly via phosphorylation-dependent ancillary proteins. With regard to the latter, there is some in vitro evidence that PKA-mediated inhibition of the rabbit renal apical Na$^+$/H$^+$ exchanger requires the involvement of a regulatory protein that is separate from the kinase and transporter.[89,90] Cell-specific expression of this protein or other factors could account for the variable responsiveness of NHE3 to individual protein kinases.

Interestingly, the activity of rat NHE3 in AP-1 cells is also inhibited when acutely exposed to hyperosmotic medium, whereas NHE1 and NHE2 are stimulated.[45] The hyperosmotic-induced inhibition of NHE3 is mimicked by the native apical NHE3 isoform in renal proximal tubule OK cells[91] and the apically-located Na$^+$/H$^+$ exchanger (possibly NHE3) in rat medullary thick ascending limb.[92] Although the mechanism is not fully understood, osmotic regulation of the exchangers is ATP-dependent, but does not involve direct phosphorylation of the exchanger, at least in the case of NHE1.[93] This process may involve G-proteins that are independent of the PKA and PKC pathways.[94,95] Potential intermediary targets for these G-proteins may include other protein kinases that have been implicated in the osmosensing pathway, such as myosin light-chain kinase,[96] tyrosine kinases,[97] p38 protein kinase[98] and Jun kinase/stress-activated protein kinase.[99]

F.2 CHRONIC REGULATION

In addition to rapid regulation, prolonged treatment of cells to various stimuli also modulates the activities of the NHE isoforms. For example, chronic exposure of renal epithelial cells to acidic medium elevates the activities and mRNA abundance of NHE1[100,101] and NHE3,[28] but inhibits those of NHE2.[43] In contrast, NHE1 mRNA and activity are reduced in NIH 3T3 fibroblasts following prolonged exposure to acidic medium.[100] Interestingly, acid-mediated stimulation of renal NHE1 is associated with activation of protein kinase C and the transcription factor AP-1,[102,103] whereas acid induction of renal NHE3 is linked to the *c-src* family of nonreceptor protein tyrosine kinases.[104] Thus, the acid-mediated regulation of Na$^+$/H$^+$ exchanger gene activity is complex, involving multiple signaling pathways that influence specific isoform genes in a cell type-dependent manner.

Similarly, hormones and high osmolarity also regulate the Na$^+$/H$^+$ exchanger in an isoform-specific manner. Retinoic acid treatment of mouse P19 embryonal carcinoma cells increases NHE1 transcription by 10-fold.[37] This response appears to be mediated by activation and direct binding of the transcription factor AP-2 to the *Nhe1* promoter region. Likewise, long-term administration of glucocorticoids to rabbits selectively elevates NHE3 mRNA and activity in ileum[105] and renal proximal convoluted tubules,[106] but has no effect on NHE1 or NHE2. Last, incubation of IMCD cells in hyperosmotic media for 72 hours stimulates total cellular Na$^+$/H$^+$ exchanger activity, which is associated with increased NHE2 and reduced NHE1 mRNA abundances.[43] At present, the underlying mechanisms for glucocorticoid and hyperosmolarity regulation have yet to be resolved, but are likely to involve altered rates of gene transcription and/or mRNA stability.

G. STRUCTURAL COMPONENTS OF THE Na$^+$/H$^+$ EXCHANGER

Biochemical and pharmacological studies have identified several discrete structural components of the Na$^+$/H$^+$ exchanger that influence Na$^+$ and H$^+$ transport, including glycosylation sites, the amiloride binding site, the H$^+$-modifier site, and potential regulatory sites that respond to ATP, osmolarity and various protein kinases. Recent structure-function studies are now beginning to delineate some of these elements.

G.1 GLYCOSYLATION

Examination of the primary structures of the NHE isoforms reveals several potential glycosylation sites. Recent studies with human[75] and rabbit[14] NHE1 stably transfected in fibroblasts (PS120 cells) show that the transporter is glycosylated. The protein contains both N- and O-glycosylated residues within the first exoplasmic loop and mutation of Asn75 abolishes the N-linked glycosylation.[107] In contrast, rabbit NHE2 expressed in the same cell type exhibits only O-linked glycosylation.[108] Unlike NHE1 or NHE2, NHE3 does not appear to be glycosylated.[46,107] At present, it is not known whether NHE4 or NHE5 are glycosylated.

Although glycosylation has been implicated in the proper biosynthetic processing and transport of ion transport proteins to the membrane surface,[109,110] the importance of this feature to the functioning of the Na$^+$/H$^+$ exchanger is unclear. Removal of N-linked oligosaccharide chains reduces the transport activity of the Na$^+$/H$^+$ exchanger in rat renal brush border membranes.[111] However, removal of the carbohydrate moieties of NHE1 and NHE2 in transfected PS120 fibroblasts has no effect on their rates of transport.[107,108,112]

G.2 DETERMINANTS OF AMILORIDE SENSITIVITY

Amiloride and its analogs exhibit simple competitive inhibition at the external Na$^+$ transport site,[113-116] implying that they occupy the same binding site. However, under different anionic buffer conditions, these compounds can also inhibit transport noncompetitively, suggesting that the external Na$^+$ and amiloride binding sites may not be identical.[117,118] Furthermore, the extracellular Na$^+$- and amiloride-binding sites can be altered independently of each other using genetic selection techniques.[119] Overall, these data indicate that amiloride may bind to multiple regions of the exchanger.

Recent data from molecular studies provide support for the above postulation. Pouysségur and colleagues, using a H$^+$-killing selection technique, isolated mutant cells that express an amiloride-resistant variant of the Chinese hamster NHE1 isoform.[63] Sequencing and site-directed mutagenesis of this variant cDNA demonstrate that a single amino acid substitution, Leu167 to Phe167, in the fourth transmembrane region of NHE1 is responsible for modulating amiloride sensitivity and possibly binding. Likewise, mutagenesis of the equivalent residue in rabbit NHE2 (Leu143 to Phe143) also reduces its sensitivity to amiloride compounds.[62] In addition, Wang et al[120] recently reported that mutation of His349 to Tyr, Phe, Gly or Leu in the putative ninth transmembrane segment of human NHE1 produces a modest alteration in amiloride sensitivity, although other amino acid substitutions are without effect. Thus, both the fourth and ninth transmembrane segments appear to contribute to amiloride sensitivity. However, mutations at the above mentioned sites do not confer the full degree of resistance observed for the amiloride low affinity NHE3 isoform,[56,63,64] implicating

G.3 H⁺-MODIFIER SITE

One of the most interesting kinetic features of the Na^+/H^+ exchanger is its allosteric activation by intracellular H^+.[59] This apparent allosteric activation by H^+_i can be interpreted most simply by assuming the presence of one or more ionizable groups, possibly those present on histidine residues that, upon protonation, alter the conformation of the protein and cause activation.

Some structural evidence supporting this paradigm is provided from studies of mammalian and prokaryotic Na^+/H^+ exchangers. Wakabayashi et al,[51] using deletion mutagenesis, demonstrated that the N-terminal transmembranous region of human NHE1 most likely contains the H^+-modifier site while the C-terminal cytoplasmic domain modulates the pH_i set-point value. In a related study, Gerchman et al used site-directed mutagenesis to identify a histidine residue (His226) in the intracellular loop between transmembrane segments 7 and 8 as a component of the pH sensor of the NhaA isoform of the *Escherichia coli* Na^+/H^+ exchanger.[121] Although the bacterial transporter shares little similarity in primary structure to its mammalian homologs, presumably an analogous histidine residue(s) in the plasma membrane Na^+/H^+ exchangers fulfills a similar role.

G.4 QUATERNARY STRUCTURE

Although little is known about the tertiary or quaternary structure of the Na^+/H^+ exchangers, it is generally believed that the enzyme functions as a monomer.[4] However, recent biochemical and immunological analyses indicate that the exchanger exists in the membrane as a homodimer[122,123] and does not appear to form a heterodimer, at least in the cases of NHE1 and NHE3.[123] The site of interaction between the monomers resides in the putative transmembranous region,[123] possibly linked by disulfide bonding,[122] although the precise location(s) has yet to be defined. Whether these homodimers function in higher orders in cells is unknown, although pre-steady-state kinetic analyses[124] and radiation inactivation studies[125] support the possibility of a tetramer organization.

H. CONCLUDING REMARKS

Significant advances in our understanding of the structure-function relationships of the Na^+/H^+ exchanger gene family have been made over the last several years using molecular-genetic approaches. The primary structures of distinct members of this class of cation transporter have been deduced by cDNA cloning and sequencing. Computer analyses of the hydrophobicity and potential secondary structure and sequence comparisons with homologs from various species have identified common structural features. This has provided the molecular foundation for the design of models and hypotheses aimed at understanding the relationships between NHE structure and function.

Development of a hypothetical transmembrane organization for the Na+/H+ exchanger and application of genetic selection and site-directed mutagenesis techniques has allowed significant progress to be made toward identifying amino acids involved in amiloride sensitivity and possibly binding. Likewise, deletion analyses have provided significant insight into the structural regions of the transporter that are involved in acute regulation by various protein kinases.

At present, some progress has been made in characterizing the regulatory complexity of the different isoform genes in various cell types during development and in response to chronic exposure to acid, high osmolarity and hormones. The regulatory mechanisms are likely to involve transcriptional as well as post-transcriptional processes in a cell-specific manner.

REFERENCES

1. Mahnensmith RL, Aronson PS. The plasma membrane sodium-hydrogen exchanger and its role in physiological and pathophysiological processes. Circ Res 1985; 56:773-788.
2. Grinstein S, Rotin D, Mason MJ. Na/H exchange and growth factor-induced cytosolic pH changes. Role in cellular proliferation. Biochim Biophys Acta 1989; 988:73-97.
3. Garlid KD. Sodium/proton antiporters in the mitochondrial inner membrane. Adv Exp Med Biol 1988; 232:37-46.
4. Sardet C, Franchi A, Pouysségur J. Molecular cloning, primary structure, and expression of the human growth factor-activatable Na/H antiporter. Cell 1989; 56:271-280.
5. Moolenaar WH. Effects of growth factors on intracellular pH regulation. Annu Rev Physiol 1986; 48:363-376.
6. Grinstein S, Rothstein A. Mechanisms of regulation of the Na+/H+ exchanger. J Membr Biol 1986; 90:1-12.
7. Clark JD, Limbird LE. Na+-H+ exchanger subtypes: a predictive review. Am J Physiol 1991; 261:C945-C953.
8. Fliegel L, Fröhlich O. The Na+/H+ exchanger: an update on structure, regulation and cardiac physiology. Biochem J 1993; 296:273-285.
9. Counillon L, Pouysségur J. Structure-function studies and molecular regulation of the growth factor activatable sodium-hydrogen exchanger (NHE-1). Cardiovasc Res 1995; 29:147-154.
10. Yun CHC, Tse C-M, Nath SK, et al. Mammalian Na+/H+ exchanger gene family: Structure and function studies. Am J Physiol 1995; 269:G1-G11.
11. Takaichi K, Wang D, Balkovetz DF, et al. Cloning, sequencing, and expression of Na+-H+ antiporter cDNAs from human tissues. Am J Physiol 1992; 262:C1069-C1076.
12. Fliegel L, Dyck JRB, Wang H, et al. Cloning and analysis of the human myocardial Na+/H+ exchanger. Mol Cell Biochem 1993; 125:137-143.

13. Orlowski J, Kandasamy RA, Shull GE. Molecular cloning of putative members of the Na/H exchanger gene family. cDNA cloning, deduced amino acid sequence, and mRNA tissue expression of the rat Na/H exchanger NHE-1 and two structurally related proteins. J Biol Chem 1992; 267:9331-9339.
14. Tse C-M, Ma AI, Yang VW, et al. Molecular cloning and expression of a cDNA encoding the rabbit ileal villus cell basolateral membrane Na$^+$/H$^+$ exchanger. EMBO J 1991; 10:1957-1967.
15. Hildebrandt F, Pizzonia JH, Reilly RF, et al. Cloning, sequence, and tissue distribution of a rabbit renal Na$^+$/H$^+$ exchanger transcript. Biochim Biophys Acta Gene Struct Expression 1991; 1129:105-108.
16. Reilly RF, Hildebrandt F, Biemesderfer D, et al. cDNA Cloning and immunolocalization of a Na$^+$-H$^+$ exchanger in LLC-PK$_1$ renal epithelial cells. Am J Physiol 1991; 261:F1088-F1094.
17. Counillon L, Pouysségur J. Nucleotide sequence of the Chinese hamster Na$^+$/H$^+$ exchanger NHE1. Biochim Biophys Acta Gene Struct Expression 1993; 1172:343-345.
18. Borgese F, Sardet C, Cappadoro M, et al. Cloning and expression of a cAMP-activated Na$^+$/H$^+$ exchanger: Evidence that the cytoplasmic domain mediates hormonal regulation. Proc Natl Acad Sci USA 1992; 89:6765-6769.
19. Marra MA, Prasad SS, Baillie DL. Molecular analysis of two genes between let-653 and let-56 in the unc-22 (IV) region of *Caenorhabditis elegans*. Mol Gen Genet 1993; 236:289-298.
20. Jia Z-P, McCullough N, Martel R, et al. Gene amplification at a locus encoding a putative Na$^+$/H$^+$ antiporter confers sodium and lithium tolerance in fission yeast. EMBO J 1992; 11:1631-1640.
21. Wang Z, Orlowski J, Shull GE. Primary structure and functional expression of a novel gastrointestinal isoform of the rat Na/H exchanger. J Biol Chem 1993; 268:11925-11928.
22. Tse C-M, Levine SA, Yun CHC, et al. Cloning and expression of a rabbit cDNA encoding a serum-activated ethylisopropylamiloride-resistant epithelial Na$^+$/H$^+$ exchanger isoform (NHE-2). J Biol Chem 1993; 268:11917-11924.
23. Collins JF, Honda T, Knobel S, et al. Molecular cloning, sequencing, tissue distribution, and functional expression of a Na$^+$/H$^+$ exchanger (NHE-2). Proc Natl Acad Sci USA 1993; 90:3938-3942.
24. Honda T, Knobel SM, Bulus NM, et al. Kinetic characterization of stably expressed novel Na$^+$/H$^+$ exchanger (NHE-2). Biochim Biophys Acta Bio-Membr 1993; 1150:199-202.
25. Mount SM. A catalogue of splice junction sequences. Nucleic Acids Res 1982; 10:459-472.
26. Miller RT, Counillon L, Pages G, et al. Structure of the 5'-flanking regulatory region and gene for the human growth factor-activatable Na/H exchanger NHE-1. J Biol Chem 1991; 266:10813-10819.
27. Tse C-M, Brant SR, Walker MS, et al. Cloning and sequencing of a rabbit cDNA encoding an intestinal and kidney-specific Na$^+$/H$^+$ exchanger isoform (NHE-3). J Biol Chem 1992; 267:9340-9346.

28. Amemiya M, Yamaji Y, Cano A, et al. Acid incubation increases NHE-3 mRNA abundance in OKP cells. Am J Physiol 1995; 269:C126-C133.
29. Brant SR, Yun CHC, Donowitz M, et al. Cloning, tissue distribution, and functional analysis of the human Na$^+$/H$^+$ exchanger isoform, NHE3. Am J Physiol 1995; 269:C198-C206.
30. Klanke CA, Su YR, Callen DF, et al. Molecular cloning, physical and genetic mapping of a novel human Na$^+$/H$^+$ exchanger (NHE5/SLC9A5) to chromosome 16q22.1. Genomics 1995; 25:615-622.
31. Mattei MG, Sardet C, Franchi A, et al. The human amiloride-sensitive Na+/H+ antiporter: localization to chromosome 1 by in situ hybridization. Cytogenet Cell Genet 1988; 48:6-8.
32. Lifton RP, Sardet C, Pouysségur J, et al. Cloning of the human genomic amiloride-sensitive Na$^+$/H$^+$ antiporter gene, identification of genetic polymorphisms, and localization on the genetic map of chromosome 1p. Genomics 1990; 7:131-135.
33. Szpirer C, Szpirer J, Rivière M, et al. Chromosomal assignment of four genes encoding Na/H exchanger isoforms in human and rat. Mamm Genome 1994; 5:153-159.
34. Kolyada AY, Lebedeva TV, Johns CA, et al. Proximal regulatory elements and nuclear activities required for transcription of the human Na$^+$/H$^+$ exchanger (NHE-1) gene. Biochim Biophys Acta Gene Struct Expression 1994; 1217:54-64.
35. Dyck JRB, Silva NLCL, Fliegel L. Activation of the Na$^+$/H$^+$ exchanger gene by the transcription factor AP-2. J Biol Chem 1995; 270:1375-1381.
36. Blaurock MC, Rebouças NA, Kusnezov JL, et al. Phylogenetically conserved sequences in the promoter of the rabbit sodium-hydrogen exchanger isoform 1 gene (*NHE1/SLC9A1*). Biochim Biophys Acta Gene Struct Expression 1995; 1262:159-163.
37. Dyck JRB, Fliegel L. Specific activation of the Na$^+$/H$^+$ exchanger gene during neuronal differentiation of embryonal carcinoma cells. J Biol Chem 1995; 270:10420-10427.
38. Krapf R, Solioz M, Fehlmann C. Na/H antiporter mRNA expression in single nephron segments of rat kidney cortex. J Clin Invest 1991; 88:783-788.
39. Azarani A, Goltzman D, Orlowski J. Parathyroid hormone and parathyroid hormone-related peptide inhibit the apical Na$^+$/H$^+$ exchanger NHE-3 isoform in renal cells (OK) via a dual signalling cascade involving protein kinase A and C. J Biol Chem 1995; 270:20004-20010.
40. Biemesderfer D, Reilly RF, Exner M, et al. Immunocytochemical characterization of Na$^+$-H$^+$ exchanger isoform NHE-1 in rabbit kidney. Am J Physiol 1992; 263:F833-F840.
41. Bookstein C, DePaoli AM, Xie Y, et al. Na$^+$/H$^+$ exchangers, NHE-1 and NHE-3, of rat intestine. Expression and localization. J Clin Invest 1994; 93:106-113.

42. Kulanthaivel P, Furesz TC, Moe AJ, et al. Human placental syncytiotrophoblast expresses two pharmacologically distinguishable types of Na$^+$-H$^+$ exchangers, NHE-1 in the maternal-facing (brush border) membrane and NHE-2 in the fetal-facing (basal) membrane. Biochem J 1992; 284:33-38.

43. Soleimani M, Singh G, Bizal GL, et al. Na$^+$/H$^+$ exchanger isoforms NHE-2 and NHE-1 in inner medullary collecting duct cells. Expression, functional localization, and differential regulation. J Biol Chem 1994; 269:27973-27978.

44. Mrkic B, Tse C-M, Forgo J, et al. Identification of PTH-responsive Na/H-exchanger isoforms in a rabbit proximal tubule cell line (RKPC-2). Pflügers Arch 1993; 424:377-384.

45. Kapus A, Grinstein S, Wasan S, et al. Functional characterization of three isoforms of the Na$^+$/H$^+$ exchanger stably expressed in Chinese hamster ovary cells: ATP dependence, osmotic sensitivity and role in cell proliferation. J Biol Chem 1994; 269:23544-23552.

46. Biemesderfer D, Pizzonia J, Abu-Alfa A, et al. NHE3: A Na$^+$/H$^+$ exchanger isoform of renal brush border. Am J Physiol 1993; 265:F736-F742.

47. Soleimani M, Bookstein C, Bizal GL, et al. Localization of the Na$^+$/H$^+$ exchanger isoform NHE-3 in rabbit and canine kidney. Biochim Biophys Acta 1994; 1195:89-95.

48. Ives HE, Yee VJ, Warnock DG. Asymmetric distribution of the Na+/H+ antiporter in the renal proximal tubule epithelial cell. J Biol Chem 1983; 258:13513-13516.

49. Preisig PA, Ives HE, Cragoe Jr EJ, et al. Role of the Na+/H+ antiporter in rat proximal tubule bicarbonate absorption. J Clin Invest 1987; 80:970-978.

50. Bookstein C, Musch MW, DePaoli A, et al. A unique sodium-hydrogen exchange isoform (NHE-4) of the inner medulla of the rat kidney is induced by hyperosmolarity. J Biol Chem 1994; 269:29704-29709.

51. Wakabayashi S, Fafournoux P, Sardet C, et al. The Na$^+$/H$^+$ antiporter cytoplasmic domain mediates growth factor signals and controls "H$^+$-sensing." Proc Natl Acad Sci USA 1992; 89:2424-2428.

52. Wakabayashi S, Bertrand B, Shigekawa M, et al. Growth factor activation and "H$^+$-sensing" of the Na$^+$/H$^+$ exchanger Isoform 1 (NHE1). Evidence for an additional mechanism not requiring direct phosphorylation. J Biol Chem 1994; 269:5583-5588.

53. Levine SA, Nath SK, Yun CHC, et al. Separate C-terminal domains of the epithelial specific brush border Na$^+$/H$^+$ exchanger isoform NHE3 are involved in stimulation and inhibition by protein kinases/growth factors. J Biol Chem 1995; 270:13716-13725.

54. Winkel GK, Sardet C, Pouysségur J, et al. Role of cytoplasmic domain of the Na$^+$/H$^+$ exchanger in hormonal activation. J Biol Chem 1993; 268:3396-3400.

55. Aronson PS. Kinetic properties of the plasma membrane Na-H exchanger. Annu Rev Physiol 1985; 47:545-560.

56. Orlowski J. Heterologous expression and functional properties of the amiloride high affinity (NHE-1) and low affinity (NHE-3) isoforms of the rat Na/H exchanger. J Biol Chem 1993; 268:16369-16377.
57. Yu FH, Shull GE, Orlowski J. Functional properties of the rat Na/H exchanger NHE-2 isoform expressed in Na/H exchanger-deficient Chinese hamster ovary cells. J Biol Chem 1993; 268:25536-25541.
58. Levine SA, Montrose MH, Tse C-M, et al. Kinetics and regulation of three cloned mammalian Na^+/H^+ exchangers stably expressed in a fibroblast cell line. J Biol Chem 1993; 268:25527-25535.
59. Aronson PS, Nee J, Suhm MA. Modifier role of internal H^+ in activating the Na^+-H^+ exchanger in renal microvillus membrane vesciles. Nature 1982; 299:161-163.
60. Grinstein S, Cohen S, Rothstein A. Cytoplasmic pH regulation in thymic lymphocytes by an amiloride-sensitive Na^+/H^+ antiport. J Gen Physiol 1984; 83:341-369.
61. Paris S, Pouysségur J. Growth factors activate the Na^+/H^+ antiporter in quiescent fibroblasts by increasing its affinity for intracellular H^+. J Biol Chem 1984; 259:10989-10994.
62. Yun CHC, Little PJ, Nath SK, et al. Leu143 in the putative fourth membrane spanning domain is critical for amiloride inhibition of an epithelial Na^+/H^+ exchanger isoform (NHE-2). Biochem Biophys Res Commun 1993; 193:532-539.
63. Counillon L, Franchi A, Pouysségur J. A point mutation of the Na^+/H^+ exchanger gene (*NHE1*) and amplification of the mutated allele confer amiloride resistance upon chronic acidosis. Proc Natl Acad Sci USA 1993; 90:4508-4512.
64. Counillon L, Scholz W, Lang HJ, et al. Pharmacological characterization of stably transfected Na^+/H^+ antiporter isoforms using amiloride analogs and a new inhibitor exhibiting anti-ischemic properties. Mol Pharmacol 1993; 44:1041-1045.
65. Scholz W, Albus U, Lang HJ, et al. Hoe 694, a new Na^+/H^+ exchange inhibitor and its effects in cardiac ischaemia. Br J Pharmacol 1993; 109:562-568.
66. Ganapathy V, Balkovetz DF, Miyamoto Y, et al. Inhibition of human placental Na^+-H^+ exchanger by cimetidine. J Pharmacol Exp Ther 1986; 239:192-197.
67. Kulanthaivel P, Leibach FH, Mahesh VB, et al. The Na^+-H^+ exchanger of the placental brush-border membrane is pharmacologically distinct from that of the renal brush-border membrane. J Biol Chem 1990; 265: 1249-1252.
68. Ganapathy ME, Leibach FH, Mahesh VB, et al. Interaction of clonidine with human placental Na+-H+ exchanger. Biochem Pharmacol 1986; 35:3989-3994.
69. Aronson PS, Bounds SE. Harmaline inhibition of Na-dependent transport in renal microvillus membrane vesicles. Am J Physiol 1980; 238:F210-F217.

70. Knickelbein R, Aronson PS, Atherton W, et al. Sodium and chloride transport across rabbit ileal brush border. I. Evidence for Na-H exchange. Am J Physiol 1983; 245:G504-G510.
71. Balkovetz DF, Miyamoto Y, Tiruppathi C, et al. Inhibition of Brush-border membrane Na$^+$-H$^+$ exchanger by loperamide. J Pharmacol Exp Ther 1987; 243:150-154.
72. Frelin C, Vigne P, Barbry P, et al. Interaction of guanidinium and guanidinium derivatives with the Na$^+$/H$^+$ exchange system. Eur J Biochem 1986; 154:241-245.
73. Sardet C, Fafournoux P, Pouysségur J. α-Thrombin, epidermal growth factor, and okadaic acid activate the Na$^+$/H$^+$ exchanger, NHE-1, by phosphorylating a set of common sites. J Biol Chem 1991; 266:19166-19171.
74. Kandasamy RA, Yu FH, Harris R, et al. Plasma membrane Na$^+$/H$^+$ exchanger isoforms (NHE-1, -2, and -3) are differentially responsive to second messenger agonists of the protein kinase A and C pathways. J Biol Chem 1995; 270:29209-29216.
75. Sardet C, Counillon L, Franchi A, et al. Growth factors induce phosphorylation of the Na/H antiporter, a glycoprotein of 110 kD. Science 1990; 247:723-725.
76. Bertrand B, Wakabayashi S, Ikeda T, et al. The Na$^+$/H$^+$ exchanger isoform 1 (NHE1) is a novel member of the calmodulin-binding proteins. Identification and characterization of calmodulin-binding sites. J Biol Chem 1994; 269:13703-13709.
77. Wakabayashi S, Bertrand B, Ikeda T, et al. Mutation of calmodulin-binding site renders the Na$^+$/H$^+$ exchanger (NHE1) highly H$^+$-sensitive and Ca^{2+} regulation-defective. J Biol Chem 1994; 269:13710-13715.
78. Silva NLCL, Haworth RS, Singh D, et al. The carboxyl-terminal region of the Na$^+$/H$^+$ exchanger interacts with mammalian heat shock protein. Biochemistry 1995; 34:10412-10420.
79. Helmle-Kolb C, Counillon L, Roux D, et al. Na/H exchange activities in NHE1-transfected OK-cells: Cell polarity and regulation. Pflügers Arch 1993; 425:34-40.
80. Kong SK, Choy YM, Fung KP, et al. cAMP activates Na$^+$/H$^+$ antiporter in murine macrophages. Biochem Biophys Res Commun 1989; 165:131-137.
81. Moule SK, McGivan JD. Epidermal growth factor and cyclic AMP stimulate Na$^+$/H$^+$ exchange in isolated rat hepatocytes. Eur J Biochem 1990; 187:677-682.
82. Gupta A, Schwiening CJ, Boron WF. Effects of CGRP, forskolin, PMA, and ionomycin on pH$_i$ dependence of Na-H exchange in UMR-106 cells. Am J Physiol 1994; 266:C1083-C1092.
83. Demaurex N, Orlowski J, Brisseau G, et al. The mammalian Na$^+$/H$^+$ antiporters NHE-1, NHE-2, and NHE-3 are electroneutral and voltage independent, but can couple to a H$^+$ conductance. J Gen Physiol 1995; 106:85-111.

84. Azarani A, Orlowski J, Goltzman D. Parathyroid hormone and parathyroid hormone-related peptide activate the Na$^+$/H$^+$ exchanger NHE-1 isoform in osteoblastic cells (UMR-106) via a cAMP-dependent pathway. J Biol Chem 1995; in press.
85. Borgese F, Malapert M, Fievet B, et al. The cytoplasmic domain of the Na$^+$/H$^+$ exchangers (NHEs) dictates the nature of the hormonal response: Behavior of a chimeric human NHE1/trout βNHE antiporter. Proc Natl Acad Sci USA 1994; 91:5431-5435.
86. Kennelly PJ, Krebs EG. Consensus sequences as substrate specificity determinants for protein kinases and protein phosphatases. J Biol Chem 1991; 266:15555-15558.
87. Helmle-Kolb C, Montrose MH, Murer H. Parathyroid hormone regulation of Na$^+$/H$^+$ exchange in oppossum kidney cells: polarity and mechanisms. Pflügers Arch 1990; 416:615-623.
88. Weinman EJ, Shenolikar S, Kahn AM. cAMP-associated inhibition of Na$^+$-H$^+$ exchanger in rabbit kidney brush-border membranes. Am J Physiol 1987; 252:F19-F25.
89. Weinman EJ, Steplock D, Shenolikar S. cAMP-mediated inhibition of the renal brush border membrane Na$^+$-H$^+$ exchanger requires a dissociable phosphoprotein cofactor. J Clin Invest 1993; 92:1781-1786.
90. Weinman EJ, Steplock D, Wang Y, et al. Characterization of a protein cofactor that mediates protein kinase A regulation of the renal brush border membrane Na$^+$-H$^+$ exchanger. J Clin Invest 1995; 95:2143-2149.
91. Harada K, Franklin A, Johnson RG, et al. Acidemia and hypernatremia enhance postischemic recovery of excitation-contraction coupling. Circ Res 1994; 74:1197-1209.
92. Watts BA, III, Good DW. Apical membrane Na$^+$/H$^+$ exchange in rat medullary thick ascending limb. pH$_i$-dependence and inhibition by hyperosmolality. J Biol Chem 1994; 269:20250-20255.
93. Grinstein S, Woodside M, Sardet C, et al. Activation of the Na$^+$/H$^+$ antiporter during cell volume regulation. Evidence for a phosphorylation-independent mechanism. J Biol Chem 1992; 267:23823-23828.
94. Davis BA, Hogan EM, Boron WF. Role of G proteins in stimulation of Na-H exchange by cell shrinkage. Am J Physiol 1992; 262:C533-C536.
95. Voyno-Yasenetskaya T, Conklin BR, Gilbert RL, et al. Gα13 stimulates Na-H exchange. J Biol Chem 1994; 269:4721-4724.
96. Shrode LD, Klein JD, O'Neill WC, et al. Shrinkage-induced activation of Na$^+$/H$^+$ exchange in primary rat astrocytes: role of myosin light-chain kinase. Am J Physiol 1995; 269:C257-C266.
97. Good DW. Hyperosmolality inhibits bicarbonate absorption in rat medullary thick ascending limb via a protein-tyrosine kinase-dependent pathway. J Biol Chem 1995; 270:9883-9889.
98. Han J, Lee J-D, Bibbs L, et al. A MAP kinase targeted by endotoxin and hyperosmolarity in mammalian cells. Science 1994; 265:808-811.
99. Galcheva-Gargova Z, Dérijard B, Wu I-H, et al. An osmosensing signal transduction pathway in mammalian cells. Science 1994; 265:806-808.

100. Moe OW, Miller RT, Horie S, et al. Differential regulation of Na/H antiporter by acid in renal epithelial cells and fibroblasts. J Clin Invest 1991; 88:1703-1708.
101. Igarashi P, Freed MI, Ganz MB, et al. Effects of chronic metabolic acidosis on Na^+-H^+ exchangers in LLC-PK_1 renal epithelial cells. Am J Physiol 1992; 263:F83-F88.
102. Horie S, Moe O, Miller RT, et al. Long-term activation of protein kinase C causes chronic Na/H antiporter stimulation in cultured proximal tubule cells. J Clin Invest 1992; 89:365-372.
103. Horie S, Moe O, Yamaji Y, et al. Role of protein kinase C and transcription factor AP-1 in the acid-induced increase in Na/H antiporter activity. Proc Natl Acad Sci USA 1992; 89:5236-5240.
104. Yamaji Y, Amemiya M, Cano A, et al. Overexpression of csk inhibits acid-induced activation of NHE-3. Proc Natl Acad Sci USA 1995; 92:6274-6278.
105. Yun CHC, Gurubhagavatula S, Levine SA, et al. Glucocorticoid stimulation of ileal Na^+ absorptive cell brush border Na^+/H^+ exchange and association with an increase in message for NHE-3, an epithelial Na^+/H^+ exchanger isoform. J Biol Chem 1993; 268:206-211.
106. Baum M, Moe OW, Gentry DL, et al. Effect of glucocorticoids on renal cortical NHE-3 and NHE-1 mRNA. Am J Physiol 1994; 267:F437-F442.
107. Counillon L, Pouysségur J, Reithmeier RAF. The Na^+/H^+ exchanger NHE-1 possesses N- and O-linked glycosylation restricted to the first N-terminal extracellular domain. Biochemistry 1994; 33:10463-10469.
108. Tse C-M, Levine SA, Yun CHC, et al. Na^+/H^+ exchanger-2 is an O-linked but not an N-linked sialoglycoprotein. Biochemistry 1994; 33:12954-12961.
109. Geering K. Subunit assembly and functional maturation of Na,K-ATPase. J Membr Biol 1990; 115:109-121.
110. Reithmeier RAF. Mammalian exchangers and co-transporters. Curr Opin Cell Biol 1994; 6:583-594.
111. Yusufi ANK, Szczepanska-Konkel M, Dousa TP. Role of N-linked oligosaccharides in the transport activity of the Na/H antiporter in rat renal brush-border membrane. J Biol Chem 1988; 263:13683-13691.
112. Haworth RS, Fröhlich O, Fliegel L. Multiple carbohydrate moieties on the Na^+/H^+ exchanger. Biochem J 1993; 289:637-640.
113. Kinsella JL, Aronson PS. Amiloride inhibtion of the Na^+-H^+ exchanger in renal microvillus membrane vesicles. Am J Physiol 1981; 241:F374-F379.
114. Mahnensmith RL, Aronson PS. Interrelationships among quinidine, amiloride, and lithium as inhibitors of the renal Na^+-H^+ exchanger. J Biol Chem 1985; 260:12586-12592.
115. Paris S, Pouysségur J. Biochemical characterization of the amiloride-sensitive Na^+/H^+ antiport in Chinese hamster lung fibroblasts. J Biol Chem 1983; 258:3503-3508.

116. L'Allemain G, Franchi A, Cragoe Jr EJ et al. Blockade of the Na$^+$/H$^+$ antiport abolishes growth factor-induced DNA synthesis in fibroblasts. Structure-activity relationships in the amiloride series. J Biol Chem 1984; 259:4313-4319.
117. Ives HE, Yee VJ, Warnock DG. Mixed type inhibition of the renal Na$^+$/H$^+$ antiporter by Li$^+$ and amiloride. Evidence for a modifier site. J Biol Chem 1983; 258:9710-9716.
118. Warnock DG, Yang W, Huang Z, et al. Interactions of chloride and amiloride with the renal Na$^+$/H$^+$ antiporter. J Biol Chem 1988; 263:7216-7221.
119. Franchi A, Cragoe Jr E Pouysségur J. Isolation and properties of fibroblast mutants overexpressing an altered Na$^+$/H$^+$ antiporter. J Biol Chem 1986; 261:14614-14620.
120. Wang D, Balkovetz DF, Warnock DG. Mutational analysis of transmembrane histidines in the amiloride-sensitive Na$^+$/H$^+$ exchanger. Am J Physiol 1995; 269:C392-C402.
121. Gerchman Y, Olami Y, Rimon A, et al. Histidine-226 is part of the pH sensor of NhaA, a Na$^+$/H$^+$ antiporter in *Escherichia coli*. Proc Natl Acad Sci USA 1993; 90:1212-1216.
122. Fliegel L, Haworth RS, Dyck JRB. Characterization of the placental brush border membrane Na$^+$/H$^+$ exchanger: identification of thiol-dependent transitions in apparent molecular size. Biochem J 1993; 289:101-107.
123. Fafournoux P, Noël J, Pouysségur J. Evidence that Na$^+$/H$^+$ exchanger isoforms NHE1 and NHE3 exist as stable dimers in membranes with a high degree of specificity for homodimers. J Biol Chem 1994; 269:2589-2596.
124. Otsu K, Kinsella JL, Heller P, et al. Sodium dependence of the Na$^+$-H$^+$ exchanger in the pre-steady state. Implications for the exchange mechanism. J Biol Chem 1993; 268:3184-3193.
125. Beliveau R, Demeule M, Potier M. Molecular size of the Na$^+$/H$^+$ antiport in renal brush border membranes, as estimated by radiation inactivation. Biochem Biophys Res Commun 1988; 152:484-489.
126. Thompson JD, Higgins DG, Gibson TJ, et al. Improving the sensitivity of progressive multiple sequence alignment through sequence weighting, positions-specific gap penalties and weight matrix choice. Nucleic Acids Res 1994; 22:4673-4680.
127. Kumir S, Tamura K, Nei M. MEGA: Molecular Evolutionary Genetics Analysis. The Pennsylvania State University 1993.

CHAPTER 8

THE REGULATORY CYTOPLASMIC DOMAIN OF THE Na$^+$/H$^+$ EXCHANGER

Bernhard M. Schmitt, Toshitaro Ikeda,
Munekazu Shigekawa and Shigeo Wakabayashi

A. INTRODUCTION

This review will describe briefly the main structural and functional features of the cytoplasmic tail of vertebrate sodium proton exchangers according to currently available data. Particular attention is given to the human isoform NHE1 which has been studied intensively by many groups, including ours.

Sodium proton exchangers (Na$^+$/H$^+$ exchanger or NHE) participate in several physiological processes. Some of these processes occur in every mammalian cell, such as control of intracellular pH and cell volume, while others are only observed in specialized tissues, e.g., the unidirectional transport of sodium or bicarbonate across epithelia. The contribution of Na$^+$/H$^+$ exchangers to the various processes is always the electroneutral countertransport of sodium and protons. At least five different isoforms have been found to carry out this basic function. These isoforms are expressed in a tissue and developmental specific pattern and subject to specific targeting in polarized cells. Furthermore, the responses to various signaling molecules differ greatly between the isoforms.

From a first glance at a single cell, the only consequence of Na$^+$/H$^+$ exchange activity seems to be a tendency towards lower H^+_i and higher Na^+_i. In polarized epithelial cells, however, the consequences on the ion fluxes and concentrations can be very diverse, depending on whether Na$^+$/H$^+$ exchange occurs on the apical or on the basolateral membrane. For instance, in proximal tubular cells only the apically located Na$^+$/H$^+$

The Na$^+$/H$^+$ Exchanger, edited by Larry Fliegel. © 1996 R.G. Landes Company.

exchangers participate in sodium and bicarbonate reabsorption, while basolaterally located Na$^+$/H$^+$ exchangers are somewhat privileged in controlling pH$_i$ and volume since their activity is not coupled to an influx of CO_2. Na$^+$/H$^+$ exchangers are thus involved in the homeostasis of both cytoplasm and extracellular space. Without the possibility to regulate apical and basolateral Na$^+$/H$^+$ exchangers up and down separately the organism could not adequately counteract deviations in one compartment without affecting the balance of the other. However, such regulatory conflicts do not arise since Na$^+$/H$^+$ exchangers can be targeted and regulated in an isoform-specific way. According to present knowledge, this isoform-specific responsiveness to growth factors, hormones, or second messengers is mainly provided by the cytoplasmic portion of the NHE protein.

A wealth of information describing the effects of numerous agents on Na$^+$/H$^+$ exchanger activity, notably NHE1 and NHE3, has been accumulated (see refs. 1-17 for reviews), with data on the other isoforms becoming increasingly available. Yet the signal transduction pathways are not well understood. The action of most agents cannot be traced down to the NHE molecule since crucial elements are missing or hypothetical, and the sodium proton exchanger protein itself has largely remained a "black box" when it comes to explain in what way it is modified in response to various stimuli and how these modifications control the rate of ion transport.

B. GROSS FUNCTIONAL ANATOMY OF THE CYTOPLASMIC DOMAIN

No direct experimental information on the three-dimensional structure of Na$^+$/H$^+$ exchangers, e.g., by X-ray crystallography, immunofluorescence microscopy, partial proteinase digestion etc., is available. In the absence of such data which would advance the understanding of Na$^+$/H$^+$ exchange substantially our ideas of the secondary, tertiary and quaternary structures of the various sodium proton exchangers had to be derived mostly from indirect evidence. This includes accessibility or inaccessibility of certain epitopes by specific antibodies from the extracellular site, digestion of sugar moieties by externally added glycosidases, functional characterization of mutant and chimeric exchangers, modification by cytoplasmic factors, and analysis of the primary amino acid sequence according to several computerized algorithms.

It appears to be a common structural feature of all vertebrate Na$^+$/H$^+$ exchangers that they consist of two major portions, an N-terminal transmembrane domain and a large C-terminal cytoplasmic domain of ~300 amino acids. The N-terminal portions of all known isoforms are predicted to comprise 10-12 transmembrane segments, the exact number depending on the specific algorithm used in the hydropathy plot analysis. In NHE1, the isoform characterized in most detail, this transmembrane domain was shown to contain the ion translocation site,[18] the amiloride binding site,[19,20] sites for N-linked and O-linked glycosylation,[21-23] and a putative N-terminal signaling sequence. Furthermore, unknown segments of the transmembrane segment allow dimerization of NHE1 to occur.[24] Possibly the transmembrane segment also harbors a defined structural entity, termed "proton modifier site" or "pH sensor," which gives rise to the allosteric behavior of the exchanger at varying internal pH. The amino acid homology is higher between the transmembrane segments of various Na$^+$/H$^+$ exchanger isoforms than between the cytoplasmic segments (50-60% vs. 20-25%). Nevertheless, major differences of the transmembrane segments affecting the sensitivities to inhibition by amiloride analogs[25-27] or the setpoint of the H^+_i activation curve[18,26-29] should not be overlooked.

The first thing to be noted about the cytoplasmic domain is the remarkable fact that all eukaryotic Na$^+$/H$^+$ exchangers are bearing such a cytoplasmic segment, of considerable size, at their C-termini. Bacterial Na$^+$/H$^+$ exchangers, in contrast, contain only short cytoplasmic C-termini with no known function (see ref. 30 for review). In all species, exchanger activity is regulated effectively and quickly in response to changing ion concentrations, notably pH$_i$, but only in mammals, fish, and amphibia, i.e., in multicellular organisms with an obvious need to coordinate the function of its individual cells, Na$^+$/H$^+$ exchange has been found to be regulated by circulating or local signaling molecules. While the transmembrane segment contains the complete molecular machinery required for Na$^+$/H$^+$ exchange,[28,31] the principal functions of the cytoplasmic domain described so far are all related to sensing of intracellular signals and transformation of this input into activation or inhibition of the exchanger.

As discussed in more detail below, several sites within the cytoplasmic portion of NHE seem to exist, each specialized to respond to a particular subset of signals. Thus, several residues have been identified in NHE1 which undergo protein kinase C-dependent phosphorylation; one well-characterized protein domain can bind calmodulin in the presence of calcium, and a further domain is thought to interact with a regulatory cofactor. Interestingly, whatever the stimulus and whatever domain mediating its effect, NHE1 activity is generally upregulated. Apparently the mechanism by which the exchanger activity is increased is the same for all stimulators of NHE1, namely an increased sensitivity to allosteric activation by internal protons.

NHE3, the sodium proton exchanger isoform found in the apical membranes of reabsorptive epithelia as intestine and kidney tubule cells, is also a calmodulin binding protein,[32] and a regulatory protein cofactor has been cloned recently.[33] This isoform can be stimulated, e.g., by angiotensin II, growth factors, serum or phosphatase inhibitor okadaic acid (OA), but unlike NHE1, there are also several inhibitory agents, e.g., parathyroid hormone, cAMP or phorbol ester PMA. Successive truncation of the NHE3 cytoplasmic tail revealed several distinct domains for both activation and inhibition.[28]

β–NHE, the Na⁺/H⁺ exchanger found in trout red blood cells,[29] can be activated by phorbol esters, β-adrenergic agents or cAMP analogs. Since its cytoplasmic portion contains consensus sites for phosphorylation by protein kinase C (PKC) as well as for protein kinase A (PKA) and mutation of PKA consensus sites blunted the effect of PKA,[34] direct phosphorylation of β-NHE after activation of these pathways may occur and may be responsible for the altered activity. However, this remains to be demonstrated directly.

C. PHOSPHORYLATION

Sardet and coworkers first demonstrated that the NHE1 is a phosphorylated glycoprotein under in vivo conditions.[23] Increased phosphorylation was observed after stimulation by thrombin or EGF and was associated with cytoplasmic alkalinization, suggesting enhanced sodium proton exchange. Phosphorylation was restricted to serine residues, and purification and trypsin digestion of NHE1 yielded similar phosphopeptides for both stimuli.[35] NHE1 was reported to be equally phosphorylated when expressed in other cells, e.g., human platelets, granulocytes or Chinese hamster lung fibroblasts.[23,35-37]

Phosphopeptide mapping of NHE1 and derived deletion mutants revealed that the major phosphorylation sites all mapped to the cytoplasmic segment while the transmembrane segment did not appear to be phosphorylated;[31] growth factors and phosphatase inhibitor OA enhanced phosphorylation of all phosphopeptides, and one common additional phosphopeptide emerged. Surprisingly, removal of these phosphorylation sites individually or in pairs by mutating their serine residues to alanine did not affect the response to growth factor stimulation. In addition, a truncation mutant of NHE1 in which the large cytoplasmic portion C-terminal of amino acid 636 (Δ635) was missing together with all major phosphorylation sites was still significantly stimulated by growth factors.[31] The extent of this activation amounted to about 50% of the activation observed in wild-type NHE1, and it was thought that the difference is due to lacking stimulation by protein phosphorylation.

This clearly demonstrates that phosphorylation-independent pathways exist for the activation of NHE1 by growth factors. It remains unclear, however, to what extent and by what molecular mechanisms the phosphorylation of cytoplasmic amino acid residues is involved in the regulation of NHE1. Recently, it was reported[38] that the increased Na⁺/H⁺ exchanger activity in primary hypertensive humans was associated with an increased phosphorylation of NHE1; this illustrates that a clearer picture of NHE1 phosphorylation and its functional relevance is needed to interpret such findings correctly.

While the relevance of direct phosphorylation of NHE1, NHE3 and β–NHE remains elusive and little information is available on NHE2, NHE4 and NHE5, the initial finding that protein kinase-dependent steps are involved in the response to many agents is well documented.[3,5,39,40] Sodium proton exchangers of several isoforms have been found to be activated or inhibited by stimulation of tyrosine kinase receptors, e.g., EGF, FGF, PDGF, CSF, insulin, IGF, or by stimulating protein kinases PKA, PKC or PKG either directly or via upstream receptors coupled to them by heterotrimeric G-proteins. This includes receptors for PTH and beta-adrenergic agents which are coupled to protein kinase A, angiotensin II and several drugs acting on PKC, and atrial natriuretic peptide which leads to stimulation of PKG. PKG seems to be essential for the action of atrial natriuretic peptides (ANP) on sodium proton exchange since the cGMP, generated from GTP by particulate guanylate cyclase coupled to the ANP receptor, has no direct effect on sodium proton exchange, in contrast to an apical Na^+-channel which is directly regulated by cGMP.[41] However, NHE1 was not phosphorylated by PKG in vitro (Schmitt et al, unpublished results), although it contains PKG consensus sites.

Two different models have been suggested to explain how such a great number of regulating agents finally modifies the Na^+/H^+ exchanger. First, a putative "NHE1 kinase" could integrate the various upstream signals and regulate NHE1 activity by direct phosphorylation.[3] Second, it has been proposed[31] that a cofactor exists which interacts directly with the Na^+/H^+ exchanger protein and modifies its activity by mechanisms other than direct phosphorylation, while the cofactor itself is regulated by direct phosphorylation in response to multiple upstream stimuli. While there is no direct evidence for a common "NHE1 kinase," a protein factor binding to the exchanger and containing consensus sites for PKA phosphorylation has been identified and cloned recently in the case of NHE3[33,42] (see section 5); this approach could prove fruitful for other isoforms, too.

D. CALCIUM/CALMODULIN

A role for calcium in the regulation of Na^+/H^+ exchange has been suggested early by several studies. Villereal and coworkers found that Ca^{2+} ionophore A23187 stimulated Na^+-influx into human foreskin fibroblasts.[43] These data did not allow to distinguish between direct Ca^{2+} effects and those secondary to cell acidification. Cell acidification by Ca^{2+} is a phenomenon observed in several cell types including 3T3 fibroblasts,[44] HSWP human foreskin fibroblasts,[45] vascular smooth muscle cells,[46] and adrenal glomerulosa cells,[47] and was explained by displacement of H^+ from intracellular buffers by Ca^{2+}[48] or by activation of Ca^{2+}/H^+ exchange via the plasma membrane Ca^{2+}-pump.[49] Other cell types responded to increased intracellular Ca^{2+} by an increased cytoplasmic pH, including thymic lymphocytes,[50] WS-1 human foreskin fibroblasts[45] and human platelets.[51-53]

Recent studies have assessed Na^+/H^+ exchange activity after mobilization of intracellular Ca^{2+} more specifically as an EIPA-sensitive acid extrusion rate in the rat osteosarcoma cell line UMR-106[54] or as an EIPA-sensitive ^{22}Na-uptake into exchanger-deficient fibroblasts transfected with a plasmid encoding NHE1.[55] In both studies, Ca^{2+} was found to shift the H^+_i-activation curve of NHE1 towards more alkaline values; thus Na^+/H^+ exchange appears to be stimulated by increased Ca^{2+}_i, regardless of the change in pH_i produced by the integrated effect of Ca^{2+} on the other proton loading and extrusion mechanisms present in the studied cell type.

In principle, this effect might be brought about by direct phosphorylation of NHE1 by protein kinases which are activated by calcium, e.g., Ca^{2+}/CaM-dependent Protein Kinase (CAMK) or PKC; however, phosphorylation of NHE1 was not increased after Ca^{2+} activation.[55] Alternatively, calmodulin could affect NHE1 activity in a Ca^{2+}-dependent manner.

Calmodulin, a monomeric 17 kD protein present in virtually all cells, contains four binding sites for Ca^{2+}; upon binding of Ca^{2+} it undergoes large conformational changes. Thus, binding of Ca^{2+} can induce conformational and functional changes in several cytoplasmic enzymes in which calmodulin is either a permanently bound subunit, e.g., phosphorylase kinase, or becomes bound at increased cytoplasmic Ca^{2+} e.g., CAMKII. Our group recently demonstrated that NHE1 is a calmodulin binding protein,[56] showed that Ca^{2+}/calmodulin is a major pathway of NHE1 regulation in vivo,[32,55] and presented a simple model of the molecular mechanism of Ca^{2+} activation.[55] (See also Fig. 8.5.)

Binding experiments with CaM-sepharose and dansylated calmodulin[56] showed that NHE1 binds CaM in a strictly Ca^{2+}-dependent manner (Fig. 8.1). Two different binding sites, one with a high affinity (K_d ~20 nM) and one with a relatively low affinity (K_d ~350 nM), could be identified and located to amino acid 636-656, termed "region A," and amino acids 664-684, termed "region B," respectively. These regions are rich in positively charged amino acids (Arg, Lys, His) and lack almost completely negatively charged amino acids (Asp or Glu), a common feature of many calmodulin binding proteins. According to the algorithm of Chou & Fasman,[57] however, the high-affinity region A will rather fold into hydrophilic β-strands than into the amphiphilic α-helix generally held to be typical for high affinity, calmodulin binding regions.

Several experimental approaches unambiguously supported the notion that calmodulin is indeed essentially involved in NHE1 activation by Ca^{2+}. Firstly,[55] mutants of NHE1 were constructed in which region A was deleted (amino acids 637-656) or all its positive charges reversed by substituting glutamic acids for Lys641 and Arg643/Arg645/Arg647, the respective mutants being termed Δ 637-656 and 1K3R4E (Fig. 8.1). Expression in exchanger-deficient PS120 fibroblasts and measurement of EIPA-sensitive ^{22}Na-uptake showed that both mutant exchangers could be activated by lower H^+_i, i.e., at higher pH_i, compared to wild-type NHE1, while the maximum activities were similar in all forms; thus, the H^+_i-

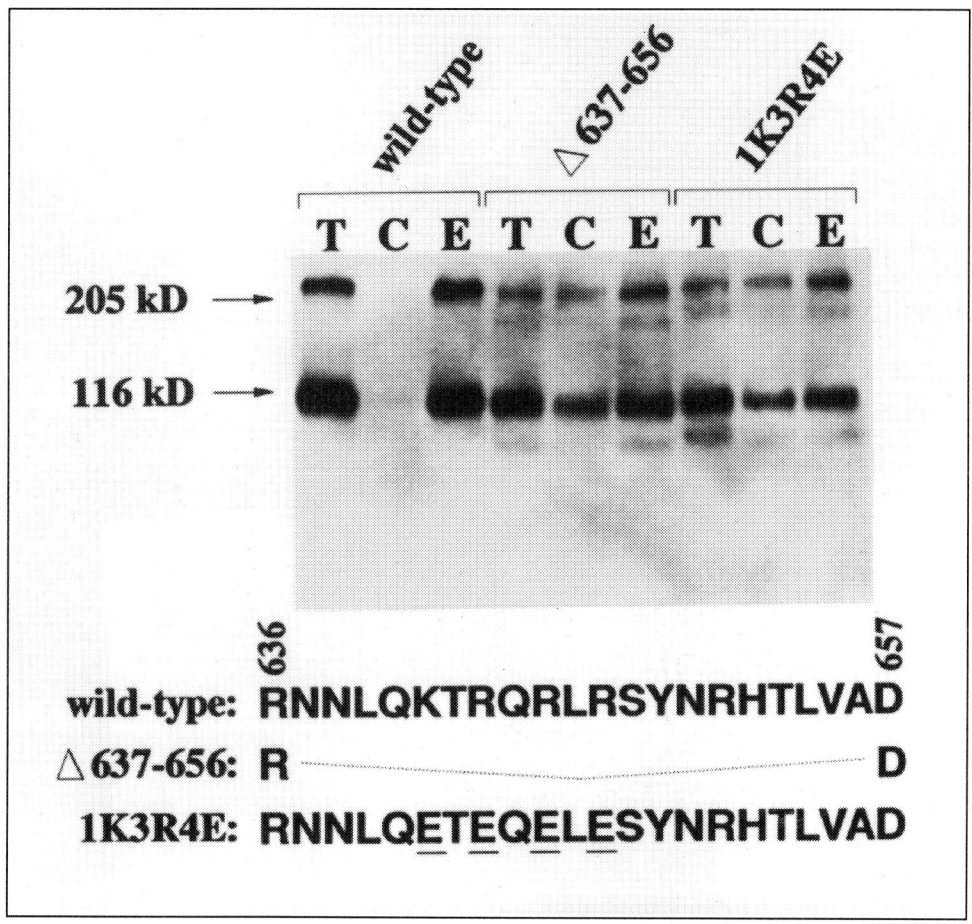

Fig. 8.1. In vitro binding of calmodulin to NHE1. Immunoprecipitated wild-type NHE1 and mutants of NHE1 in which the high-affinity calmodulin binding site was deleted (Δ 647-656) or altered by charge reversal (1K3R4E) were incubated with calmodulin-sepharose beads in the presence of Ca^{2+} (C) or of the Ca^{2+}-chelator EGTA (E); T indicates controls without beads. Supernatants were visualized by immunoblot. Lower panel shows the sequences of the different exchanger types used. Wild-type NHE1 was bound to calmodulin-sepharose in strictly Ca^{2+}-dependent manner. Calmodulin binding was greatly reduced in both mutants; due to second, low-affinity calmodulin binding site still present this effect was not complete. Upper band corresponds to dimeric exchanger.

activation curve was shifted towards more alkaline values by approximately 0.3 pH units. Consistent with these observations, resting pH_i was higher in fibroblasts expressing these mutants compared to cells expressing wild type NHE1 (Fig. 8.2). In contrast to wild-type NHE1, elevated Ca^{2+}_i did not further affect this pH-dependency curve, i.e., the mutants had lost their calcium responsiveness (Fig. 8.3).

This prompted the hypothesis that the presence of region A decreases the activity of NHE1 by interfering with H^+_i-activation of NHE1,

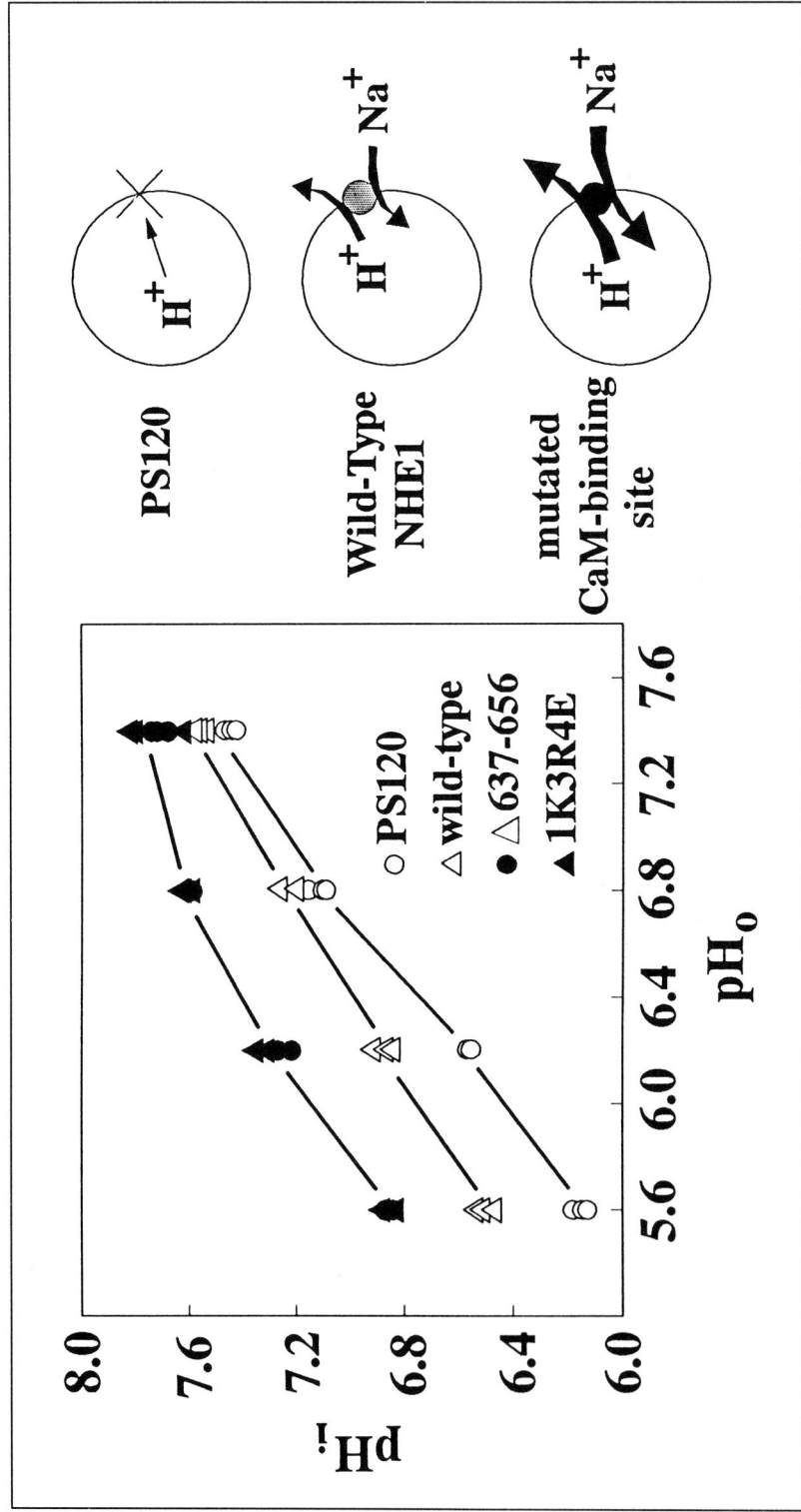

Fig. 8.2. Cytoplasmic pH in PS120 fibroblasts at different extracellular pH. PS120 cells expressing no Na^+/H^+ exchanger (O), wild-type NHE1 (△), or mutants of NHE1 in which the CaM-binding/autoinhibitory domain was deleted (•) or inactivated by charge reversal (▲) were exposed media of different pH in the absence of bicarbonate; cytoplasmic pH was determined by the weak acid-equilibration method using ^{14}C-benzoic acid. In all cells, changes in pH_o caused a parallel shift of the steady-state levels of cytoplasmic pH. The mutants established a higher intracellular pH at all values of pH_o a consequence of abolished autoinhibition.

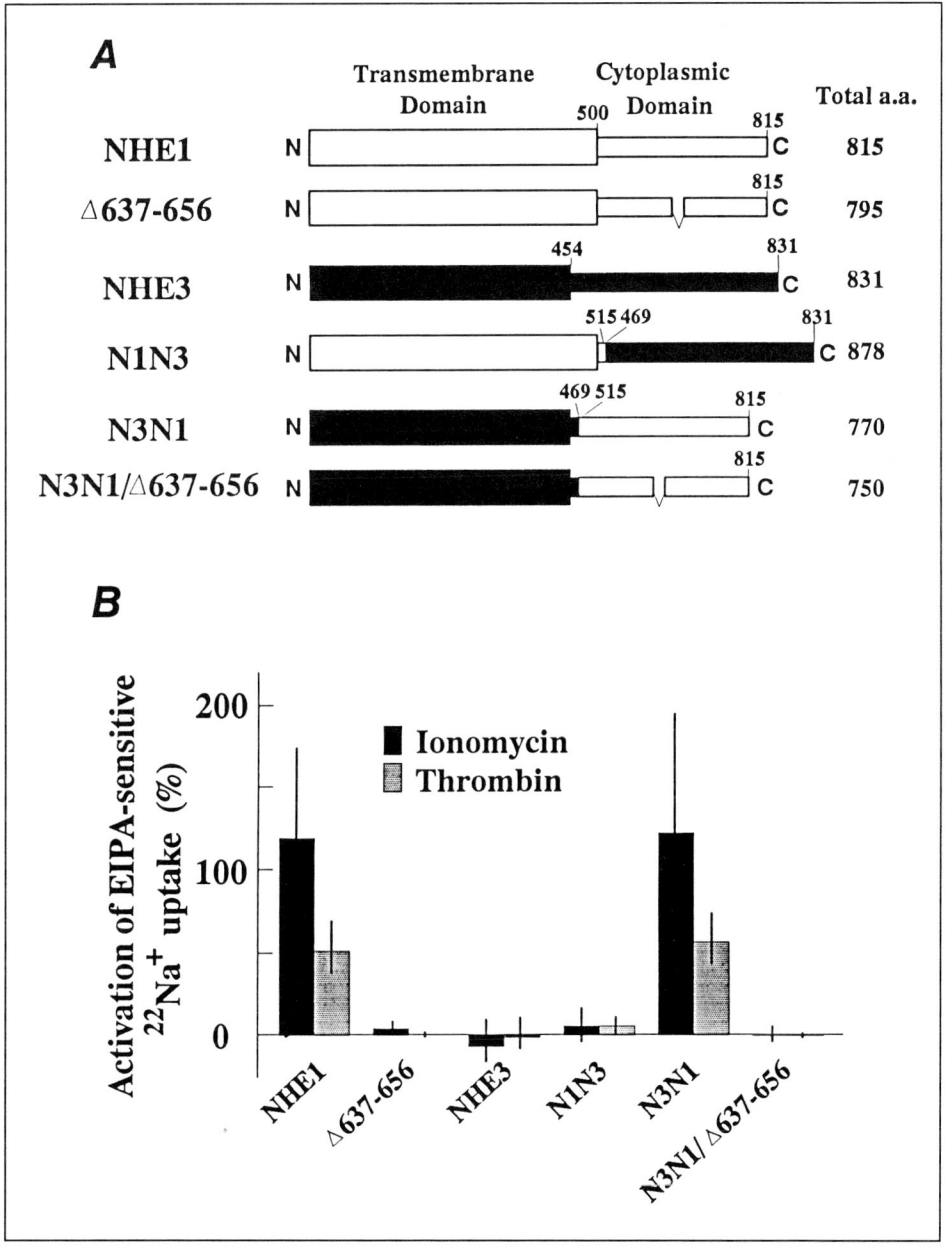

Fig. 8.3. Panel A: Na+/H+ exchanger isoforms, deletion mutants, and chimeric exchangers used to study the role of the calmodulin binding/autoinhibitory domain in NHE1 and NHE3. Constructs were cloned in pECE vector, transfected into exchanger-deficient fibroblasts, and functionally characterized. Panel B: Effect of thrombin and ionomycin on Na+/H+ exchanger activity, assessed as EIPA sensitive ^{22}Na-uptake. Wild-type NHE1 was activated by increased cytoplasmic Ca^{2+}; calcium-resistant NHE3 became responsive to Ca^{2+} when its cytoplasmic domain was replaced by the NHE1-cytoplasmic domain. Deletion of the calmodulin binding/autoinhibitory domain resulted in a loss of calcium responsiveness.

thus acting as an autoinhibitory domain. Furthermore, analogous to the activation brought about by the deletion of region A and consequently the abolishment of its autoinhibitory interaction with other exchanger domains, the Ca^{2+}-induced activation could be explained by binding of calmodulin to NHE1 in a way that will also disrupt this interaction.[55] Similar models have been proposed for Ca^{2+}-induced activation of other calmodulin-binding proteins, e.g., plasma membrane Ca^{2+} pump or CAMKII.[58,59] The calmodulin binding sites in these enzymes partially overlap with the respective autoinhibitory domains; our data suggest that this pattern also applies to NHE1.

Second,[32] we could confirm and extend this model of NHE1 activation by Ca^{2+} studying wild-types NHE1 and NHE3 and their chimeras expressed in exchanger-deficient PS120 fibroblasts. Chimeric exchanger proteins were constructed in which the cytoplasmic domain of NHE1 (amino acids 515-815) was replaced by the cytoplasmic domain of NHE3 (amino acids 469-831; termed N1N3) and vice versa, i.e., the cytoplasmic domain of NHE3 was replaced by the cytoplasmic domain of NHE1 (termed N3N1). Addition of ionomycin or thrombin to the medium elicited fast calcium transients in these cells, as confirmed by fluorometric determination of Ca^{2+}_i. These transients were similar in all PS120 fibroblasts expressing the various isoforms and chimeras. In contrast to well known stimulation of NHE1 by ionomycin or thrombin, wild-type NHE3 was not activated within 2 minutes, the period in which the calcium transients occurred (Fig. 8.4).

Phorbol ester PMA or hyperosmotic medium containing sucrose (final osmolarity: 400 mosmol), which are both known to stimulate NHE1 after prolonged exposure, did not stimulate any isoform or chimera within this short interval. The lack of effect was associated with the absence of calcium transients after PMA and hyperosmotic shock. This was compatible with our hypothesis that the short-term regulation of NHE1 is essentially mediated by an increase in Ca^{2+}_i.

Interestingly, when the cytoplasmic domain of the Ca^{2+}-activatable exchanger NHE1 was removed and replaced by the cytoplasmic domain of the Ca^{2+} resistant isoform NHE3, the resulting chimeric exchanger N1N3 did not further respond to ionomycin or thrombin. On the other hand, when the cytoplasmic domain of NHE3 was substituted by the cytoplasmic domain of NHE1, the resulting exchanger N3N1 responded to both thrombin and ionomycin with an increased uptake of ^{22}Na similar to that observed in wild-type NHE1 (Fig. 8.4).

This showed that calcium responsiveness of NHE1 can be conveyed to NHE3 by transferring the cytoplasmic segment of NHE1. Furthermore, this implies that the relevant cytoplasmic domains of NHE1 are able to interact efficiently with the transmembrane segment of NHE3. In a similar way, NHE1 had been reported to become responsive to PKA stimulation after replacing its cytoplasmic domain by that of the trout red blood cell Na^+/H^+ exchanger.[34]

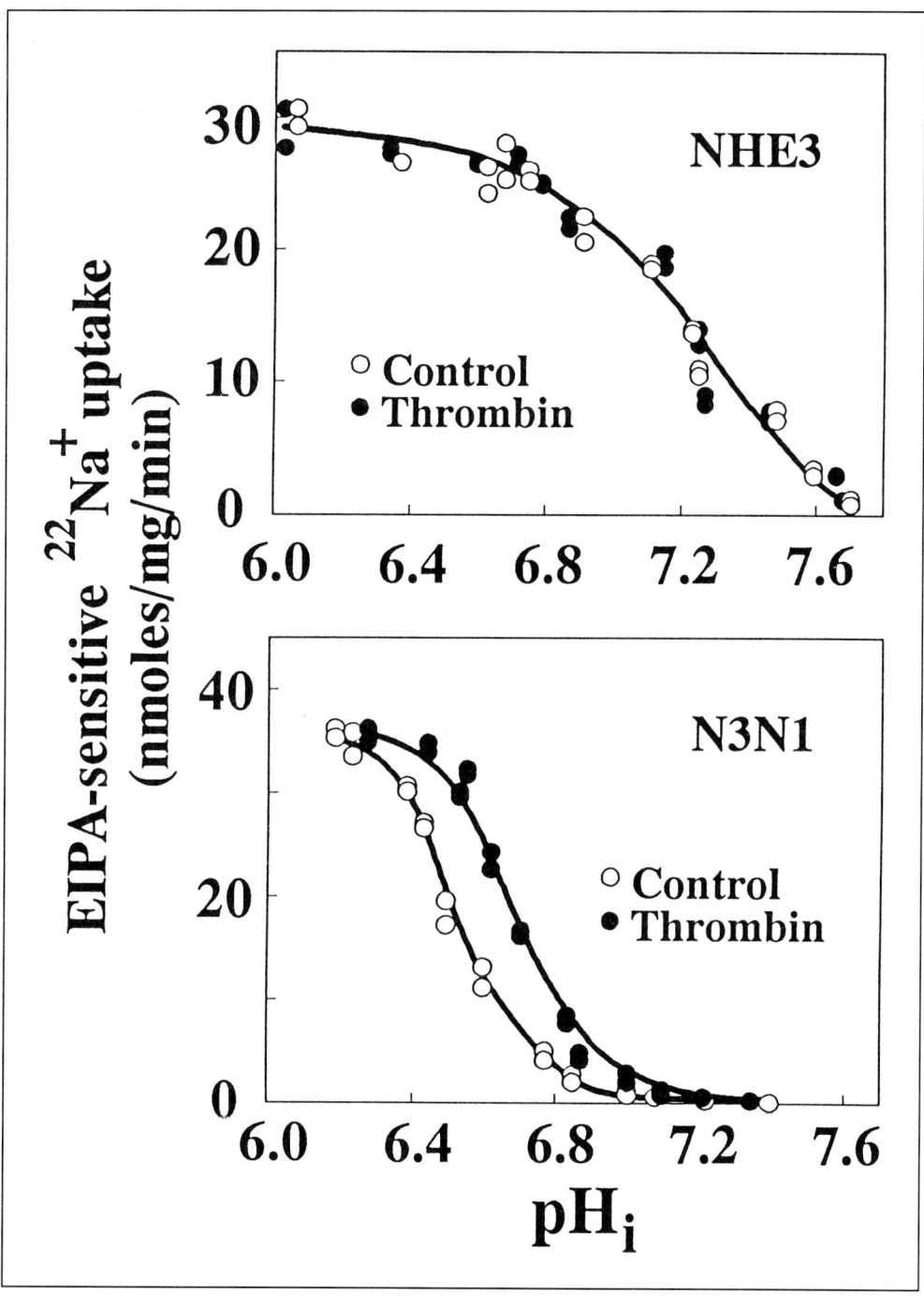

Fig. 8.4. Effect of thrombin on pH-dependent ^{22}Na uptake in fibroblasts expressing wild-type NHE3 or the N3N1 chimera in which the cytoplasmic domain of NHE3 was replaced by the cytoplasmic tail of NHE1. Thrombin had no effect on NHE3, but shifted the H^+_i activation curve of N3N1 to more alkaline values.

The deletion of the NHE1 calmodulin binding/autoinhibitory domain within the grafted cytoplasmic segment of the N3N1 chimera led to complete abolishment of calcium activation; this paralleled the previous finding that wild-type NHE1 was no longer activated by calcium when this region was deleted from its cytoplasmic segment. This demonstrates that the short-term activation of NHE1 and the N3N1 chimera by calcium is completely dependent on this calmodulin binding region.

The structural changes afflicted to the exchanger molecule by the deletion of these 20 amino acids were relatively small; thus, other mechanisms of Ca^{2+} action on NHE1 can almost been ruled out, including phosphorylation by Ca^{2+}/calmodulin dependent protein kinases and Ca^{2+}-induced cell shrinkage.[50,60] It is not clear, however, whether other cytoplasmic domains participate in this Ca^{2+}-mediated regulation. There should be at least some flexible "spacer" to meet the minimum steric requirements of the interacting NHE1 domains, but more domains could be involved. Interestingly, when only the NHE1 calmodulin binding/autoinhibitory domain (amino acids 637-656) was transferred to NHE3, replacing amino acids 591-636 of NHE3, this was not sufficient to convey Ca^{2+}-responsiveness (S. Wakabayashi et al, unpublished results).

The short term response of NHE3 to calcium had not been characterized previously. The cytoplasmic domain of NHE3 contains a site which shares some homology with the NHE1 calmodulin binding site (amino acids 588-644, 9 conserved amino acids). However, an intervening stretch of 27 unrelated amino acids disrupts this domain. Thus, it was not clear whether this domain can function as a calmodulin binding and/or autoinhibitory domain.

Unexpectedly, we found that NHE3 was bound to CaM-sepharose in a strictly Ca^{2+}-dependent manner, as we had previously shown for NHE1.[32,56] While this demonstrates that NHE3, too, is a calmodulin binding protein, it is recalled that this isoform could not be activated by Ca^{2+} in the aforementioned experiments. Moreover, the binding site and the consequences of calmodulin binding for the function of NHE3 are still unknown. An indirect and inhibitory effect of Ca^{2+} and calmodulin on NHE3 has been found in studies where the more prolonged effects were examined.[28,61,62]

E. REGULATORY COFACTORS AND "pH MAINTENANCE DOMAIN"

From deletion mutant studies of NHE1 it has become clear that phosphorylation can only partially explain the exchanger activation upon stimulation by growth factors.[31] Deletion of the cytoplasmic domain carrying all major phosphorylation sites (amino acids 636-815) did not abolish the increase of pH_i after the addition of serum, PDGF-BB, PMA, or α-thrombin, but only decreased its extent by approximately 50%. This prompted the speculation that a regulatory cofactor might be involved in the growth-factor mediated stimulation, which might itself be phosphorylated, but could act independently from direct phosphorylation of NHE1.[31]

The binding site of this regulatory factor has been tentatively assigned to somewhere in between amino acids 566-635, since growth factor activation was completely lost in a mutant of NHE1 truncated at amino acids 566.[18] Intriguingly, this mutant had simultaneously become considerably less sensitive to protons, i.e., its H^+_i-activation curve was moved towards more acidic values. A more detailed analysis was performed using additional truncation mutants and revealed that amino acids 515-595 are required for maintenance of '"normal" H^+_i sensitivity (T. Ikeda et al, manuscript in preparation). The exact role of this "pH maintenance domain" in sensing cytoplasmic pH and/or modifying the rate of ion transport across the transmembrane segment is not known to date. Furthermore, it is unknown whether this domain is related to the NHE1 site interacting with a putative regulatory cofactor. So far, such a factor has neither been characterized experimentally nor been isolated, structurally identified or cloned.

In NHE3 the concept of a regulatory cofactor had emerged earlier, based on studies in solubilized membranes and reconstitution experiments indicating the participation of an intermediary protein in the cAMP/PKA mediated inhibition of this Na^+/H^+ exchanger isoform.[63] Weinman and coworkers succeeded to obtain a partial amino acid sequence of such a protein,[64] which enabled them to raise an antibody against an analogous synthetic peptide, immunoprecipitate a single protein from renal brush border membranes and characterize it, and finally to isolate a corresponding cDNA from a rabbit kidney cDNA-library.[33]

The sequence determined from the 1.9-kb cDNA predicts a novel protein of 358 amino acids containing potential sites for phosphorylation by PKA; its mRNA could be detected in proximal small intestine, liver, and distinct nephron segments. This regulatory cofactor has not yet been characterized on the protein level, so its binding site and the mechanisms by which the transport rate of the NHE3 molecule is altered remain to be elucidated. However, the successful identification of this factor should make it possible to tackle such questions; the results of such studies could also be valuable in promoting the search for cofactors in other isoforms, not to mention the possible existence of several cofactors for a single isoform.

F. G-PROTEINS AND OTHERS

Several signal transduction pathways involved in the regulation of sodium proton exchangers include guanine-nucleotide-binding proteins (G-proteins), e.g., the activation of NHE1 by α-thrombin whose receptor couples to G_9, leading to activation of a phosphoinositide-specific phosphodiesterase (PLC), breakdown of phosphoinositol-4,5-bisphosphate (PIP_2) into 1,2-diacylglycerol (1,2-DAG) and inositol-1,4,5-trisphosphate (IP_3), and finally to Ca^{2+} release and PKC activation. Another well

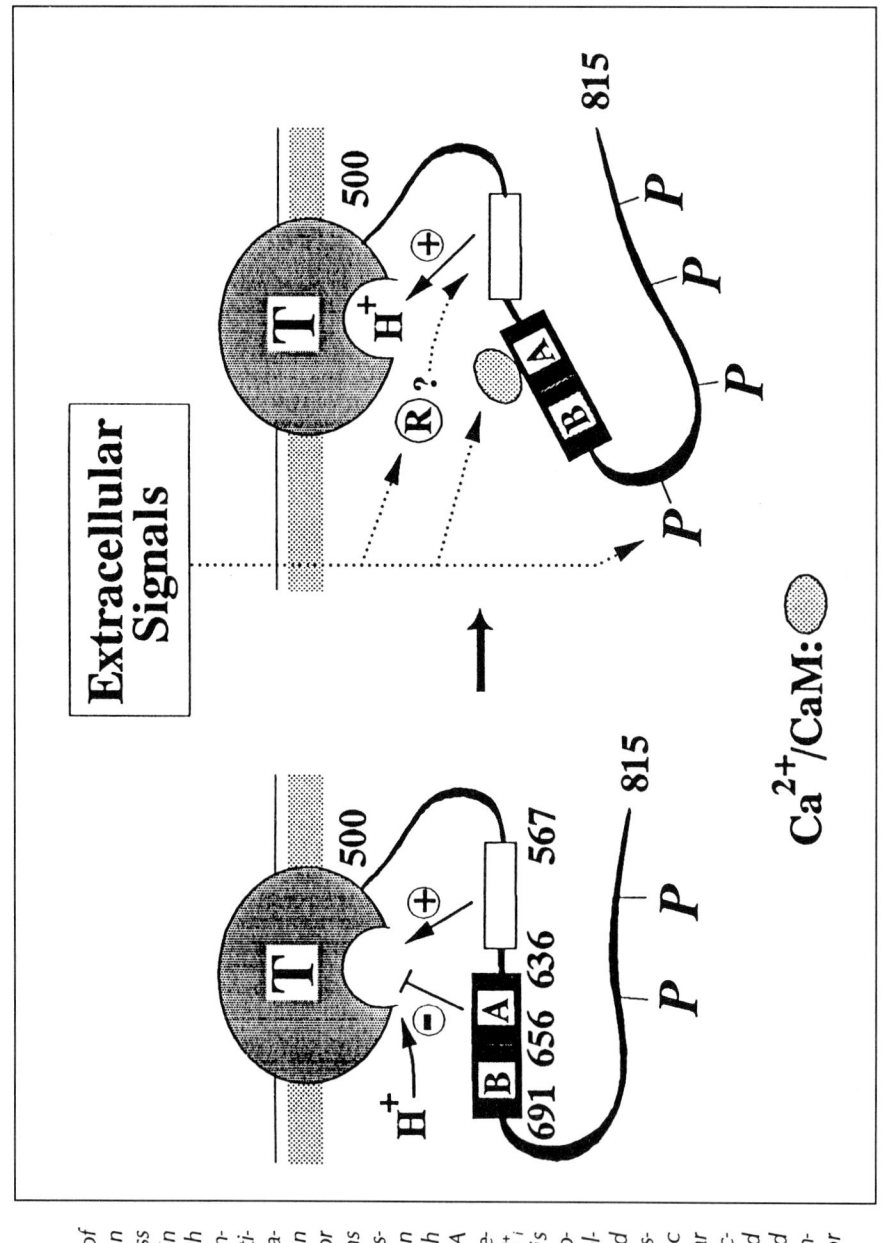

Fig. 8.5. Model drawing of the regulation of NHE1. Ion translocation occurs across the transmembrane domain (T) and is activated by high H^+_i. Cytoplasmic domains interfere with this proton activation positively and negatively: a proximal domain (white box) is required for NHE1 activation by protons or growth factors; more distally located domains can bind calmodulin with high and low affinity (regions A and B, respectively) or decrease the affinity for H^+_i when no calmodulin is bound ("autoinhibitory domain"). The functional relevance of baseline and growth factor-induced phosphorylation at cytoplasmic serine residues (P) is not clear to date. A regulatory cofactor R has been postulated since NHE1 can be activated by several stimuli independent from calmodulin or phosphorylation effects.

established example is the inhibition of NHE3 by parathyroid hormone which involves a heterotrimeric G-protein comprising the stimulatory $G\alpha_s$ and the activation of PKA.

The notion that G-proteins might directly interact with the sodium proton exchanger arose from findings hinting at a role for other heterotrimeric G-proteins and the so-called small G-proteins in exchanger regulation. Thus, Ganz and coworkers[65] transfected mouse L cells with a deletion mutant of the β_2-adrenergic receptor that could not couple to the cholera toxin-sensitive G_s subunit which is required to transduce the signal to PKA; β_2-adrenergic receptor agonist isoproterenol, sodium fluoride, and the nonhydrolyzable GTP analog GTPγS (guanosine 5'-O-[3-thiotriphosphate]) were nonetheless all found to accelerate pH_i recovery from an acute acid load. A study by Neve and colleagues[66] indicated that the activation of Na^+/H^+ exchange via the dopamine D_2 receptor in C6 glioma cells and L fibroblasts does not involve adenylate cyclase or pertussis-sensitive G-proteins. Cano and coworkers[67] recently reported that the stimulation of Na^+/H^+ exchange observed after a low dose of angiotensin II in opossum kidney cells was mediated by an AT_1 receptor and a pertussis toxin-sensitive G-protein but did not involve changes of cAMP levels. In a study conducted in internally dialyzed barnacle muscle fibers,[68] shrinkage-induced stimulation of Na^+/H^+ exchange was blunted by guanosine 5'-O-(2-thiodiphosphate), an antagonist of G-Protein activation, while cholera toxin and general G-Protein activator GTPγS stimulated Na^+/H^+ exchange even in unshrunken cells. Neither injection of cAMP nor exposure to PMA stimulated the exchanger substantially, suggesting a pathway involving unknown G-proteins. $G\alpha_{13}$, member of the recently discovered G_{12} class of G-Proteins with largely unknown functions, was found to stimulate Na^+/H^+ exchange in HEK293 embryonic kidney cells; intriguingly, neither adenylate cyclase levels nor phosphoinositide hydrolysis were changed.[69]

Whether $G\alpha_{13}$ or other G-Proteins can interact directly with the NHE molecules is not clear to date. There is evidence from another recent report[70] that $G\alpha_{13}$ is not acting directly on the Na^+/H^+ exchanger, but via a Ras-dependent pathway involving Jun kinase/stress-activated protein kinase (JNK/SAPK). In contrast to the results obtained in barnacle,[68] GTPγS stimulation of the Na^+/H^+ exchanger in *X. laevis* oocytes was found to be mediated by PKA- and PKC-dependent pathways and could be fully blocked by nonselective protein kinase inhibition.[71] Thus, a direct effect of the respective G-Proteins involved in that system seems unlikely. Evidently, trying to establish the role of G-proteins in the regulation of mammalian Na^+/H^+ exchangers will require further studies.

Apart from regulation of its activity by soluble factors, the cytoplasmic domain of Na^+/H^+ exchangers is likely to interact directly with the cytoskeleton under certain physiological conditions. Though experimental data are still limited these emerging aspects should be mentioned briefly.

NHE1 was reported to be concentrated within focal adhesions, together with F-actin, vinculin and talin;[72] this suggested that NHE1 is directed to and kept attached to the focal adhesions by these cytoskeletal elements. Such interactions were suggested to underlie the increase of pH_i in cells growing on fibronectin-coated supports[73-76] and prompted the speculation that integrin-mediated regulatory pathways of NHE1 exist. Principally, both activation of NHE1 and recruitment of new exchanger molecules could explain this effect. Recruitment of Na^+/H^+ antiporters from cytoplasmic stores is one way by which cells can control the rate of sodium proton exchange across their plasma membrane. It seems to occur in trout red blood cells[77] and could participate in the adaptation of renal tubular cells to chronic acidosis, both respiratory and metabolic, and to chronic increases in single nephron glomerular filtration rate.[78,79] Finally, interactions with the cytoskeleton could play a role in the activation of NHE1 in response to cell shrinkage upon exposure to hyperosmotic media which cannot be explained by changes of Ca^{2+}_i, phosphorylation or protein expression.[50,73,80,81] In all these processes, the cytoplasmic portion of the exchanger very likely would play an essential role.

G. OUTLOOK

In this chapter we outlined briefly the multiple involvement of the NHE cytoplasmic domain in the regulation of Na^+/H^+ exchange.

The concept of the autoinhibitory domain and the mechanism of Ca^{2+} activation have been discussed in detail not because of a particular biological significance, but rather because among the many mechanisms involved in the regulation of mammalian Na^+/H^+ exchangers the Ca^{2+} activation of NHE1 is understood completely at least in its basic aspects; no unknown factors are involved, and the events can be traced from the first messengers (e.g., thrombin) via the second messenger Ca^{2+} down to the exchanger molecule itself (Ca^{2+}/CaM binding and displacement of the autoinhibitory domain from a site controlling the sensitivity to H^+_i). This knowledge of the mechanisms of Ca^{2+} activation allowed to explain experimental findings whose interpretation had been impaired by confounding factors or conflicting results, such as activation of CAMKII or pH_i increases and decreases under different conditions.

However, the other important signaling events involved in the regulation of Na^+/H^+ exchangers are less well understood; available data are not only incomplete, but often conflicting, too. For instance, ANP was reported to stimulate NHE1, to inhibit it, or to be without any effect;[82-84] and hyperosmotic media stimulated NHE2 activity in one experiment but inhibited it in another.[4,85] Apart from different methodologies to assess Na^+/H^+ exchanger activity, different cells are used to study it; they stem from very different tissues, they are used ex vivo or cultured, cultures may be primary cultures or permanent cell lines, and may vary with respect to culture conditions, passage, confluency etc. Furthermore, Na$^+$/H$^+$ exchangers are often expressed in heterologous systems in which they might exhibit other properties than in the cells from which they were originally derived.

As a consequence, the receptors expressed on the cell membrane and the availability of molecules required in the cytoplasm for various signal transduction pathways will vary extremely. Thus, regarding the cells as a "black box," exposing them to a drug or hormone, and recording the net effect on Na$^+$/H$^+$ exchange will probably not by itself lead to a deeper understanding of how Na$^+$/H$^+$ exchangers are regulated.

In contrast to the large and still growing number of hormones, growth factors and drugs reported to affect its activity, the number of cytoplasmic domains involved in the regulation of Na$^+$/H$^+$ exchanger activity is probably very small; in other words, the signal transduction pathways seem to converge. Ca^{2+}/calmodulin is one example of such a final common route shared by several messengers, and probably the regulatory protein cofactors of NHE1 and NHE3 are analogous candidates. Thus, characterizing structure and functions of the Na$^+$/H$^+$ exchanger molecule and then departing from the events at the cytoplasmic domain in an upstream direction could be a successful approach to unravel the regulation of Na$^+$/H$^+$ exchangers.

ACKNOWLEDGMENTS

The authors wish to thank the colleagues who contributed to this work, including the laboratory of J. Pouysségur.

REFERENCES

1. Fliegel L, Dyck JRB. Molecular biology of the cardiac sodium/hydrogen exchanger. Cardiovasc Res 1995; 29:155-159.
2. Lucchesi PA, Berk BC. Regulation of sodium-hydrogen exchange in vascular smooth muscle. Cardiovasc Res 1995; 29:172-177.
3. Nöel J, Pouysségur J. Hormonal regulation, pharmacology, and membrane sorting of vertebrate Na$^+$/H$^+$ exchanger isoforms. Am J Physiol (Cell Physiol 37) 1995; 268:C283-C296.
4. Yun CH, Tse C-M, Nath S, et al. Structure/function studies of mammalian Na-H exchangers—an update. J Physiol Lond 1995; 482:1S-6S.
5. Murer H, Krapf R, Helmle Kolb C. Regulation of renal proximal tubular Na/H exchange: a tissue culture approach. Kidney Int Suppl 1994; 44:S23-S31.
6. Alpern RJ, Yamaji Y, Cano A, et al. Chronic regulation of the Na/H antiporter. J Lab Clin Med 1993; 122:137-140.
7. Fliegel L, Fröhlich O. The Na$^+$/H$^+$ exchanger: an update on structure, regulation and cardiac physiology. Biochem J 1993; 296:273-285.
8. Tse M, Levine S, Yun C, et al. Structure/function studies of the epithelial isoforms of the mammalian Na$^+$/H$^+$ exchanger gene family. J Membr Biol 1993; 135:93-108.

9. Weinman EJ, Shenolikar S. Regulation of the renal brush border membrane Na^+-H^+ exchanger. Annu Rev Physiol 1993; 55:289-304.
10. Motais R, Borgese F, Fievet B, et al. Regulation of Na^+/H^+ exchange and pH in erythrocytes of fish. Comp Biochem Physiol Comp Physiol 1992; 102:597-602.
11. Barber DL. Mechanisms of receptor-mediated regulation of Na-H exchange. Cell Signal 1991; 3:387-397.
12. Igarashi P, Reilly RF, Hildebrandt F, et al. Molecular biology of renal Na^+-H^+ exchangers. Kidney Int Suppl 1991; 33:S84-S89.
13. Frelin C, Vigne P, Lazdunski M. The Na^+/H^+ exchange system in vascular smooth muscle cells. Adv Nephrol Necker Hosp 1990; 19:17-29.
14. Grinstein S, Goetz JD, Cohen S, et al. Regulation of Na^+/H^+ exchange in lymphocytes. Ann N Y Acad Sci 1985; 456:207-219.
15. Wakabayashi S, Shigekawa M, Pouysségur J. Molecular physiology of vertebrate Na^+/H^+ exchangers. Physiol Rev 1995; (in press).
16. Counillon L, Pouysségur J. Structure-function studies and molecular regulation of the growth factor activatable sodium-hydrogen exchanger (NHE-1). Cardiovasc Res 1995; 29:147-154.
17. Demaurex N, Grinstein S. Na^+/H^+ antiport: modulation by ATP and role in cell volume regulation. J Exp Biol 1994; 196:389-404.
18. Wakabayashi S, Fafournoux P, Sardet C, et al. The Na^+/H^+ antiporter cytoplasmic domain mediates growth factor signals and controls "H^+-sensing." Proc Natl Acad Sci USA 1992; 89:2424-2428.
19. Counillon L, Franchi A, Pouysségur J. A point mutation of the Na^+/H^+ exchanger gene (NHE1) and amplification of the mutated allele confer amiloride resistance upon chronic acidosis. Proc Natl Acad Sci USA 1993; 90 (10):4508-4512.
20. Franchi A, Cragoe EJ, Pouysségur J. Isolation and properties of fibroblast mutants overexpressing an altered Na^+/H^+ antiporter. J Biol Chem 1986; 261:14614-14620.
21. Counillon L, Pouysségur J, Reithmeier RA. The Na^+/H^+ exchanger NHE-1 possesses N- and O-linked glycosylation restricted to the first N-terminal extracellular domain. Biochemistry 1994; 33:10463-10469.
22. Haworth RS, Frohlich O, Fliegel L. Multiple carbohydrate moieties on the Na^+/H^+ exchanger. Biochem J 1993; 289:637-640.
23. Sardet C, Counillon L, Franchi A, et al. Growth factors induce phosphorylation of the Na^+/H^+ antiporter, glycoprotein of 110 kD. Science 1990; 247:723-726.
24. Fafournoux P, Nöel J, Pouysségur J. Evidence that Na^+/H^+ exchanger isoforms NHE1 and NHE3 exist as stable dimers in membranes with a high degree of specificity for homodimers. J Biol Chem 1994; 269: 2589-2596.
25. Tse C-M, Levine SA, Yun CH, et al. Functional characteristics of a cloned epithelial Na^+/H^+ exchanger (NHE3): resistance to amiloride and inhibition by protein kinase C. Proc Natl Acad Sci USA 1993; 90:9110-9114.
26. Orlowski J. Heterologous expression and functional properties of amiloride high affinity (NHE-1) and low affinity (NHE-3) isoforms of the rat Na/H exchanger. J Biol Chem 1993; 268:16369-16377.

27. Yu FH, Shull GE, Orlowski J. Functional properties of the rat Na/H exchanger NHE-2 isoform expressed in Na/H exchanger-deficient Chinese hamster ovary cells. J Biol Chem 1993; 268:25536-25541.
28. Levine SA, Montrose MH, Tse CM, et al. Kinetics and regulation of three cloned mammalian Na^+/H^+ exchangers stably expressed in a fibroblast cell line. J Biol Chem 1993; 268:25527-25535.
29. Borgese F, Sardet C, Cappadoro M, et al. Cloning and expression of a cAMP-activated Na^+/H^+ exchanger: evidence that the cytoplasmic domain mediates hormonal regulation. Proc Natl Acad Sci USA 1992; 89: 6765-6769.
30. Padan E, Schuldiner S. Molecular physiology of Na^+/H^+ antiporters, key transporters in circulation of Na^+ and H^+ in cells. Biochim Biophys Acta 1994; 1185:129-151.
31. Wakabayashi S, Bertrand B, Shigekawa M, et al. Growth factor activation and "H^+-sensing" of the Na^+/H^+ exchanger isoform 1 (NHE1). Evidence for an additional mechanism not requiring direct phosphorylation. J Biol Chem 1994; 269:5583-5588.
32. Wakabayashi S, Ikeda T, Noel J, et al. Cytoplasmic domain of the ubiquitous Na^+/H^+ exchanger NHE1 can confer Ca^{2+} responsiveness to the apical isoform NHE3. J Biol Chem 1995; 270:26460-26465.
33. Weinman EJ, Steplock D, Wang Y, et al. Characterization of a protein cofactor that mediates protein kinase A regulation of the renal brush border membrane Na^+-H^+ exchanger. J Clin Invest 1995; 95:2143-2149.
34. Borgese F, Malapert M, Fievet B, et al. The cytoplasmic domain of the Na^+/H^+ exchangers (NHEs) dictates the nature of the hormonal response: behavior of a chimeric human NHE1/trout beta NHE antiporter. Proc Natl Acad Sci USA 1994; 91:5431-5435.
35. Sardet C, Fafournoux P, Pouysségur J. Alpha-thrombin, epidermal growth factor, and okadaic acid activate the Na^+/H^+ exchanger, NHE-1, by phosphorylating a set of common sites. J Biol Chem 1991; 266:19166-19171.
36. Livne AA, Sardet C, Pouysségur J. The Na^+/H^+ exchanger is phosphorylated in human platelets in response to activating agents. FEBS Lett 1991; 284:219-222.
37. Rao GN, Sardet C, Pouysségur J, et al. Phosphorylation of Na^+-H^+ antiporter is not stimulated by phorbol ester and acidification in granulocytic HL-60 cells. Am J Physiol 1993; 264:C1278-C1284.
38. Ng LL, Sweeney FP, Siczkowski M, et al. Na^+-H^+ antiporter phenotype, abundance, and phosphorylation of immortalized lymphoblasts from humans with hypertension. Hypertension 1995; 25:971-977.
39. Wakabayashi S, Sardet C, Fafournoux P, et al. Structure function of the growth factor-activatable Na^+/H^+ exchanger (NHE1). Rev Physiol Biochem Pharmacol 1992; 119:157-186.
40. Alpern RJ. Cell mechanisms of proximal tubule acidification. Physiol Rev 1990; 70:79-114.
41. Light DG, Schwiebert EM, Karlson KH, et al. Atrial natriuretic peptide inhibits a cation channel in renal inner medullary collecting duct cells. Science 1989; 243:383-385.

42. Weinman EJ, Steplock D, Corry D, et al. Identification of the human NHE-1 form of Na$^+$-H$^+$ exchanger in rabbit renal brush border membranes. J Clin Invest 1993; 91:2097-2102.
43. Villereal ML. Sodium fluxes in human fibroblasts: effect of serum, Ca^{2+}, and amiloride. J Cell Physiol 1981; 107:359-369.
44. Ives HE, Daniel TO. Interrelationship between growth factor-induced pH changes and intracellular Ca^{2+}. Proc Natl Acad Sci USA 1987; 84:1950-1954.
45. Moolenaar WH, Tertoolen LGJ, de Laat SW. Phorbol ester and diacylglycerol mimic growth factors in raising cytoplasmic pH. Nature 1984; 312:371-374.
46. Berk BC, Brock TA, Gimbrone MAJ, et al. Early agonist-mediated ionic events in cultured vascular smooth muscle cells. Calcium mobilization is associated with intracellular acidification. J Biol Chem 1987; 262:5065-5072.
47. Conlin PR, Cirillo M, Zerbini G, et al. Calcium-mediated intracellular acidification and activation of Na$^+$/H$^+$ exchange in adrenal glomerulosa cells stimulated with potassium. Endocrinology 1993; 132:1345-1352.
48. Meech RW, Thomas RC. The effect of calcium injection on the intracellular sodium and pH of snail neurons. J Physiol Lond 1977; 265:867-879.
49. Niggli VE, Sigel E, Carafoli E. The purified Ca^{2+} pump of human erythrocyte membranes catalyzes an electroneutral Ca^{2+}-H$^+$ exchange in reconstituted liposomal systems. J Biol Chem 1982; 257:2350-2356.
50. Grinstein S, Cohen S. Cytoplasmic Ca^{2+} and intracellular pH in lymphocytes. Role of membrane potential and volume-activated Na$^+$/H$^+$ exchange. J Gen Physiol 1987; 89:185-213.
51. Kimura M, Gardner JP, Aviv A. Agonist-evoked alkaline shift in the cytosolic pH set point for activation of Na$^+$/H$^+$ antiport in human platelets. The role of cytosolic Ca^{2+} and protein kinase C. J Biol Chem 1990; 265:21068-21074.
52. Kimura M, Aviv A. Regulation of the cytosolic pH set point for activation of the Na$^+$/H$^+$ antiport in human platelets: the roles of the Na$^+$/Ca^{2+} exchange, the Na$^+$-K$^+$-2Cl-co-transport and cellular volume. Pflugers Arch 1993; 422:585-590.
53. Pouysségur J, Sardet C, Franchi A, et al. Cytoplasmic pH, a key determinant of growth factor-induced DNA synthesis in quiescent fibroblasts. FEBS Lett 1985; 190:115-119.
54. Gupta A, Schwiening CJ, Boron WF. Effects of CGRP, forskolin, PMA, and ionomycin on pH$_i$ dependence of Na-H exchange in UMR-106 cells. Am J Physiol 1994; 266:C1088-C1092.
55. Wakabayashi S, Bertrand B, Ikeda T, et al. Mutation of calmodulin-binding site renders the Na$^+$/H$^+$ exchanger (NHE1) highly H$^+$-sensitive and Ca^{2+} regulation-defective. J Biol Chem 1994; 269:13710-13715.
56. Bertrand B, Wakabayashi S, Ikeda T, et al. The Na$^+$/H$^+$ exchanger isoform 1 (NHE1) is a novel member of the calmodulin-binding proteins. Identification and characterization of calmodulin-binding sites. J Biol Chem 1994; 269:13703-13709.
57. Chou PY, Fasman GD. Conformational parameters for amino acids in heli-

cal, beta-sheet, and random coil regions calculated from proteins. Biochemistry 1974; 13:211-222.
58. Soderling TR. Protein kinases. Regulation by autoinhibitory domains. J Biol Chem 1990; 265:1823-1826.
59. Carafoli E. Calcium pump of the plasma membrane. Physiol Rev 1991; 71:129-153.
60. Wöll E, Ritter M, Scholz W, et al. The role of calcium in cell shrinkage and intracellular alkalinization by bradykinin in Ha-ras oncogene expressing cells. FEBS Lett 1993; 322:261-265.
61. Chakraborty M, Chatterjee D, Gorelick FS, et al. Cell cycle-dependent and kinase-specific regulation of the apical Na/H exchanger and the Na,K-ATP-ase in the kidney cell line LLC-PK1 by calcitonin. Proc Natl Acad Sci USA 1994; 91:2115-2119.
62. Cohen ME, Reinlib L, Watson AJ, et al. Rabbit ileal villus cell brush border Na^+/H^+ exchange is regulated by Ca^{2+}/calmodulin-dependent protein kinase II, a brush border membrane protein. Proc Natl Acad Sci USA 1990; 87:8990-8994.
63. Weinman EJ, Steplock D, Bui G, et al. Regulation of renal Na^+-H^+ exchanger by cAMP-dependent protein kinase. Am J Physiol 1990; 258:F1254-F1258.
64. Weinman EJ, Steplock D, Shenolikar S. cAMP-mediated inhibition of the renal brush border membrane Na^+-H^+ exchanger requires a dissociable phosphoprotein cofactor. J Clin Invest 1993; 92:1781-1786.
65. Ganz MB, Pachter JA, Barber DL. Multiple receptors coupled to adenylate cyclase regulate Na-H exchange independent of cAMP. J Biol Chem 1990; 265:8989-8992.
66. Neve KA, Kozlowski MR, Rosser MP. Dopamine D_2 receptor stimulation of Na^+/H^+ exchange assessed by quantification of extracellular acidification. J Biol Chem 1992; 267:25748-25753.
67. Cano A, Preisig P, Alpern RJ. Cyclic adenosine monophosphate acutely inhibits and chronically stimulates Na/H antiporter in OKP cells. J Clin Invest 1993; 92:1632-1638.
68. Davis BA, Hogan EM, Boron WF. Role of G-Proteins in stimulation of Na-H exchange by cell shrinkage. Am J Physiol 1992; 262:C533-C536.
69. Voyno Yasenetskaya T, Conklin BR, Gilbert RL, et al. G alpha 13 stimulates Na-H exchange. J Biol Chem 1994; 269:4721-4724.
70. Dhanasekaran Nl, Vara Prasad MVVS, Wadsworth SJ, et al. Protein kinase C-dependent and -independent activation of Na/H exchanger by Ga12 class of G-Proteins. J Biol Chem 1994; 269:11802-11806.
71. Busch S, Wieland T, Esche H, et al. G-Protein regulation of the Na^+/H^+ antiporter in *Xenopus laevis* oocytes. Involvement of protein kinases A and C. J Biol Chem 1995; 270:17898-17901.
72. Grinstein S, Woodside M, Waddell TK, et al. Focal localization of the NHE-1 isoform of the Na^+/H^+ antiport: assessment of effects on intracellular pH. EMBO J 1993; 12:5209-5218.
73. Grinstein S, Woodside M, Sardet C, et al. Activation of the Na^+/H^+ antiporter during cell volume regulation. Evidence for a phosphorylation-independent mechanism. J Biol Chem 1992; 267:23823-23828.

74. Schwartz MA, Lechene C. Adhesion is required for protein kinase C-dependent activation of the Na$^+$/H$^+$ antiporter by platelet-derived growth factor. Proc Natl Acad Sci USA 1992; 89:6138-6141.
75. Schwartz MA, Cragoe Jr EJ, Lechene CP. pH regulation in spread cells and round cells. J Biol Chem 1990; 265:1327-1332.
76. Schwartz MA, Ingber DE, Lawrence M, et al. Multiple integrins share the ability to induce elevation of intracellular pH. Exp Cell Res 1991; 195:533-535.
77. Guizouarn H, Borgese F, Pellissier B, et al. Regulation of Na$^+$/H$^+$ exchange activity by recruitment of new Na$^+$/H$^+$ antiporters: effect of calyculin A. Am J Physiol 1995; 268:C434-C441.
78. Cohn DE, Klahr S, Hammerman MR. Metabolic acidosis and parathyroidectomy increase Na$^+$-H$^+$ exchange in brush border vesicles. Am J Physiol 1983; 245:F217-F222.
79. Tsai CJ, Ives HE, Alpern RJ, et al. Increased V_{max} for Na$^+$/H$^+$ antiporter activity in proximal tubule brush border vesicles from rabbits with metabolic acidosis. Am J Physiol 1984; 247:F339-F343.
80. Soleimani M, Singh G, Dominguez JH, et al. Long-term high osmolality activates Na$^+$-H$^+$ exchange and protein kinase C in aortic smooth muscle cells. Circ Res 1995; 76:530-535.
81. Dascalu A, Nevo Z, Korenstein R. Hyperosmotic activation of the Na$^+$-H$^+$ exchange in barnacle muscle fibers. J Physiol Lond 1992; 456:503-518.
82. O'Donnell ME, Bush EN, Holleman W, et al. Biologically active atrial natriuretic peptides selectively activate Na,K,Cl co-transporter in cultured vascular smooth muscle cells. J Pharmacol Exp Ther 1987; 243:822-828.
83. Gupta S, Cragoe EJ, Deth RC. Influence of atrial natriuretic factor on 5-(N-ethyl-N-isopropyl)-amiloride-sensitive ^{22}Na$^+$ uptake in rabbit aorta. J Pharmacol Exp Ther 1989; 248:991-996.
84. Petrov V, Amery A, Lijnen P. Role of cyclic GMP in atrial-natriuretic-peptide stimulation of erythrocyte Na/H exchange. Eur J Biochem 1993; 221:195-199.
85. Kapus A, Grinstein S, Wasan S, et al. Functional characterization of three isoforms of the Na$^+$/H$^+$ exchanger stably expressed in Chinese hamster ovary cells. ATP dependence, osmotic sensitivity, and role in cell proliferation. J Biol Chem 1994; 269:23544-23552.

PART II

CHAPTER 9

ROLE OF THE SARCOLEMMAL Na^+/H^+ EXCHANGER IN ARRHYTHMOGENESIS DURING REPERFUSION OF ISCHEMIC MYOCARDIUM

Metin Avkiran and Masahiro Yasutake

A. INTRODUCTION

In many experimental studies, reperfusion of myocardium subjected to a brief period of ischemia has been shown to result in the rapid induction of severe ventricular arrhythmias (for a review, see Manning and Hearse[1]). Commonly, polymorphic ventricular tachycardia (VT) develops within a few beats after the onset of reperfusion, and in the absence of protective interventions, this frequently degenerates into ventricular fibrillation (VF).[2] Observations made in patients with Prinzmetal's angina[3] and silent myocardial ischemia[4] suggest that a similar pattern of reperfusion-induced arrhythmogenesis may occur also in man.

In several species, a "bell-shaped" relationship has been demonstrated between the severity of reperfusion-induced arrhythmias and the duration of the preceding period of ischemia.[1] The duration that results in maximum susceptibility to reperfusion-induced VF is relatively short: 5-10 minutes in the anesthetized rat,[1] 10-15 minutes in the isolated rat heart,[1,5] 20-25 minutes in the isolated rabbit heart,[6] and 20-30 minutes in the anesthetized dog.[1] With more prolonged ischemia, susceptibility to reperfusion-induced VF declines with increasing duration. Although, for obvious ethical reasons, the presence

---

The Na^+/H^+ Exchanger, edited by Larry Fliegel. © 1996 R.G. Landes Company.

of a bell-shaped relationship (between ischemia duration and susceptibility to reperfusion-induced arrhythmias) has not been directly tested in man, there is circumstantial evidence to suggest that such a relationship may exist. Thus, reperfusion during cardiac surgery after relatively short (< 60 min.) periods of ischemia has been shown to result in a 50% incidence of ventricular fibrillation.[7] In contrast, as discussed previously,[8] serious reperfusion-induced ventricular arrhythmias were uncommon in large scale clinical trials with intravenously administered thrombolytic agents. A contributory factor to this phenomenon may have been the fact that thrombolytic therapy was started hours, rather than minutes, after the onset of symptoms. Indeed, a survey of clinical trials with intracoronary thrombolysis has revealed a decreasing incidence of severe reperfusion-induced ventricular arrhythmias with increasing duration of ischemia.[9] The malignant nature of the arrhythmias observed during reperfusion in experimental studies (polymorphic VT and/or VF) has led several investigators to suggest that such arrhythmias may underlie some cases of sudden cardiac death in man.[10,11]

The cellular mechanisms of reperfusion-induced arrhythmias are likely to differ from those responsible for ischemia-induced arrhythmias.[12,13] Although the fundamental question of whether the arrhythmogenic process is initiated by the washout of substances accumulated during ischemia or by the resupply of substances absent during ischemia remains to be resolved,[14] several factors have been implicated in reperfusion arrhythmogenesis. Among these putative arrhythmogenic factors, recent evidence appears to favor a causal involvement for free oxygen radical generation (and associated oxidant stress) and, in particular, sarcolemmal Na^+/H^+ exchanger activation.

B. AN INVOLVEMENT OF FREE OXYGEN RADICALS?

A potential link between the generation of free oxygen radicals, upon the reintroduction of oxygen during the early moments of reperfusion, and the induction of arrhythmias first became evident when scavengers of such radicals were shown to inhibit reperfusion-induced arrhythmias.[15] This association was supported by subsequent studies, which not only confirmed that reperfusion arrhythmias could be inhibited by free oxygen radical scavengers but also showed that they could be exacerbated by radical generating systems.[16] Further support came from studies which showed that a burst of free radical production occurs during early reperfusion[17] and that reactive oxygen intermediates are sufficient to generate ventricular arrhythmias in the absence of ischemia and reperfusion.[18,19] However, the results of studies with radical scavenging and generating systems have not been unequivocal,[20] and some investigators have argued that free oxygen radicals are unlikely to be the prime cause of ventricular arrhythmias during early reperfusion.[20,21] In this context, Yamada et al[22] have shown that *anoxic* reperfusion delays the time-to-onset but does not significantly alter the incidence of reperfusion-induced VF. This observation led these investigators to conclude that re-establishment of flow and readmission of oxygen were independent determinants of reperfusion-induced arrhythmias and that the latter was not a prerequisite for arrhythmogenesis. It appears, therefore, that although free oxygen radicals may modulate reperfusion-induced arrhythmias significantly, their generation may not be the dominant progenitor of these arrhythmias.

C. EVIDENCE FOR A KEY ROLE FOR THE Na^+/H^+ EXCHANGER

C.1 MECHANISM OF ACTIVATION AND CONSEQUENCES ON IONIC HOMEOSTASIS

Myocardial ischemia is known to result in acidosis, due to a retention of H^+ from glycolytic adenosine triphosphate (ATP) turnover, carbon dioxide accumulation and net ATP breakdown.[23] Intracellular acidosis during early ischemia would be expected to stimulate H^+ extrusion pathways, including the Na^+/H^+ exchanger.[23] With maintained ischemia, however, extracellular acidosis has been shown to exceed intracellular acidosis, with reversal of the transmembrane H^+ gradient occurring within 10 minutes in rabbit papillary muscle.[24] Since the sarcolemmal Na^+/H^+ exchanger is inhibited by extracellular acidosis,[25,26] such an effect on extracellular pH of prolonged ischemia would be expected to result in a secondary inhibition of the exchanger.

The potential consequences of postischemic reperfusion on sarcolemmal Na^+/H^+ exchanger activity and intracellular ionic homeostasis are schematically summarized in Figure 9.1. Lazdunski et al[27] were the first to suggest that the rapid washout of extracellular H^+ upon reperfusion could reactivate the Na^+/H^+ exchanger (by alleviating the inhibition by extracellular acidosis) and create an intracellular-to-extracellular H^+ gradient, resulting in a significant influx of Na^+ via this route. Such an influx of Na^+, in the presence of Na^+/K^+ pump inhibition caused by the preceding ischemia, could then result in an increase in intracellular Na^+. Although there is controversy over whether a large increase in intracellular Na^+ concentration occurs during early reperfusion, this may be due partly to the inability of current techniques (e.g., ion-selective microelectrodes, fluorescent dyes and nuclear magnetic resonance) to measure Na^+ concentration in the relevant cellular compartment, the subsarcolemmal "fuzzy space."[28] Even a modest increase in intracellular Na^+ concentration would be expected to alter significantly the reversal potential of the Na^+/Ca^{2+} exchanger (which, under normal conditions, operates primarily to extrude Ca^{2+} from the cell), favoring intracellular Ca^{2+} accumulation.[29] The resulting disturbance of Ca^{2+} homeostasis has been proposed as a progenitor of reperfusion-induced arrhythmias,[30] probably through a mechanism that involves the oscillatory release of Ca^{2+} from the sarcoplasmic reticulum (SR) and the subsequent induction of delayed afterdepolarizations,[30,31] as discussed below.

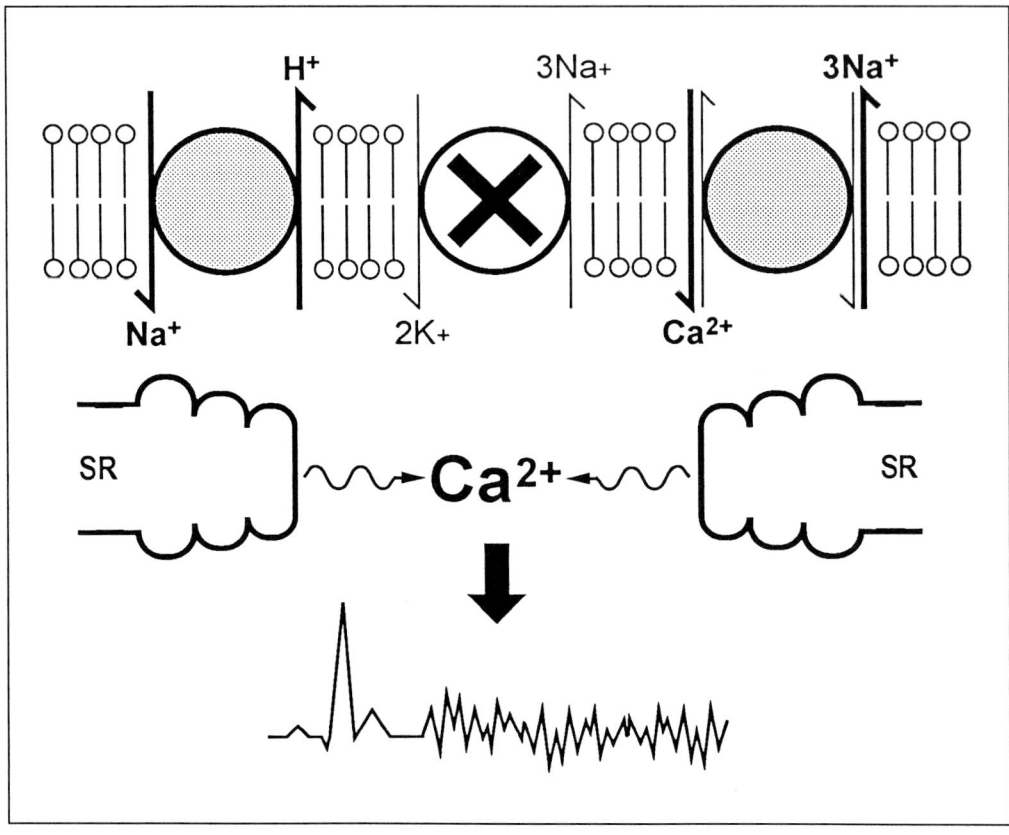

Fig. 9.1. Schematic depiction of the potential consequences of reperfusion on Na^+/H^+ exchanger activity, ionic homeostasis and cardiac rhythm. According to this scheme, an influx of Na^+ through the Na^+/H^+ exchanger results in an increased intracellular Na^+ concentration (due to inhibition of the Na^+/K^+ pump during preceding ischemia) and thereby an increased Ca^{2+} concentration (Ca^{2+} overload) via the Na^+/Ca^{2+} exchanger. The resultant oscillatory release of Ca^{2+} from the sarcoplasmic reticulum (SR) and induction of afterdepolarizations may precipitate ventricular fibrillation (see text for details).

C.2 Evidence for an Arrhythmogenic Role

Dennis et al[32] provided the first experimental evidence to suggest that activation of the Na^+/H^+ exchanger, as proposed by Lazdunski et al,[27] may contribute to arrhythmogenesis during reperfusion. This hypothesis has been supported further by studies employing acidic reperfusion in the authors' laboratory,[2,33] which have shown that slowing the rate at which extracellular pH is restored to its normal physiological value during early reperfusion inhibits the induction of VF and promotes spontaneous reversion from VT to normal sinus rhythm. This effect occurred in a pH-dependent manner,[2] with significant protection against VF observed at pH values (pH 6.6 and 6.4) that would be expected to inhibit significantly the Na^+/H^+ exchanger.[25,26] This is consistent with a major arrhythmogenic role for the rapid washout of extracellular H$^+$ during uncontrolled reperfusion, the most likely mechanism being activation of the Na$^+$/H$^+$ exchanger.[27] Furthermore, a recent study in rat hearts by the authors' group (using a novel dual coronary perfusion technique[34]) has shown that selective administration of two structurally distinct pharmacological inhibitors of the Na$^+$/H$^+$ exchanger (ethyl-isopropyl amiloride and HOE694) into the ischemic/reperfused zone reduces, in a concentration-dependent manner, the incidence of reperfusion-induced VF.[35] Interestingly, both drugs were effective not only when given before ischemia and during reperfusion, but also when given *only* during reperfusion, suggesting that the protective action of Na$^+$/H$^+$ exchanger inhibition was operative, at least in part, during the reperfusion phase. Observations consistent with these results have been obtained also with the global administration of a range of Na$^+$/H$^+$ exchanger inhibitors, such as amiloride and its analogs[32,36-38] and the newer compounds HOE694[38-40] and HOE642.[41]

C.3 Interaction with the Na$^+$/K$^+$ Pump

As discussed earlier and depicted in Figure 9.1, the magnitude of any increase in intracellular Na$^+$ concentration that occurs during early reperfusion as a consequence of Na$^+$ influx through the Na$^+$/H$^+$ exchanger may depend to a large extent on the activity of the primary Na$^+$ extrusion pathway, namely the sarcolemmal Na$^+$/K$^+$ pump. Even under normal conditions, activation of the Na$^+$/H$^+$ exchanger by the abrupt creation of an intracellular to extracellular H$^+$ gradient can result in increased intracellular Na$^+$ concentration and stimulate electrogenic Na$^+$ extrusion via the Na$^+$/K$^+$ pump.[42,43] In the presence of pharmacological inhibition of the Na$^+$/K$^+$ pump, the creation of such a H$^+$ gradient has been shown, in both isolated myocytes[44] and whole hearts,[45] to result in the rapid intracellular accumulation of Ca^{2+}, most probably through the Na$^+$/Ca^{2+} exchanger. In a similar manner, the activity of the Na$^+$/K$^+$ pump also may modulate intracellular Ca^{2+} accumulation during reperfusion.[27,46] In support of this, it has been shown that with pharmacological inhibition of the Na$^+$/K$^+$ pump reperfusion results in a greater increase in intracellular Ca^{2+} concentration relative to that observed in the absence of such inhibition.[47,48] This increase in intracellular Ca^{2+} accumulation during reperfusion has been associated with an exacerbation of contractile dysfunction,[47] as well as an increased incidence of VF.[48]

In this context, the histochemical studies of Winston et al[49] have shown that the activity of Na$^+$/K$^+$ ATPase (the biochemical correlate of the Na$^+$/K$^+$ pump) is depressed significantly during acute ischemia, while the tissue enzyme content remains unaffected. It is reasonable to propose, therefore, that the protective effect against arrhythmias of Na$^+$/H$^+$ exchanger inhibition during early reperfusion, either by pharmacological agents or by acidic reperfusion, may be mediated through inhibition of excessive Na$^+$ influx at a time when myocardial capacity to extrude Na$^+$ is significantly impaired. Interestingly, a recent study with transient acidic reperfusion in the authors' laboratory has revealed a close similarity in the duration of acidic reperfusion required to achieve: (i) a sustained protection against reperfusion-induced VF; and (ii) a significant recovery of maximal Na$^+$/K$^+$ ATPase activity from ischemia-induced inhibition.[50] This suggests that the protective mechanism of transient acidic reperfusion may involve not only a suppression of Na$^+$ influx coupled to H$^+$ extrusion via the Na$^+$/H$^+$ exchanger but also an enhanced recovery of Na$^+$ efflux via the Na$^+$/K$^+$ pump.

C.4 SELECTIVITY OF PHARMACOLOGICAL INHIBITORS

Conventional inhibitors of the Na$^+$/H$^+$ exchanger, such as amiloride and its 5-amino substituted derivatives, can interact with a number of other cation transporting proteins,[51] which may limit their value as selective pharmacological tools in delineating the arrhythmogenic role of the Na$^+$/H$^+$ exchanger in ischemia and reperfusion. Indeed, recent work by Pierce et al[52] has shown that, in the rat, amiloride and several of its derivatives may exhibit cardiodepressant effects and produce changes in action potential characteristics (particularly at high concentration and following prolonged exposure), via actions unrelated to Na$^+$/H$^+$ exchanger inhibition. Such actions may preclude attribution of the antiarrhythmic effects of these drugs to Na$^+$/H$^+$ exchanger inhibition. Indeed, it has been suggested[53] that the antiarrhythmic action of amiloride may be due partly to an inhibition of the inwardly rectifying K$^+$ current (I_{K1}).

Despite the above, amiloride derivatives still may be of value as tools in investigating the role of the Na$^+$/H$^+$ exchanger in ischemia and reperfusion, provided drug concentration and period of drug exposure are carefully controlled.[52] Furthermore, cardiodepressant actions have not been observed with Na$^+$/H$^+$ exchanger inhibitory concentrations of newer inhibitors such as HOE694,[35] which also may exhibit an enhanced selectivity for the cardiac isoform of the exchanger.[54] Nevertheless, HOE694 shares the protective efficacy of other amiloride-based exchanger inhibitors against reperfusion-induced arrhythmias.[35,38-40] Thus, it is likely that the common protection afforded by amiloride and its derivatives[32,35-38] and by the newer benzoyl guanidine compounds (such as HOE694[35,38-40] and its structural congener HOE642[41]) against reperfusion-induced arrhythmias is mediated by these agents' common pharmacological action, that is Na$^+$/H$^+$ exchanger inhibition.

C.5 IMPACT OF ACUTE NEUROHORMONAL MODULATION

Activity of the sarcolemmal Na$^+$/H$^+$ exchanger in mammalian cardiac myocytes is increased substantially by a variety of neurohormonal agents of cardiovascular relevance, including α_1-adrenergic agonists,[55-59] endothelin[60] and angiotensin II,[61] apparently through receptor-mediated mechanisms. The Na$^+$/H$^+$ exchanger-stimulatory actions of the former two agents appear to be suppressed by putative protein kinase C (PKC) inhibitors[55,56,59,60] and mimicked by PKC activators (such as phorbol esters[55,59]), although these findings have been disputed.[57]

In light of the above, it is interesting to note that exposure of isolated rabbit hearts to phorbol esters has been shown[62] to increase the incidence of VF during hypoxia and reoxygenation. In a similar manner, α_1-adrenoceptor stimulation has been shown by the authors' studies[63] to exacerbate reperfusion-induced VF in isolated rat hearts subjected to regional ischemia. The profibrillatory effect of α_1-adrenoceptor stimulation was reversed by Na$^+$/H$^+$ exchanger inhibition, implicating the exchanger in the downstream mechanism(s).[63] Since there is evidence of both norepinephrine release[64] and α_1-adrenoceptor upregulation[65] during myocardial ischemia and reperfusion, regulation of sarcolemmal Na$^+$/H$^+$ exchanger activity via endogenous catecholamines could play a significant role in modulating the severity of arrhythmias in this setting.

Another endogenous mediator that could play a role in modulation of Na$^+$/H$^+$ exchanger activity in myocardial ischemia and reperfusion is thrombin. Thrombin is a multifunctional protease which, in addition to its established role in thrombus formation, induces a variety of cellular responses through the recently cloned thrombin receptor (for review, see Coughlin[66]). Exposure to thrombin has been shown to activate the plasma membrane Na$^+$/H$^+$ exchanger in a number of cell types, including platelets,[67] endothelial cells[68] and smooth muscle cells,[69] through (at least in part) a protein kinase C-mediated pathway. In this regard, recent studies in the authors' laboratory[70] have suggested that a functional thrombin receptor may be expressed also in adult rat ventricular myocytes and may be involved in the regulation of sarcolemmal Na$^+$/H$^+$ exchanger activity. Such a regulatory mechanism could have a significant impact on susceptibility to arrhythmias in the setting of ischemia and reperfusion. Indeed, Goldstein et al[71] have shown that the incidence of malignant ventricular arrhythmias during acute ischemia is greater following thrombotic coronary occlusion than nonthrombotic balloon occlusion, implicating an arrhythmogenic role for factors (such as thrombin) that are associated with thrombus formation. Furthermore, the same group has suggested that during myocardial ischemia activation of the thrombin receptor may contribute to arrhythmogenesis by inducing an increase in intracellular Na$^+$ concentration,[72] an observation which is consistent with sarcolemmal Na$^+$/H$^+$ exchanger activation. Nevertheless, the impact of thrombin receptor activation on the severity of ischemia and reperfusion-induced arrhythmias remains to be determined. An arrhythmogenic role for thrombin during ischemia and/or reperfusion via Na$^+$/H$^+$ exchanger activation could have direct clinical implications, since intracoronary thrombosis is the commonest cause of acute ischemia in patients with coronary artery disease.

C.6 POTENTIAL IMPACT OF CHRONIC REGULATION

There is considerable experimental evidence that the severity of reperfusion-induced arrhythmias is exacerbated in hearts with pressure overload hypertrophy,[73-76] although the cellular mechanisms that are primarily responsible for this increased susceptibility are unclear. With regard to the putative arrhythmogenic scheme depicted in Figure 9.1, it is interesting to note that there is evidence of altered expression and activity of both the Na^+/K^+ pump[77,78] and the Na^+/Ca^{2+} exchanger[79,80] in left ventricular hypertrophy. In this context, mRNA expression of the ubiquitous NHE1 isoform of the Na^+/H^+ exchanger has been shown recently to be upregulated in hypertrophied rabbit myocardium.[81] If such an increase in the steady-state mRNA level is reflected by an increased NHE1 protein expression and sarcolemmal Na^+/H^+ exchanger activity (and is applicable to other species and models of hypertrophy), then this could provide a mechanism for the increased susceptibility of hypertrophied myocardium to reperfusion-induced arrhythmias.

D. CALCIUM—THE COMMON ARRHYTHMOGENIC MEDIATOR

It is apparent from the evidence presented above that the sarcolemmal Na^+/H^+ exchanger may be a key mediator of severe ventricular arrhythmias induced by reperfusion. The question that arises is how this relates to other putative arrhythmogenic processes (e.g., free oxygen radical generation and associated oxidant stress) that have been implicated in reperfusion arrhythmogenesis. The activation mapping studies of Pogwizd and Corr[82] have supported the hypothesis that the induction of VF during reperfusion is mediated by a nonreentrant mechanism that involves Ca^{2+} overload-mediated afterdepolarizations and triggered activity. Indeed, the studies of Kihara and Morgan[83] in the intact ferret heart have shown that spontaneous transitions to VF do not occur unless a state of Ca^{2+} overload is present and that diastolic Ca^{2+} oscillations precede such transitions. More recently, Brooks et al[84] from the same group have demonstrated a potentially causal association between elevated perfusate Ca^{2+} concentration, loss of intracellular Ca^{2+} homeostasis and increased susceptibility to reperfusion-induced VF in isolated rat hearts. That Ca^{2+} overload, and subsequent oscillatory Ca^{2+} release from the SR, may be a primary mechanism underlying VF induction during reperfusion is supported also by the observation that ryanodine (which inhibits Ca^{2+} release from the SR) prevents the degeneration of VT into VF.[85]

As discussed above, activation of the sarcolemmal Na^+/H^+ exchanger during early reperfusion is likely to favor Ca^{2+} overload through the Na^+/Ca^{2+} exchanger. In this regard, Ca^{2+} overload may represent a common arrhythmogenic mechanism that is precipitated by several components associated with reperfusion. Indeed, it has been suggested that the induction of oxidant stress and the activation of the Na^+/H^+ exchanger during early reperfusion may act in a synergistic manner to disrupt intracellular Ca^{2+} homeostasis.[86] Consistent with this hypothesis and the putative arrhythmogenic scheme depicted in Figure 9.1, oxidant stress has been shown to inhibit the electrogenic Na^+/K^+ pump,[87] stimulate the sarcolemmal Na^+/Ca^{2+} exchanger[88] and increase the open probability of the SR Ca^{2+} release channel.[89] Furthermore, recent studies have shown that reactive oxygen intermediates can induce arrhythmogenic oscillations in membrane potential, indicative of intracellular Ca^{2+} overload, in both isolated ventricular muscles[90] and isolated myocytes.[91] Thus, in the setting of reperfusion, the synergistic action of free oxygen radical generation and sarcolemmal Na^+/H^+ exchanger activation may result in a loss of intracellular Ca^{2+} homeostasis and thereby the induction of severe ventricular arrhythmias.

E. CONCLUDING COMMENTS

Many experimental studies have shown that reperfusion of transiently ischemic myocardium can result in the induction of severe ventricular arrhythmias, such as VF. Although incidences of such arrhythmias have been documented in man, their true clinical relevance and potential role in sudden cardiac death remain to be proven conclusively. Nevertheless, available experimental evidence suggests that during reperfusion of ischemic myocardium the intracellular accumulation of Ca^{2+} (Ca^{2+} overload) may be the final trigger responsible for the induction of VF. Ca^{2+} overload itself may arise through the synergistic interaction of more than one component associated with reperfusion, with activation of the sarcolemmal Na^+/H^+ exchanger playing a dominant role.

ACKNOWLEDGMENTS

M. Avkiran is the holder of a British Heart Foundation (Basic Science) Senior Lectureship Award. Parts of his work included in this article were funded by the St. Thomas' Hospital Heart Research Trust (STRUTH) and The David and Frederick Barclay Foundation. M. Yasutake was a Visiting Research Fellow from the Nippon Medical School, Tokyo. The collaboration of Drs. Chikao Ibuki, Yasuyuki Shimada and Peter Haddock and the support of Professor David J. Hearse are gratefully acknowledged.

REFERENCES

1. Manning AS, Hearse DJ. Reperfusion-induced arrhythmias: mechanisms and prevention. J Mol Cell Cardiol 1984; 16:497-518.
2. Avkiran M, Ibuki C. Reperfusion-induced arrhythmias: a role for washout of extracellular protons? Circ Res 1992; 71:1429-1440.
3. Tzivoni D, Keren A, Granot H, et al. Ventricular fibrillation caused by myocardial reperfusion in Prinzmetal's angina. Am Heart J 1983; 105:323-325.
4. Myerburg RJ, Kessler KM, Mallon SM, et al. Life-threatening ventricular arrhythmias in patients with silent myocardial ischemia due to coronary-artery spasm. New Eng J Med 1992; 326:1451-1455.

5. Hearse DJ. Ischemia, reperfusion, and the determinants of tissue injury. Cardiovasc Drug Ther 1990; 4:767-776.
6. Tanaka K, Hearse DJ. Reperfusion-induced arrhythmias in the isolated rabbit heart: characterization of the influence of the duration of regional ischemia and the extracellular potassium concentration. J Mol Cell Cardiol 1988; 20:201-211.
7. Kinoshita K, Mitani A, Tsuruhara Y, et al. Analysis of determinants of ventricular fibrillation induced by reperfusion: dissociation between electrical instability and myocardial damage. Ann Thorac Surg 1992; 53:999-1005.
8. Ibuki C, Hearse DJ, Avkiran M. Rate of reflow and reperfusion induced arrhythmias: studies with dual coronary perfusion. Cardiovasc Res 1992; 26:316-323.
9. Hagar JM, Kloner RA. Reperfusion arrhythmias: experimental and clinical aspects. The Age of Reperfusion 1990; 2:1-5.
10. Sheridan DJ. Reperfusion-induced arrhythmias: an experimental observation awaiting clinical discovery? In: Hearse DJ, Manning AS, Janse MJ, eds. Life Threatening Arrhythmias During Ischemia and Infarction. New York: Raven Press, 1987:49-62.
11. Corr PB, Witkowski FX. Arrhythmias associated with reperfusion: basic insights and clinical relevance. J Cardiovasc Pharmacol 1984; 6 Suppl 6:S903-S909.
12. Curtis MJ, Hearse DJ. Ischaemia-induced and reperfusion-induced arrhythmias differ in their sensitivity to potassium: implications for mechanisms of initiation and maintenance of ventricular fibrilation. J Mol Cell Cardiol 1989; 21:21-40.
13. Curtis MJ, Hearse DJ. Reperfusion-induced arrhythmias are critically dependent upon occluded zone size: relevance to the mechanism of arrhythmogenesis. J Mol Cell Cardiol 1989; 21:625-637.
14. Nakata T, Hearse DJ, Curtis MJ. Are reperfusion-induced arrhythmias caused by disinhibition of an arrhythmogenic component of ischemia? J Mol Cell Cardiol 1990; 22:843-858.
15. Woodward B, Zakaria MN. Effect of some free radical scavengers on reperfusion induced arrhythmias in the isolated rat heart. J Mol Cell Cardiol 1985; 17:485-493.
16. Bernier M, Hearse DJ, Manning AS. Reperfusion-induced arrhythmias and oxygen-derived free radicals: studies with "anti-free radical" interventions and a free radical-generating system in the isolated perfused rat heart. Circ Res 1986; 58:331-340.
17. Garlick PB, Davies MJ, Hearse DJ, et al. Direct detection of free radicals in the reperfused rat heart using electron spin resonance spectroscopy. Circ Res 1987; 61:757-760.
18. Kusama Y, Bernier M, Hearse DJ. Singlet oxygen-induced arrhythmias. Dose- and light-response studies for photoactivation of rose bengal in the rat heart. Circulation 1989; 80:1432-1448.

19. Hearse DJ, Kusama Y, Bernier M. Rapid electrophysiological changes leading to arrhythmias in the aerobic rat heart. Photosensitization studies with rose bengal-derived reactive oxygen intermediates. Circ Res 1989; 65:146-153.
20. Coetzee WA, Owen P, Dennis SC, et al. Reperfusion damage: free radicals mediate delayed membrane changes rather than early ventricular arrhythmias. Cardiovasc Res 1990; 24:156-164.
21. Curtis MJ, Pugsley MK, Walker MJA. Endogenous chemical mediators of ventricular arrhythmias in ischaemic heart disease. Cardiovasc Res 1993; 27:703-719.
22. Yamada M, Hearse DJ, Curtis MJ. Reperfusion and readmission of oxygen. Pathophysiological relevance of oxygen-derived free radicals to arrhythmogenesis. Circ Res 1990; 67:1211-1224.
23. Dennis SC, Gevers W, Opie LH. Protons in ischemia: where do they come from; where do they go to? J Mol Cell Cardiol 1991; 23:1077-1086.
24. Yan GX, Kléber AG. Changes in extracellular and intracellular pH in ischemic rabbit papillary muscle. Circ Res 1992; 71:460-470.
25. Vaughan-Jones RD, Wu ML. Extracellular H^+ inactivation of Na^+-H^+ exchange in the sheep cardiac purkinje fibre. J Physiol 1990; 428:441-466.
26. Wallert MA, Fröhlich O. Na^+-H^+ exchange in isolated myocytes from adult rat heart. Am J Physiol 1989; 257:C207-C213.
27. Lazdunski M, Frelin C, Vigne P. The sodium/hydrogen exchange system in cardiac cells: its biochemical and pharmacological properties and its role in regulating internal concentrations of sodium and internal pH. J Mol Cell Cardiol 1985; 17:1029-1042.
28. Carmeliet E. A fuzzy subsarcolemmal space for intracellular Na^+ in cardiac cells? Cardiovasc Res 1992; 26:433-442.
29. Macleod KT. Regulation and interaction of intracellular calcium, sodium and hydrogen ions in cardiac muscle. Cardioscience 1991; 2:71-85.
30. Opie LH, Coetzee WA. Role of calcium ions in reperfusion arrhythmias: relevance to pharmacologic intervention. Cardiovasc Drug Ther 1988; 2:623-636.
31. January CT, Fozzard HA. Delayed afterdepolarizations in heart muscle: mechanisms and relevance. Pharm Rev 1988; 40:219-227.
32. Dennis SC, Coetzee WA, Cragoe Jr EJ et al. Effects of proton buffering and of amiloride derivatives on reperfusion arrhythmias in isolated rat hearts: possible evidence for an arrhythmogenic role of Na^+-H^+ exchange. Circ Res 1990; 66:1156-1159.
33. Ibuki C, Hearse DJ, Avkiran M. Mechanisms of antifibrillatory effect of acidic reperfusion: role of perfusate bicarbonate concentration. Am J Physiol 1993; 264:H783-H790.
34. Avkiran M, Curtis MJ. Independent dual perfusion of left and right coronary arteries in isolated rat hearts. Am J Physiol 1991; 261:H2082-H2090.
35. Yasutake M, Ibuki C, Hearse DJ, et al. Na^+/H^+ exchange and reperfusion arrhythmias: protection by intracoronary infusion of a novel inhibitor. Am J Physiol 1994; 267:H2430-H2440.

36. Otani H, Kato Y, Tokumitsu K, et al. Effects of amiloride and an analogue on ventricular arrhythmias, contracture and cellular injury during reperfusion in isolated and perfused guinea pig heart. Jpn Circ J 1991; 55:845-856.
37. Duan J, Karmazyn M. Protective effects of amiloride on the ischemic reperfused rat heart. Relation to mitochondrial function. Eur J Pharmacol 1992; 210:149-157.
38. du Toit EF, Opie LH. Role for the Na^+/H^+ exchanger in reperfusion stunning in isolated perfused rat heart. J Cardiovasc Pharmacol 1993; 22:877-883.
39. Scholz W, Albus U, Lang HJ, et al. HOE 694, a new Na^+/H^+ exchange inhibitor and its effects in cardiac ischaemia. Brit J Pharmacol 1993; 109:562-568.
40. Sack S, Mohri M, Schwarz ER, et al. Effects of a new Na^+/H^+ antiporter inhibitor on postischemic reperfusion in pig heart. J Cardiovasc Pharmacol 1994; 23:72-78.
41. Scholz W, Albus U, Counillon L, et al. Protective effects of HOE642, a selective sodium-hydrogen exchange subtype 1 inhibitor, on cardiac ischaemia and reperfusion. Cardiovasc Res 1995; 29:260-268.
42. Piwnica-Worms D, Jacob R, Shigeto N, et al. Na/H exchange in cultured chick heart cells: secondary stimulation of electrogenic transport during recovery from intracellular acidosis. J Mol Cell Cardiol 1986; 18:1109-1116.
43. Rasmussen HH, Cragoe Jr EJ, ten Eick RE. Na^+-dependent activation of Na^+-K^+ pump in human myocardium during recovery from acidosis. Am J Physiol 1989; 256:H256-H264.
44. Nakanishi T, Seguchi M, Tsuchiya T, et al. Effect of partial Na pump and Na-H exchange inhibition on $[Ca]_i$ during acidosis in cardiac cells. Am J Physiol 1991; 261:C758-C766.
45. Anderson SE, Murphy E, Steenbergen C, et al. Na-H exchange in myocardium: effects of hypoxia and acidification on Na and Ca. Am J Physiol 1990; 259:C940-C948.
46. Tani M. Mechanisms of Ca^{2+} overload in reperfused ischemic myocardium. Annu Rev Physiol 1990; 52:543-559.
47. Meng H, Pierce GN. Involvement of sodium in the protective effect of 5-(N,N-dimethyl)-amiloride on ischemia-reperfusion injury in isolated rat ventricular wall. J Pharmacol Exp Ther 1991; 256:1094-1100.
48. Tani M, Neely JR. Deleterious effects of digitalis on reperfusion-induced arrhythmias and myocardial injury in ischemic rat hearts: possible involvements of myocardial Na^+ and Ca^{2+} imbalance. Basic Res Cardiol 1991; 86:340-354.
49. Winston DC, Spinale FG, Crawford FA, et al. Immunocytochemical and enzyme histochemical localization of Na^+/K^+-ATPase in normal and ischemic porcine myocardium. J Mol Cell Cardiol 1990; 22:1071-1082.
50. Avkiran M, Ibuki C, Shimada Y, et al. Effects of acidic reperfusion on arrhythmias and Na^+/K^+ ATPase activity in regionally ischemic rat hearts. Am J Physiol 1996; in press.

51. Kleyman TR, Cragoe Jr EJ. Amiloride and its analogs as tools in the study of ion transport. J Membrane Biol 1988; 105:1-21.

52. Pierce GN, Cole WC, Liu K, et al. Modulation of cardiac performance by amiloride and several selected derivatives of amiloride. J Pharmacol Exp Ther 1993; 265:1280-1291.

53. Duff HJ. Clinical and in vivo antiarrhythmic potential of sodium-hydrogen exchange inhibitors. Cardiovasc Res 1995; 29:189-193.

54. Counillon L, Scholz W, Lang HJ, et al. Pharmacological characterization of stably transfected Na^+/H^+ antiporter isoforms using amiloride analogs and a new inhibitor exhibiting anti-ischemic properties. Mol Pharmacol 1993; 44:1041-1045.

55. Wallert MA, Fröhlich O. α_1-adrenergic stimulation of Na-H exchange in cardiac myocytes. Am J Physiol 1992; 263:C1096-C1102.

56. Gambassi G, Spurgeon HA, Lakatta EG, et al. Different effects of α- and β-adrenergic stimulation on cytosolic pH and myofilament responsiveness to Ca^{2+} in cardiac myocytes. Circ Res 1992; 71:870-882.

57. Pucéat M, Clément-Chomienne O, Terzic A, et al. α_1-adrenoceptor and purinoceptor agonists modulate Na-H antiport in single cardiac cells. Am J Physiol 1993; 264:H310-H319.

58. Lagadic-Gossmann D, Vaughan-Jones RD. Coupling of dual acid extrusion in the guinea-pig isolated ventricular myocyte to α_1- and β-adrenoceptors. J Physiol 1993; 464:49-73.

59. Yasutake M, Coetzee WA, Avkiran M. Role of protein kinase C in α_1-adrenergic activation of sarcolemmal Na^+/H^+ exchange. J Mol Cell Cardiol 1995; 27(6):A188.

60. Kramer BK, Smith TW, Kelly RA. Endothelin and increased contractility in adult rat ventricular myocytes. Role of intracellular alkalosis induced by activation of the protein kinase C-dependent Na^+-H^+ exchanger. Circ Res 1991; 68:269-279.

61. Matsui H, Barry WH, Livsey C, et al. Angiotensin II stimulates sodium-hydrogen exchange in adult rabbit ventricular myocytes. Cardiovasc Res 1995; 29:215-221.

62. Black SC, Fagbemi SO, Chi L, et al. Phorbol ester-induced ventricular fibrillation in the Langendorff-perfused rabbit heart: antagonism by staurosporine and glibenclamide. J Mol Cell Cardiol 1993; 25:1427-1438.

63. Yasutake M, Avkiran M. Exacerbation of reperfusion arrhythmias by α_1-adrenergic stimulation: a potential role for receptor-mediated activation of sarcolemmal sodium-hydrogen exchange. Cardiovasc Res 1995; 29:222-230.

64. Schömig A, Richardt G. Cardiac sympathetic activity in myocardial ischemia: release and effects of noradrenaline. Basic Res Cardiol 1990; 85 (suppl 1):9-30.

65. Corr PB, Yamada KA, DaTorre SD. Modulation of α-adrenergic receptors and their intracellular coupling in the ischemic heart. Basic Res Cardiol 1990; 85 (suppl 1):31-45.

66. Coughlin SR. Thrombin receptor function and cardiovascular disease. Trends Cardiovasc Med 1994; 4:77-83.

67. Nieuwland R, van Willigen G, Akkerman JW. Different pathways for control of Na$^+$/H$^+$ exchange via activation of the thrombin receptor. Biochem J 1994; 297:47-52.
68. Ghigo D, Bussolino F, Garbarino G, et al. Role of Na$^+$/H$^+$ exchange in thrombin-induced platelet-activating factor production by human endothelial cells. J Biol Chem 1988; 263:19437-19446.
69. Berk BC, Taubman MB, Cragoe Jr EJ et al. Thrombin signal transduction mechanisms in rat vascular smooth muscle cells: calcium and protein kinase C-dependent and -independent pathways. J Biol Chem 1990; 265:17334-17340.
70. Yasutake M, Haworth RS, Avkiran M. Thrombin activates the sarcolemmal Na$^+$/H$^+$ exchanger in cardiac myocytes: evidence for a receptor-mediated mechanism involving protein kinase C. Circulation 1995; 92(8):I-113.
71. Goldstein JA, Butterfield MC, Ohnishi Y, et al. Arrhythmogenic influence of intracoronary thrombosis during acute myocardial ischemia. Circulation 1994; 90:139-147.
72. Yan GX, Park TH, Corr PB. Activation of thrombin receptor increases intracellular Na$^+$ during myocardial ischemia. Am J Physiol 1995; 268:H1740-H1748.
73. Taylor AL, Winter R, Thandroyen F, et al. Potentiation of reperfusion-associated ventricular fibrillation by left ventricular hypertrophy. Circ Res 1990; 67:501-509.
74. Robertson E, Hof RP, Zierhut W. Effect of hypertrophy induced by pressure overload or volume overload on reperfusion induced arrhythmias in anaesthetised rats. Cardiovasc Res 1993; 27:515-519.
75. Baxter GF, Yellon DM. Attenuation of reperfusion-induced ventricular fibrillation in the rat isolated hypertrophied heart by preischemic diltiazem treatment. Cardiovasc Drug Ther 1993; 7:225-231.
76. Shimada Y, Hearse DJ, Avkiran M. Ischaemia- and reperfusion-induced arrhythmias in LV hypertrophy. J Mol Cell Cardiol 1994; 26(6):CVIII.
77. Charlemagne D, Orlowski J, Oliviero P, et al. Alteration of Na,K-ATPase subunit mRNA and protein levels in hypertrophied rat heart. J Biol Chem 1994; 269:1541-1547.
78. Charlemagne D, Maixent JM, Preteseille M, et al. Ouabain binding sites and Na$^+$,K$^+$-ATPase activity in rat cardiac hypertrophy: expression of the neonatal forms. J Biol Chem 1986; 261:185-189.
79. Kent RL, Rozich JD, McCollam PL, et al. Rapid expression of the Na$^+$-Ca^{2+} exchanger in response to cardiac pressure overload. Am J Physiol 1993; 265:H1024-H1029.
80. Studer R, Reinecke H, Vetter R, et al. Enhanced expression and function of the Na$^+$/Ca^{2+} exchanger in rat left ventricular hypertrophy and in myocardium of neonatal rats. Circulation 1993; 88:I-86.
81. Takewaki S, Kuro O-M, Hiroi Y, et al. Activation of Na$^+$-H$^+$ antiporter (NHE-1) gene expression during growth, hypertrophy and proliferation of the rabbit cardiovascular system. J Mol Cell Cardiol 1995; 27:729-742.

82. Pogwizd SM, Corr PB. Electrophysiologic mechanisms underlying arrhythmias due to reperfusion of ischemic myocardium. Circulation 1987; 76:404-426.
83. Kihara Y, Morgan JP. Intracellular calcium and ventricular fibrillation: studies in the aequorin-loaded isovolumic ferret heart. Circ Res 1991; 68:1378-1389.
84. Brooks WW, Conrad CH, Morgan JP. Reperfusion induced arrhythmias following ischaemia in intact rat heart: role of intracellular calcium. Cardiovasc Res 1995; 29:536-542.
85. Thandroyen FT, McCarthy J, Burton KP, et al. Ryanodine and caffeine prevent ventricular arrhythmias during acute myocardial ischemia and reperfusion in rat heart. Circ Res 1988; 62:306-314.
86. Hearse DJ. Stunning: a radical re-view. Cardiovasc Drug Ther 1991; 5:853-876.
87. Shattock MJ, Matsuura H. Measurement of Na^+-K^+ pump current in isolated rabbit ventricular myocytes using the whole-cell voltage-clamp technique: inhibition of the pump by oxidant stress. Circ Res 1993; 72:91-101.
88. Reeves JP, Bailey CA, Hale CC. Redox modification of sodium-calcium exchange activity in cardiac sarcolemmal vesicles. J Biol Chem 1986; 261:4945-4955.
89. Holmberg SR, Cumming DV, Kusama Y, et al. Reactive oxygen species modify the structure and function of the cardiac sarcoplasmic reticulum calcium-release channel. Cardioscience 1991; 2:19-25.
90. Shattock MJ, Matsuura H, Hearse DJ. Functional and electrophysiological effects of oxidant stress on isolated ventricular muscle: a role for oscillatory calcium release from sarcoplasmic reticulum in arrhythmogenesis? Cardiovasc Res 1991; 25:645-651.
91. Matsuura H, Shattock MJ. Membrane potential fluctuations and transient inward currents induced by reactive oxygen intermediates in isolated rabbit ventricular cells. Circ Res 1991; 68:319-329.

CHAPTER 10

Role of Sodium-Hydrogen Exchange in Mediating Myocardial Ischemic and Reperfusion Injury. Mechanism and Therapeutic Implications

Morris Karmazyn

A. INTRODUCTION

The heart possesses distinct processes for intracellular regulation of various ionic components including Ca^{2+}, Na^+ and H^+. With respect to the latter, the two major mechanisms for restoring intracellular pH after acidosis include a Na^+-HCO_3^- symport and a Na^+/H^+ antiport. The Na^+/H^+ exchanger (NHE) is a glycoprotein with a molecular weight of approximately 110 kD and which has been identified in virtually all cells so far examined including those of the heart. There are at least five distinct isoforms differing in molecular structure and sensitivity to pharmacological inhibitors, although only the subtype 1 (NHE1) isoform has thus far been demonstrated in the heart and only in very small levels. The NHE represents an electroneutral process which renders it impossible to measure by electrophysiological means. It acts by extruding H^+ in exchange for Na^+ influx, the latter effect, as discussed below, representing a major basis for its participation in myocardial pathology associated with ischemia and reperfusion. While

The Na^+/H^+ Exchanger, edited by Larry Fliegel. © 1996 R.G. Landes Company.

the NHE is regulated by a number of processes such as phosphorylation reactions, a major activator of the antiport is intracellular acidification particularly under conditions where a transmembrane pH gradient has been established consisting of intracellular acidosis and normal pH in the extracellular environment. Conversely, the NHE is inhibited by intracellular alkalosis. Numerous pharmacological agents are available which inhibit the NHE with a good degree of specificity. Thus, the NHE can be inhibited by the K^+-sparing diuretic amiloride but more specifically by N-5 disubstituted derivatives of amiloride. Recently, a number of drugs have been developed which are unrelated to the amiloride structure and which inhibit the NHE. In particular, 3-methylsulphonyl-4-piperidinobenzoyl-guanidine methanesulphonate (HOE694) and, more recently, 4-isopropyl-3-methylsulphonylbenzoyl-guanidine methanesulphonate (HOE642) have been shown to be effective NHE inhibitors. HOE642 is of particular interest as it may represent a selective inhibitor of the cardiac-specific NHE-1 isoform rendering it particularly attractive for therapeutic interventions for cardiac disorders while minimizing the potential for sideeffects. Thus, a major advantage in the study of NHE-dependent processes is the availability of relatively selective and specific antiport pharmacological inhibitors. NHE can be activated by various kinases including calmodulin-dependent kinase although most extensive evidence relates to diacylglycerol-activated calcium-dependent protein kinase C (PKC) as an important activator of NHE. As discussed below, the latter can occur as a consequence of receptor-mediated stimulation of phosphoinositide hydrolysis or by direct activation of PKC with phorbol esters. As noted above, NHE cannot be measured electrophysiologically. A useful tool for the assessment of NHE activity is to examine recovery from intracellular acidosis following NH_4Cl pulsing in cells exposed to bicarbonate-free medium, which renders the process dependent on NHE activity. Thus, restoration of intracellular pH is either slowed or accelerated by NHE inhibitors or activators. As illustrated in Figure 10.1, the α_1 adrenergic agonists phenylephrine, endothelin-1 or phorbol myristate acetate (PMA) accelerate the recovery from acidosis produced by NH_4Cl pulsing in guinea pig ventricular myocytes, indicative of NHE activation.

As seen in Figure 10.2, one of the important consequences of phosphoinositide (PI) hydrolysis is the liberation of diacylglycerol, activation of PKC and the subsequent phosphorylation and activation of NHE. As the heart possesses a number of distinct PKC isoforms, it is not known at present which isoform(s) is responsible for NHE activation. For further discussions of the regulation of the myocardial NHE and its relevance to cardiovascular function, the reader is referred to a number of excellent reviews[1-4] as well as chapters 9 and 11 in this volume and *Cardiovascular Research* (Volume 29, February 1995).

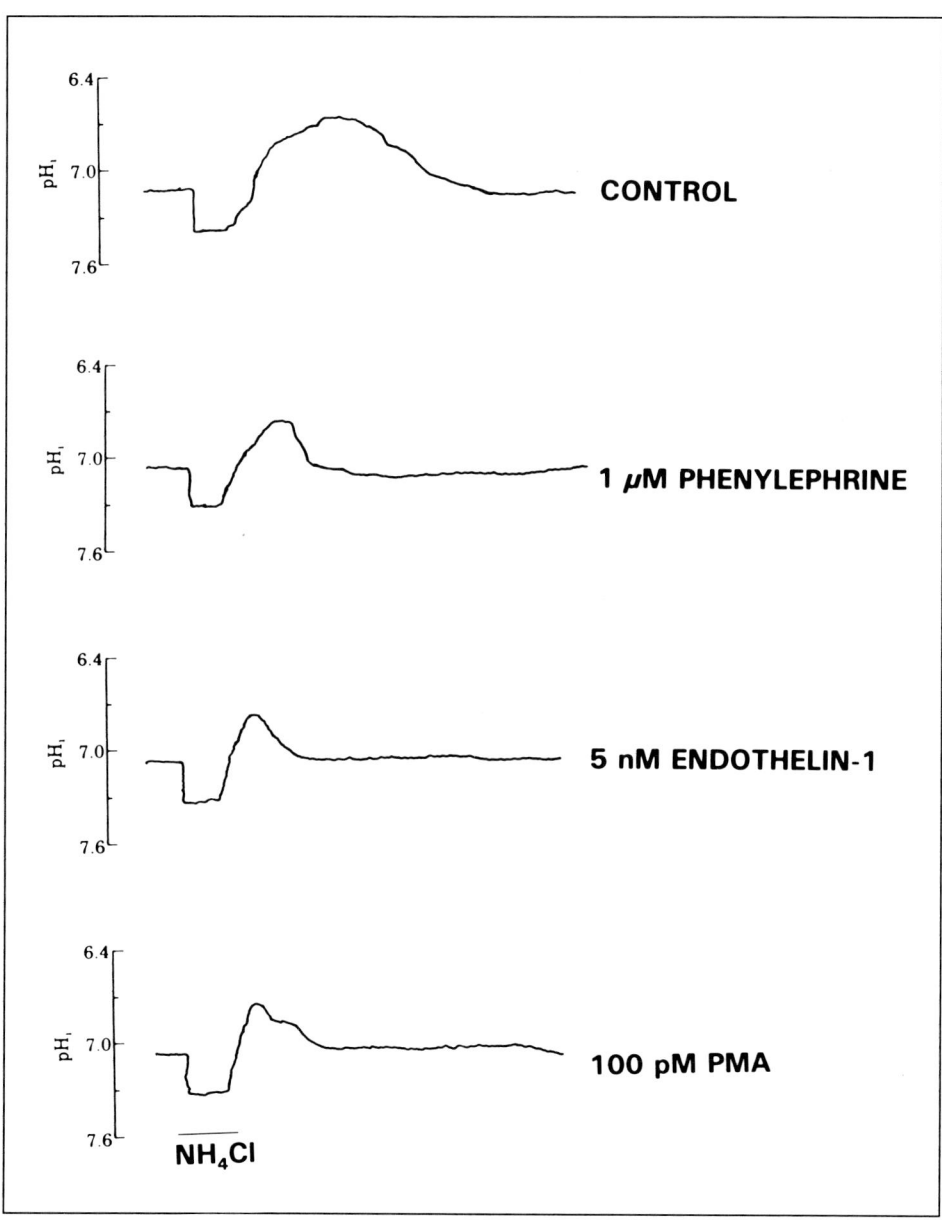

Fig. 10.1. Intracellular pH recordings demonstrating recovery of pH after acid loading produced by NH$_4$Cl pulsing in cells superfused with bicarbonate-free buffer. Note the accelerated pH recovery in cells pretreated with phenylephrine, endothelin-1 or phorbol myristate acetate (PMA) indicative of NHE activation.

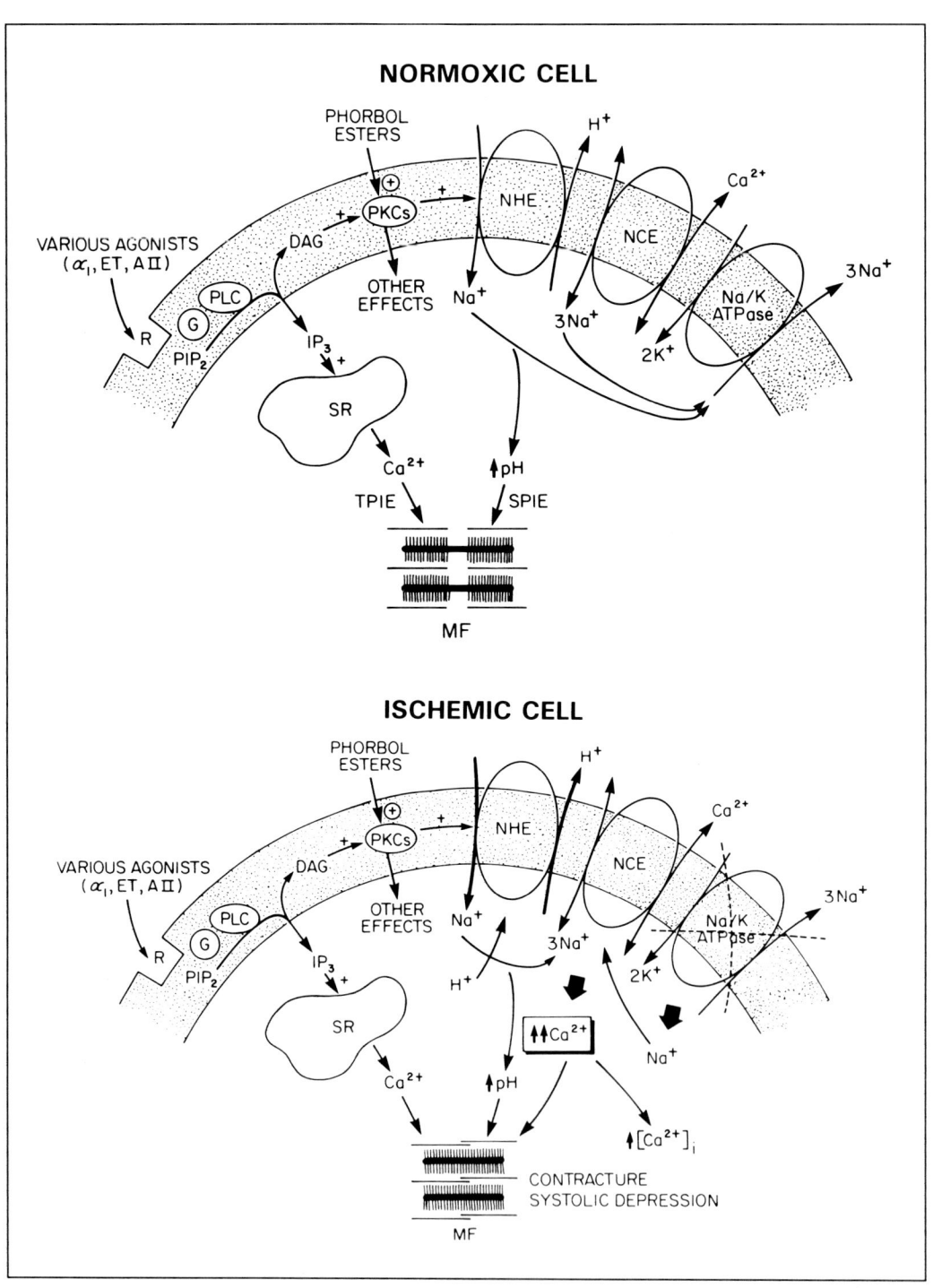

B. IMPORTANCE OF INTRACELLULAR pH REGULATORY MECHANISMS FOR MAINTENANCE OF CARDIAC FUNCTION

Changes in intracellular pH can produce substantial alterations in cardiac contractility, particularly acidosis-induced negative inotropic effects. The mechanisms involved in pH-regulated contractility are very complex but reflect the direct interfering effects of H^+ on various cellular processes involved with excitation-contraction coupling.[5] Thus, it is critical that the cell possess mechanisms by which intracellular pH is regulated especially after intracellular acidosis which occurs during myocardial ischemia. A number of such mechanisms exist including intracellular H^+ buffering as well as a variety of sarcolemmal ion exchangers. Acid extrusion is primarily accomplished by the NHE as well as a Na^+-HCO_3^- symport mechanism. There is evidence that these two systems are differentially regulated. For example, in guinea pig ventricular myocytes α_1 adrenoceptor stimulation activates NHE whereas the Na^+-HCO_3^- is inhibited; in contrast β adrenergic activation exerts the opposite effect.[6] The NHE represents one of the most important modes of extrusion of H^+ that are continually generated within the cell as a consequence of normal energy metabolism or after H^+ loading such as that which may occur during myocardial ischemia. The concomitant entry of Na^+ during NHE activation means that this process is also an important route for increasing intracellular Na^+ concentrations during various conditions.

C. EVIDENCE THAT NA^+/H^+ EXCHANGE REGULATES INTRACELLULAR NA^+ AND CA^{2+} CONCENTRATIONS IN THE HEART

Although the major function of the NHE is to maintain intracellular pH following acidosis, there is extensive evidence that by virtue of its coupling to Na^+ influx, the antiport also regulates intracellular

Fig. 10.2. (Facing page) Cellular events in normoxic or ischemic cardiac cells after Na^+/H^+ exchange (NHE) activation. In normoxic cells, receptor-stimulated G-protein dependent stimulation of phosphoinositide hydrolysis leads to the formation of diacylglycerol (DAG) which in turn stimulates the various isoforms of protein kinase C (PKCs) which in turn phosphorylates and activates NHE leading to Na^+ influx and an increase in intracellular pH. Under normoxic conditions, activity of $Na+$/$K+$ ATPase is able to extrude intracellular Na^+ which enters via either the NHE or the Na^+/Ca^{2+} exchanger (NCE). The production of inositol (1,4,5) trisphosphate (IP_3) results in the transient release of Ca^{2+} from sarcoplasmic reticulum (SR) which produces a transient positive inotropic effect (TPIE) whereas the intracellular alkalinization due to NHE activation sensitizes the myofibrils (MF) to Ca^{2+} causing a sustained positive inotropic response (SPIE). In contrast, in the ischemic cell, NHE activity is stimulated (as depicted by bolder arrows) because of H^+ buildup. Moreover, Na^+/K^+ ATPase is inhibited (as indicated by dashed X) thus precluding the efficient removal of Na^+ following NHE activation. The increase in intracellular Na^+ concentrations will thus result in a NCE-mediated elevation in intracellular Ca^{2+} concentrations which, coupled with pH-dependent myofibrillar sensitization results in defective diastolic and systolic functions.

Na^+ concentrations and therefore Ca^{2+} levels in the cardiac cell. These effects may have bearing on both normal cardiac function as well as the cardiac response to ischemia and reperfusion. In rat and chick cultured cardiac cells, administration of nigericin and ouabain produces a decrease in intracellular pH and inhibition of Na^+ efflux via Na^+/K^+ ATPase, respectively. Activation of the NHE has been shown to account for up to 50% of the basal membrane permeability to Na^+, an effect which was inhibited by amiloride.[7] It is probable that this property of amiloride likely represents the mechanistic basis for its ability to decrease the cardiac effects of digitalis glycosides.[8,9] In cardiac cells, which possess both the NHE and Na^+/Ca^{2+} exchange (NCE) systems, changes in intracellular pH and in cytosolic Ca^{2+} have been shown to be closely related,[10,11] suggesting a close coupling between the two exchangers through changes in intracellular Na^+ concentrations.[7] Thus, Na^+ accumulated via NHE activation may be exchanged for Ca^{2+} via NCE leading to an increase in intracellular Ca^{2+} concentrations. In normal cardiac tissue this would produce a positive inotropic effect, whereas toxicity would be evident in the ischemic myocardium (see below).

The concept of a NHE-mediated increase in intracellular Ca^{2+} concentrations through the NCE has been studied by various investigators.[10,12-14] Inhibition of Na^+ influx with amiloride attenuated the ability of ouabain to elevate intracellular Na^+ concentrations and to increase the rate and extent of Ca^{2+} entry through the NCE.[7] As amiloride on its own did not markedly alter intracellular Na^+ concentrations, it was reasoned that a balance must exist between the rate of Na^+ entry via NHE and the rate of Na^+ efflux via the Na^+/K^+ ATPase.[7] Piwnica-Worms and coworkers[15] measured total cell Ca^{2+} during and after an NH_4^+-induced acid loading in chick heart muscle cells. During exposure to NH_4^+, intracellular Ca^{2+} concentrations declined by about 30%. Changing to a NH_4^+-free solution, which results in intracellular acidification and stimulation of the NHE, produced an increase in intracellular Ca^{2+} concentrations back to control values. The net uptake of Ca^{2+} and net Na^+ extrusion were temporally correlated leading the authors to suggest that both the NCE and the Na^+/K^+ ATPase were important in re-establishing the Na^+ gradient subsequent to intracellular pH regulation. Duan and Moffat have previously demonstrated that accumulation of intracellular Ca^{2+} by isolated ventricular myocytes during realkalinization following a brief period of lactate acidosis was inhibited by hexamethylene amiloride in a concentration-dependent fashion.[16] However, the increase in intracellular Ca^{2+} concentrations was completely dependent on cell stimulation and did not occur if the cells were allowed to remain quiescent during the initial four minutes of realkalinization.[16] This observation suggested that the Na^+/K^+ ATPase by itself was able to dissipate the acidosis-induced Na^+ load in the absence of Na^+ entry through the voltage-dependent channels.

The association between intracellular Ca^{2+} and NHE activation may explain, at least in part, the basis for the cardiotoxic effects of tumor-promoting phorbol esters. These diacylglycerol analogs have been shown to stimulate NHE probably through PKC activation.[11,17-19] Phorbol ester-induced negative inotropic responses are attenuated by NHE inhibitors as well as Ca^{2+} channel blockers suggestive of a link between NHE and Ca^{2+} on the contractile depression produced by phorbol esters.[20,21] This is supported by studies in ventricular myocytes which have shown that PKC activation with phorbol esters produced increased intracellular Ca^{2+} concentrations.[22,23] It must be stressed however that the cardiac effects of phorbol esters are very complex, and many of their actions could be mediated by processes unrelated to either PKC or NHE activation particularly at high concentrations where the effects are insensitive to NHE inhibitors and can indeed be mimicked by phorbols devoid of PKC-stimulating activity.[21]

D. NA^+/H^+ EXCHANGE AND MYCARDIAL ISCHEMIC AND REPERFUSION INJURY

Myocardial reperfusion injury occurs when blood flow to the myocardium is reintroduced following periods of myocardial ischemia. While procedures aimed at restoration of blood flow are critical for myocardial salvage following infarction, there is ample evidence that this results in exacerbation of tissue injury. Myocardial reperfusion injury may represent tissue dysfunction associated with cell necrosis as well as reversible contractile depression occurring in the absence of tissue injury, a phenomenon known as myocardial stunning. The mechanisms for such injury are very complex and most likely represent numerous cellular processes acting in concert.[24]

D.1 THEORETICAL CONCEPTS FOR NHE INVOLVEMENT IN MYOCARDIAL REPERFUSION INJURY

There is now increasing evidence as initially proposed a number of years ago[10] that activation of the NHE at the time of reflow represents a major component of reperfusion-associated dysfunction and cell injury. Theoretically, such injury can occur when flow is introduced to the previously ischemic, and thereby acidic, cardiac cell thus establishing a rapid transarcolemmal H^+ gradient resulting in activation of NHE. As discussed below, this has the rapid effect of restoring intracellular pH, however the concomitant Na^+ influx could result in increases in intracellular Ca^{2+} concentrations through the NCE likely by reduced Ca^{2+} efflux because of the reduction of the Na^+ gradient driving the NCE. The cellular bases for these actions are summarized in Figure 10.2 which also demonstrates potential differences in the cellular responses to NHE activation between those observed in the normoxic and ischemic cardiomyocyte. In the normal cell, intracellular ion homeostasis is regulated by a number of ion transporting systems such as the NHE, NCE and the Na$^+$/K$^+$ ATPase which maintains intracellular Na$^+$ levels by extruding Na$^+$ in exchange for K$^+$. Substantial evidence has demonstrated that the NHE may in fact represent an important mediator for the positive inotropic effects of various agents such as α_1 adrenergic agonists, endothelins and angiotensin II which stimulate NHE through receptor mediated activation of phosphinositide hydrolysis and the production of diacylglycerol which in turn stimulates PKC resulting in activation of the NHE. This series of events can be summarized according to the following pathway: receptor (e.g., α_1 adrenergic, endothelin, angiotensin II) activation → phosphoinositide breakdown → increased production of diacylglycerol → activation of PKC (or distinct PKC isoforms) → activation of NHE → H$^+$ extrusion → increase in intracellular pH → myofibrillar sensitization to Ca^{2+} → sustained positive inotropic effect. Thus, activation of the NHE by various agents may represent the basis for their positive inotropic actions, and indeed studies have demonstrated that the positive inotropic effect of both endothelin-1 as well α_1 adrenergic agonists are dependent on NHE activation.[25,26] It has also recently been shown that angiotensin II activation of the L-type Ca^{2+} channel in rabbit ventricular myocytes involves stimulation of the NHE[27] suggesting that under some situations the antiporter could be an important regulator of ion channel activity. It should be added that the above sequence of events also results in the production of inositol (1,4,5) trisphosphate (IP$_3$) which releases Ca^{2+} from the sarcoplasmic reticulum; this short-lasting enhanced availability of Ca^{2+} results in a transient positive inotropic effect in response to various agonists which stimulate PI hydrolysis.

The scenario is markedly different in the ischemic myocardial cell particularly one subjected to reperfusion. As shown in Figure 10.2, an important consequence of myocardial ischemia is inhibition of the Na$^+$/K$^+$ ATPase and the resultant elevation in intracellular Na$^+$ concentrations due to reduced ability to pump out Na$^+$ through this route of extrusion. Thus, activation of NHE upon reperfusion, which in itself is greater due to the prior ischemia-induced accumulation of H$^+$, provokes a greater elevation in intracellular Na$^+$- and Ca^{2+} concentrations particularly under conditions of defective ion regulatory mechanisms such as depressed Na$^+$/K$^+$ ATPase activity. The net result of large elevations in intracellular Ca^{2+} levels coupled with intracellular alkalosis due to NHE-mediated H$^+$ extrusion results in tissue damage manifested by intracellular Ca^{2+} overload, contracture and depressed systolic function. Moreover, pharmacological activation of the NHE could be further deleterious under ischemic conditions based on this scenario. For example, endothelin-1 (ET-1) production is increased in myocardial ischemia, and it has been shown to produce deleterious effects on the reperfused myocardium in terms of inducing both diastolic and systolic dysfunction.[28] The deleterious effects of ET-1 on the ischemic and reperfused heart in terms of compromised ventricular recovery and contracture formation can be effectively attenuated by the NHE inhibitor methylisobutyl amiloride suggesting an important role of the antiporter in mediating the detrimental effect of the peptide on the ischemic and reperfused myocardium.[29] Similarly, it has recently been shown that α_1 adrenergic agonists enhance ventricular arrhythmias in the reperfused ischemic myocardium which was attenuated by NHE inhibition with HOE694.[30]

D.2 Experimental Evidence for NHE Involvement in Myocardial Reperfusion Injury

The theoretical basis for the critical role of the NHE in mediating myocardial reperfusion injury has been overwhelmingly substantiated in numerous studies which have demonstrated a protective effect of NHE inhibition on a wide variety of parameters. A large number of these studies have been compiled and summarized in Table 10.1. As shown in this table, NHE inhibition can be accomplished either pharmacologically or through early transient reperfusion with acidotic medium which blunts NHE activation by diminution of the transarcolemmal H^+ gradient. Both approaches have indeed been effective in exerting cardioprotective actions. Moreover, as shown in the table these salutary effects have been demonstrated on numerous parameters of cardiac function including enhanced contractility, reduced contracture and a decrease in the incidence of arrhythmias. In addition, improvements in biochemical and ultrastructural indices have been extensively demonstrated with NHE inhibition.

D.3 Pharmacological Studies

The first study demonstrating a cardioprotective effect of NHE inhibition was reported by Karmazyn[31] who showed that amiloride enhanced ventricular recovery and diminished enzyme efflux from reperfused ischemic isolated rat hearts. More recently, we have demonstrated that the protective effect of amiloride is also associated with reduced incidence of arrhythmias and preservation of ultrastructural integrity.[32] Further evidence that this protective action may be due to NHE inhibition stems from a number of studies. For example, Tani and Neely[33] have shown that improved postischemic recovery with amiloride is associated with diminished tissue contents of Na^+ and Ca^{2+}, a result compatible with the concept of NHE- and NCE-linked mechanisms as discussed above. More selective and potent NHE inhibitors including hexamethylene amiloride, dimethyl amiloride and methylisobutyl amiloride (MIA) have also been demonstrated to provide substantial protection against reperfusion injury (see table). In addition to ischemia and reperfusion, we and others have shown that amiloride and amiloride analogs reduce injury in hearts subjected to normal flow hypoxia followed by reoxygenation suggesting NHE involvement in this form of injury.[39,54] Figure 10.3 illustrates the potent ability of the

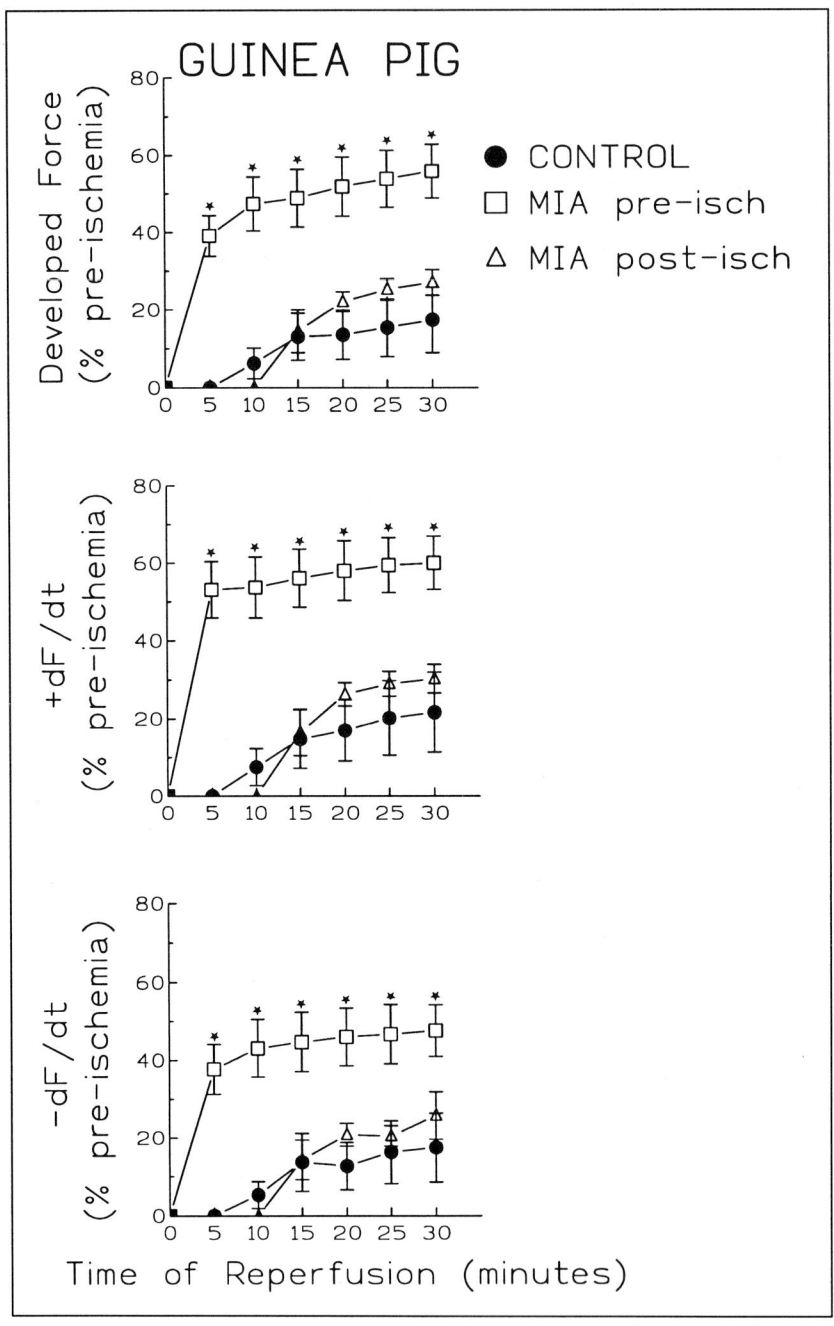

Fig. 10.3. Contractile recovery of isolated guinea pig hearts during reperfusion following 45 minutes of global total ischemia in the presence or absence of the NHE inhibitor methylisobutyl amiloride (MIA). MIA was added either before ischemia as well as during reperfusion (pre-isch) or during reperfusion only (post-isch). *$P < 0.05$ from either the control or post-isch group. From Moffat and Karmazyn,[52] with permission. +dF/dt; rate of force development; -dF/dt; rate of relaxation.

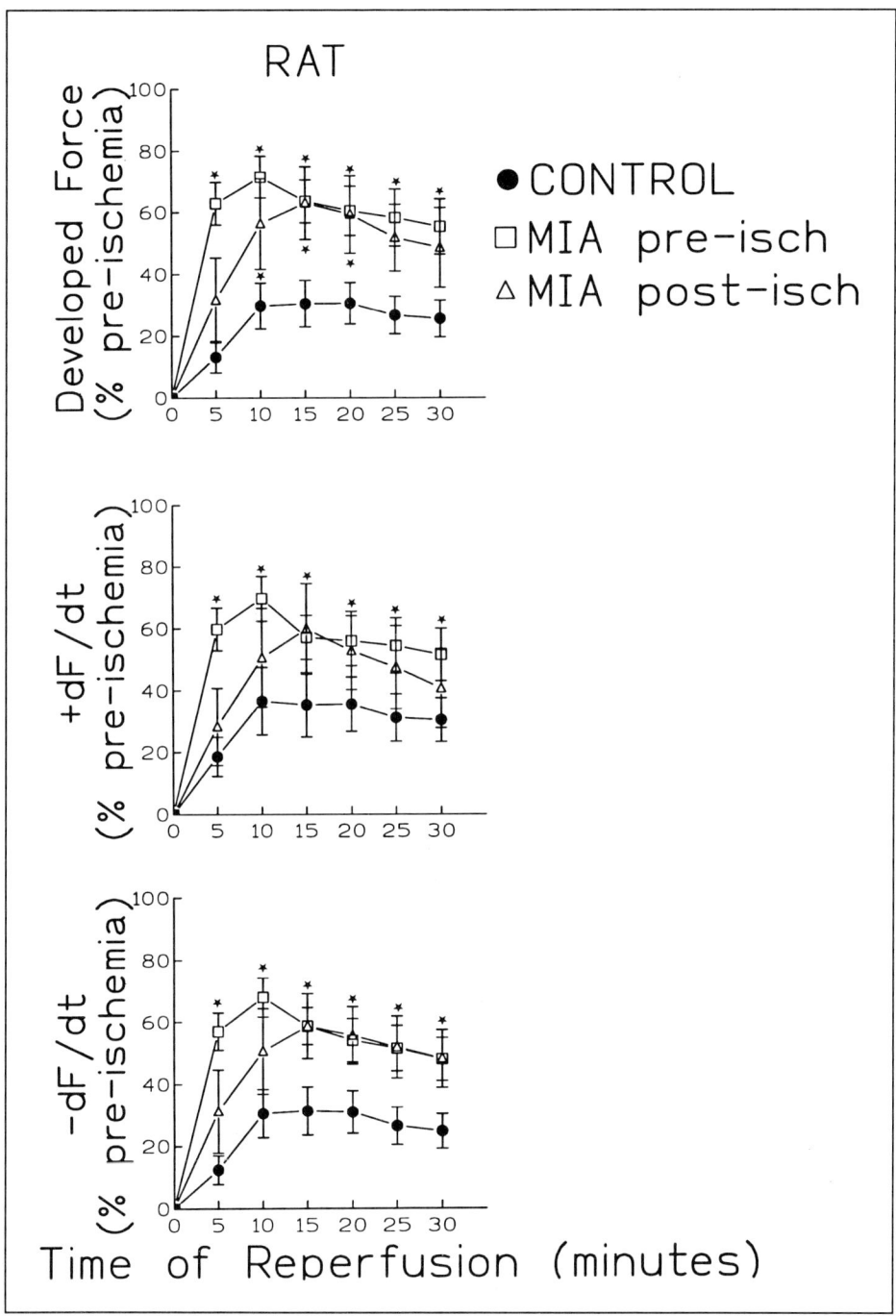

Fig. 10.4. Contractile recovery of isolated rat hearts during reperfusion following 45 minutes of global total ischemia in the presence or absence of MIA. Description as for Figure 10.3. $^*P < 0.05$ from control. From Moffat and Karmazyn, J Moll Cell Cardio 1993; 25:959-971 with permission.

NHE inhibitor MIA to enhance recovery in contractility of isolated reperfused guinea pig hearts after 45 minutes total ischemia. Markedly improved recovery of contractility was evident when MIA was present during both the ischemic and reperfusion periods with no protection when the drug was administered during reperfusion only. In contrast, as shown in Figure 10.4, some protective effects of MIA were observed in rat hearts with MIA postischemic treatment. Moreover, such protection in terms of magnitude of ventricular recovery was also associated with reduced contracture development (Fig. 10.5) as well as accelerated recovery time following reperfusion (Fig. 10.6). The controversy regarding ideal administration protocols of NHE inhibitors to produce cardioprotection is further discussed below.

As discussed in detail in chapter 9 of this volume, extensive evidence has demonstrated a role of NHE in arrhythmogenesis associated with reperfusion. Most in vitro studies have concluded that the antiarrhythmic effects of NHE inhibitors were due to inhibition of NHE and prevention of the sequence of events which are described above for the beneficial effects of these drugs on functional recovery after reperfusion. Duff et al proposed however, that at least part of the antiarrhythmic effect of amiloride in the infarcted myocardium could be attributed to inhibition of the inward rectifier potassium current and a prolongation of repolarization in the border zone.[35,36] Nonetheless, these studies could not exclude the participation of inhibition of NHE and NCE in the effects of amiloride and its analogs. Duan and Moffat[16] have reported that blockade of NHE during realkalinization of guinea pig ventricular myocytes exposed to 20 mM d, l-lactate for five minutes prevented the increase in intracellular Ca^{2+} concentrations and the development of both delayed afterdepolarizations and depolarization-induced automaticity.

The advent of compounds produced by Hoechst (HOE694 and HOE642) which are potent NHE inhibitors but unrelated to the amiloride structure have further advanced the concept of NHE involvement in myocardial reperfusion injury. The protective effects of HOE694 in terms of improved recovery after ischemia, reduced incidence of arrhythmias and ultrastructural preservation have been shown in a variety of experimental models as well as in species including rat, rabbit and pig (Table 10.1).

D.4 Locus of the Protective Effects of NHE Inhibitors in the Ischemic and Reperfused Heart

As noted above, there is some uncertainly regarding the locus of the protective actions of NHE inhibitors on the ischemic and reperfused myocardium and the necessity for preischemic treatment to demonstrate cardioprotective actions. Thus, a wide degree of responses have been reported in studies in which NHE inhibitors were added only during reperfusion including a total lack of protection, diminished

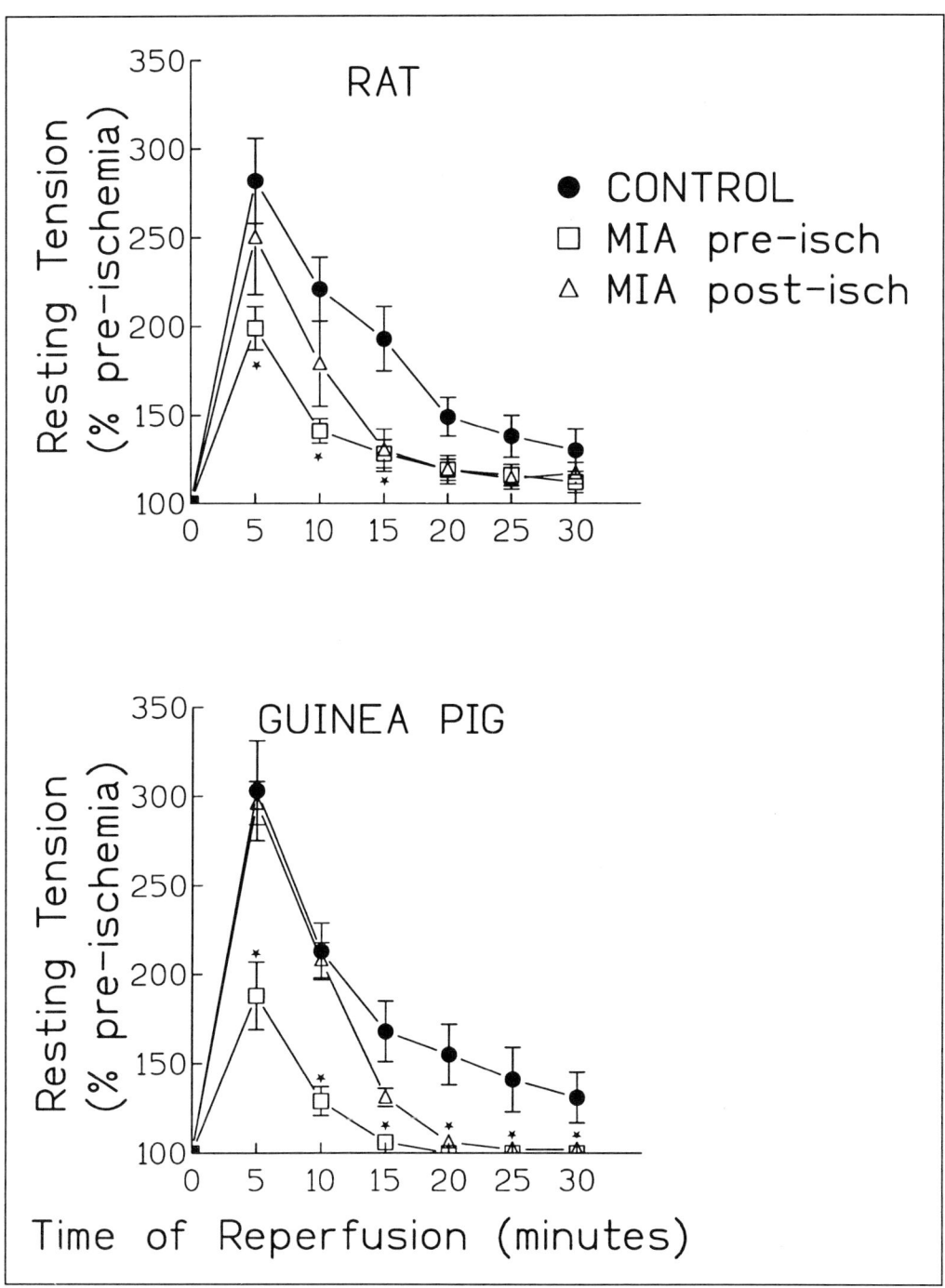

Fig. 10.5. Resting tension of isolated rat or guinea pig hearts during reperfusion following 45 minutes of global total ischemia in the presence or absence of MIA. Drug addition protocols as in Figure 10.3. *$P < 0.05$ from control. From Moffat and Karmazyn, J Moll Cell Cardio 1993; 25:959-971 with permission.

Table 10.1. *Summary of studies in chronological order demonstrating protective effects of sodium-hydrogen exchange inhibition in the ischemic reperfused heart*

Authors and reference #	Drug or Manipulation	Model	Result
Karmazyn, 1988[31]	Amiloride	Isolated rat heart	↑function, ↑CK efflux
Kitakaze et al, 1988[34]	Acidic reoxygenation	Isolated ferret heart	↓stunning
Duff et al, 1988,[35] 1989[36]	Amiloride	Dog, post infarct	antiarrhythmic
Tani and Neely, 1989[33]	Amiloride	Isolated rat heart	↑function, ↓Ca^{2+} and Na^{2+}
Meng and Pierce, 1990[37]	DMA	Rat right ventricular wall	↑function, ↓CK release
Khandoudi et al, 1990[38]	Amiloride	Isolated rat heart	↑function
Weiss et al, 1990[39]	Amiloride	Isolated rat heart	↑function, ↓Ca^{2+} gain
Dennis et al, 1990[40]	DMA, HMA	Isolated rat heart	antiarrhythmic
Bond et al, 1991[41]	Acidic reoxygenation	Rat cardiomyocytes	↓lethal cell injury
Murphy et al, 1991[42]	Amiloride	Isolated rat heart	↑function, ↓Ca^{2+} and Na^+ gain during ischemia
Hori et al, 1991[43]	Acidic reperfusion	Dog in vivo	↑function
Meng et al, 1991[44]	Acidic reperfusion	Rat ventricular wall	↑function
Matsuda et al, 1991[45]	Acidic reperfusion	Isolated rat heart	↑function
Avkiran and Ibuki, 1992[46]	Acidic reperfusion	Isolated rat heart	antifibrillatory
Duan and Moffat, 1992[16]	HMA	Guinea pig cardiomyocytes	↓Ca^{2+} gain
Duan and Karmazyn, 1992[32]	Amiloride	Isolated rat heart	↑function, antiarrhythmic, mitochondrial preservation
Ibuki et al, 1993[47]	Acidic reperfusion	Isolated rat heart	antifibrillatory
Bond et al, 1993[48]	DMA	Rat cardiomyocytes	↓lethal cell injury
Scholz et al, 1993[49]	HOE694	Isolated rat heart and in vivo CA ligation	↑function, ↓CK & LDH release, antiarrhythmic
Mochizuki et al, 1993[50]	Amiloride	Isolated rat heart	↓reperfusion arrhythmias
du Toit and Opie, 1993[51]	HOE694	Isolated rat heart	↑function, antiarrhythmic
Moffat and Karmazyn, 1993[52]	Amiloride, MIA	Rat and guinea pig heart	↑function, accelerated recovery
Meng et al, 1993[53]	Amiloride, HMA, DMA, EIPA	Rat right ventricular wall	↑function
Karmazyn et al, 1993[54]	Amiloride, HMA	Isolated rat heart	↑function
Karmazyn, 1993[55]	Amiloride, HMA	Isolated rat heart	↓lactate-induced toxicity
Khandoudi et al, 1994[29]	MIA	Isolated rat heart	↓endothelin toxicity
Sack et al, 1994[56]	HOE694	Pig in vivo	↓stunning
Hendrikx et al, 1994[57]	HOE694	Isolated rabbit heart	↑function and HEPs, ↓Ca^{2+}
Hata et al, 1994[58]	DMA	Isolated canine heart	↑postacidosis dysfunction
Myers et al, 1995[59]	Amiloride, MIA	Isolated rabbit heart	↑function, ↓CK efflux
Scholz et al, 1995[60]	HOE642	Isolated rat heart and in vivo rat	↓reperfusion arrhythmias preserved metabolites
Yasutake and Avkiran, 1995[30]	HOE694	Isolated rat heart	↓α_1-mediated reperfusion arrhythmias

Authors and reference #	Drug or Manipulation	Model	Result
Ward & Moffat, 1995[61]	MIA	Guinea pig cardiomyocytes	↓Ca^{2+}, ↓contracture, ↑recovery
Bugge and Ytrehus, 1995[62]	EIPA	Isolated rat heart	↓infarct size
Ladilov et al, 1995[63]	HOE694	Rat cardiomyocytes	↓contracture, ↓Ca^{2+}
Klein et al, 1995[64]	HOE694	Pig in vivo	↑recovery, ↓infarct size
Faes et al, 1995[65]	MIA	Isolated rat heart	Reversal of neutrophil-induced inhibition of recovery
Kaplan et al, 1995[66]	DMA or acidic reperfusion	Rabbit papillary	↓muscle cell death

Abbreviations used: CK, Creatine kinase; DMA, dimethylamiloride; EIPA, ethylisopropyl amiloride; HMA, hexamethylene amiloride; LDH, lactate dehysdrogenase; MIA methylisobutyl amiloride

protection as well as marked protection identical to that observed if the drug was administered prior to ischemia. Some investigators have suggested that pretreatment during myocardial ischemia is necessary for cardioprotection, or at least for *maximum* protection to be evident. For example Karmazyn[31] reported that amiloride failed to improve recovery of isolated ischemic rat hearts when administered only at reflow. Similar results were demonstrated by du Toit and Opie[51] with respect to amiloride, however these investigators did show beneficial effects of HOE694 when it was administered at the time of reperfusion. However, HOE694 was ineffective in blood-perfused isolated ischemic rabbit hearts[57] as well as in porcine hearts[64] when it was administered solely at reperfusion. In other studies, amiloride, or amiloride analogs, exerted diminished protection but was not completely ineffective in protecting the reperfused heart when added only during reperfusion compared to protection seen with drug pretreatment prior to ischemia.[33,52] Pierce's group has demonstrated equal effective protection of reperfused right ventricular tissue with a variety of NHE inhibitors irrespective of whether the agents were administered prior to ischemia or upon reperfusion.[37,53] This phenomenon precluding the necessity for pretreatment has important clinical advantages. Moreover, such a property would be more in line with the theoretical concept that maximum activation of NHE occurs at reperfusion, and hence inhibition of the antiporter at that time should be conducive to cardioprotection.

The apparent discrepancy between the above cited studies is difficult to resolve at present, although such differences may be related to differences in experimental design or the choice of NHE inhibitor. Moreover, it is possible that differences in drug administration protocols may have resulted in insufficient NHE inhibition at reperfusion

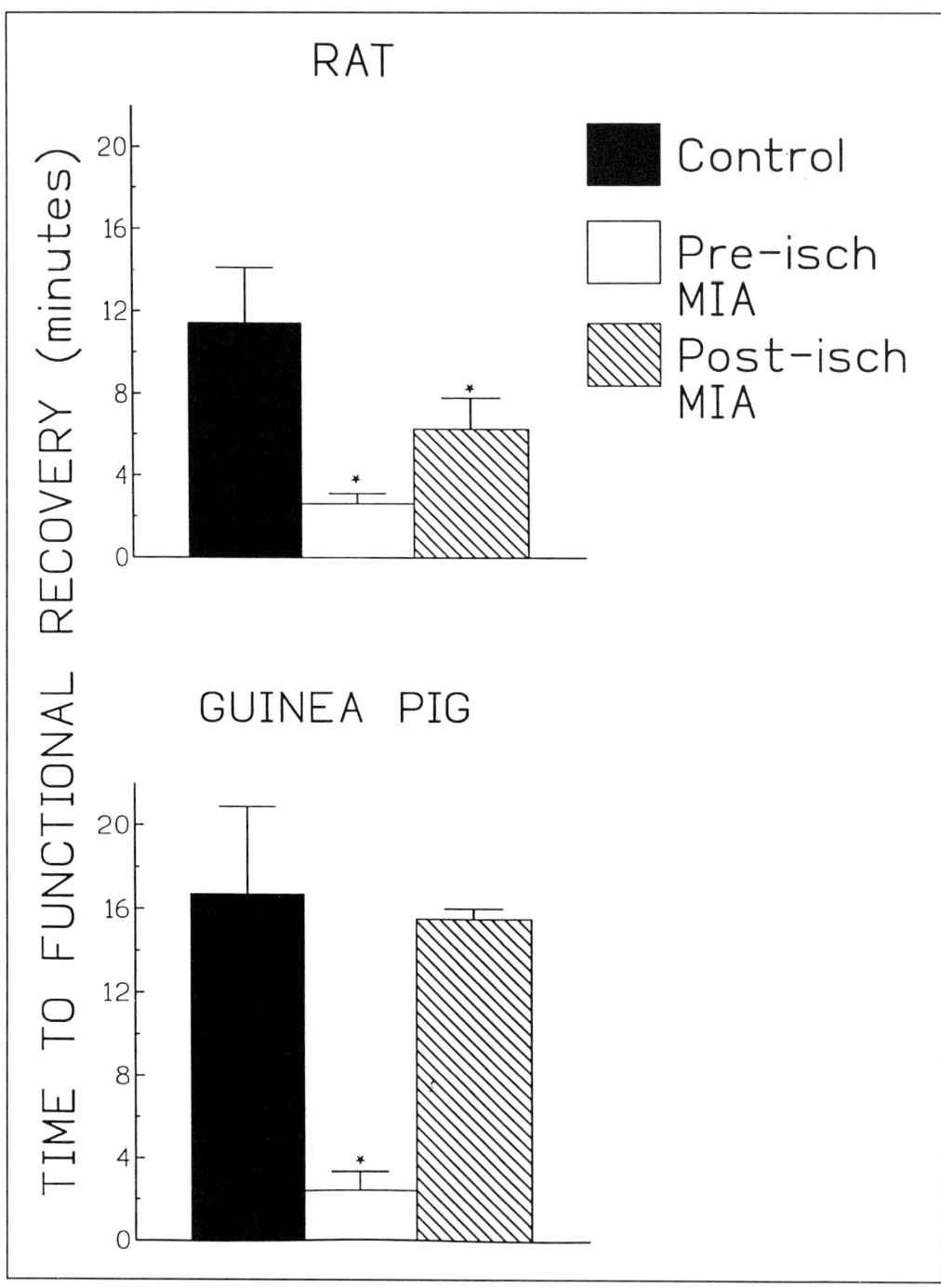

Fig. 10.6. Time to restoration of sustained ventricular contractility of isolated rat or guinea pig hearts after reperfusion following 45 minutes of global total ischemia in the presence or absence of MIA. Drug addition protocols as in Figure 10.3. *P < 0.05 from control. From Moffat and Karmazyn, J Moll Cell Cardio 1993; 25:959-971 with permission.

in some studies thus precluding cardioprotective effects. However, some evidence suggests that NHE activity during ischemia may also contribute to tissue injury, and it accordingly has been proposed that antiport inhibition during ischemia or hypoxia could represent the basis for the beneficial effects of these drugs. For example, Anderson et al[67] reported that amiloride and ethyl-isopropyl amiloride selectively prevented the elevation in tissue Na^+ and Ca^{2+} contents during hypoxic perfusion, an effect associated with diminished creatine phosphate depletion. Moreover, investigators from the same laboratory have demonstrated that amiloride attenuates the elevation in tissue Ca^{2+} and Na^+ contents in the globally ischemic rat heart in the absence of diminished reperfusion-induced creatine kinase efflux.[42] The above two studies suggest that the inhibition of NHE activity during hypoxia or ischemia per se is the mechanism of the protective effects of amiloride or its analogs. This issue is difficult to resolve at present, particularly as it has been suggested by one group that the antiport is not operative under either ischemic or hypoxic conditions.[67] The latter scenario is unlikely in view of existing evidence. It is possible that, depending on experimental conditions, NHE inhibition during both ischemia and reperfusion offers optimal cardioprotection by blunting cellular ionic derangement during both phases of perfusion.

D.5 NON-PHARMACOLOGICAL PROTECTION OF THE ISCHEMIC AND REPERFUSED MYOCARDIUM: ROLE OF NHE INHIBITION

Nonpharmacological approaches implicating NHE activation in reperfusion injury have also been presented. This has generally been accomplished by initial reperfusion with an acidotic medium in which the pH is gradually returned to normal thus blunting NHE activation during the critical period of early reflow. For example, initial reperfusion of isolated ferret hearts with acidotic (pH 6.6) buffer produced a greater recovery of ventricular function which was dissociated from improved contents of high energy phosphastes.[34] Identical results were obtained using an in vivo model of myocardial stunning in dogs in which initial reperfusion with acidotic coronary venous blood after 15 minutes of coronary occlusion resulted in markedly enhanced ventricular recovery three hours after reperfusion.[43] While these studies were not carried out to assess the role of NHE per se, it could be suggested that early acidotic reperfusion exerts cardioprotective actions by attenuating NHE activation during reflow in a manner akin to pharmacological inhibition of the antiporter. This concept has recently been strengthened by the finding that reperfusion of ischemic isolated papillary muscle with alkaline (pH 7.6) medium enhances cell death, whereas treatment with the NHE inhibitor dimethylamiloride or reperfusion with acidotic medium both significantly exerted identical effects.[66] While the above studies are suggestive of NHE involvement in the pH-dependent modulation of cardiac responses to reperfusion, other possible effects of acidosis, particularly on Ca^{2+} regulatory processes which modulate ventricular recovery after reperfusion, cannot be excluded in providing a mechanism for the cardioprotective actions of early acidotic reperfusion.

D.6 ROLE OF NHE IN LACTATE-INDUCED DEPRESSION OF POSTISCHEMIC VENTRICULAR RECOVERY

The NHE may also serve to explain the observation, initially made by Neely and Grotyohann,[69] that inhibition of glycolytic flux by glycogen depletion results in improved postischemic ventricular recovery presumably as a result of reduction in the lactate burden at the end of ischemia. These authors also showed that exogenous lactate was able to reverse the salutary effects of inhibition of glycolytic flux, further implicating lactate as a principal modulator of postischemic recovery.[69] Whether the beneficial effects of lactate depletion are due to attenuation of H^+ load is uncertain, particularly as the role of lactate in producing intracellular acidosis in the ischemic myocardium remains controversial. Nonetheless, indirect evidence does support a role for lactate in this phenomenon. For example, inhibition of lactate accumulation in the ischemic myocardium can mimic the effects of amiloride in terms of postischemic contractile recovery as well as diminution of Na^+ and Ca^{2+} loading after reperfusion.[33] Moreover, Karmazyn[55] recently reported that the lactate-induced reduction in contractile recovery after reperfusion is reversed by NHE inhibitors. In these studies, lactic acid was administered only during ischemia and was not present during the reperfusion period indicating that the reduced recovery in function did not reflect a direct cardiodepressant effect of lactate per se. While these observations are suggestive of NHE involvement in the lactate-induced depression in postischemic recovery, it should be noted that in both of the above studies[33,65] intracellular pH, particularly at the end of ischemia, was not assessed to confirm enhanced ischemia-induced acidosis by lactate. Therefore it still remains to be precisely determined what the relationship is between lactate-induced attenuation of postischemic recovery, intracellular pH and NHE activity.

D.7 A POSSIBLE NON-Ca^{2+}-DEPENDENT COMPONENT OF NHE-MEDIATED MYOCARDIAL REPERFUSION INJURY

The above discussions have centered on an important role of Ca^{2+} in mediating reperfusion injury as a consequence of NHE activation. Work from Lemasters' group has proposed the concept of NHE-mediated injury following reperfusion which is unrelated to elevation in intracellular Ca^{2+} concentrations, a phenomenon termed by this group as the "pH paradox." For example, neonatal cultured rat cardiomyocytes exposed to ischemia mimetic conditions demonstrated extensive injury when re-exposed to normal but not acidotic pH buffer, or buffer containing NHE inhibitors, which was dissociated from elevations in intracellular Ca^{2+} concentration and which also occurred in the absence of extracellular Ca^{2+}.[41,48] These observations, while further strengthening the concept of NHE involvement in reperfusion injury, are suggestive of an additional component of this phenomenon which is possibly distinct from intracellular Ca^{2+} elevations at least under certain conditions.

D.8 POSSIBLE ROLE OF THE NHE IN MYOCARDIAL PRECONDITIONING

Brief multiple periods of myocardial ischemia bestows protection against prolonged ischemia in terms of reduced infarct size and improved recovery of contractilty after reperfusion, a phenomenon termed myocardial preconditioning.[70] The mechanisms underlying this phenomenon are very complex and, as for reperfusion injury in general, most likely involve numerous cellular processes. A number of recent studies have demonstrated altered intracellular pH profiles in the preconditioned myocardium which may be a consequence of changes in NHE activity or which modulate NHE activity during ischemia or reperfusion. The major finding in the majority of these studies is diminished acidosis during ischemia, although the precise relationship to NHE remains to be determined. Asimakis and coworkers[71] have demonstrated that preconditioned rat hearts better maintain intracellular pH during prolonged ischemia in terms of reduced acidosis during that period. Similar observations were shown by other investigators. Steenbergen et al[72] demonstrated that the reduced acidosis during ischemia was associated with diminished intracellular Na^+ and Ca^{2+} content. These investigators also proposed that the lowered H^+ loading during ischemia was not a reflection of changes in NHE activity but rather that it caused diminished NHE activity during ischemia which resulted in diminished Ca^{2+} loading. Other investigators have demonstrated a good correlation between intracellular pH during ischemia and subsequent recovery of function, that is, as intracellular pH during ischemia is markedly reduced, recovery is diminished.[73] Taken together, the results of these studies are suggestive of a critical role of intracellular pH maintenance in enhancing recovery of the preconditioned myocardium.

As mentioned previously, the precise role of the NHE in myocardial preconditioning needs to be determined. It is possible that the reduced H^+ accumulation in the preconditioned myocardium may reflect diminished glycolytic flux due to reduced glycogen breakdown,[70] which would result in decreased H^+ and lactate production. This in turn would attenuate NHE activity particularly upon reperfusion because of reduced transarcolemmal H^+ gradient. However, in a recent study, it was shown that NHE activity in the preconditioned myocardium is increased by virtue of the ability of these hearts to recover much quicker from acid loading produced by NH_4Cl pulsing, a phenomenon which was prevented by the NHE inhibitor ethyl-isopropyl amiloride.[74] These authors also found that the protective effect of preconditioning was attenuated by ethyl-isopropyl amiloride thus implicating NHE activation as a mediating protection by preconditioning, a finding which is difficult to reconcile in view of the extensive evidence implicating NHE as a mediator of reperfusion-induced dysfunction.

Recently, mechanisms for myocardial preconditioning dissimilar from NHE inhibition have been proposed based on the observation that NHE inhibition produced additive protective effects in preconditioned rat hearts.[62] While this finding may indeed support the notion of distinct mechanisms for each form of protection, an alternate explanation could be that NHE inhibition was not maximum in the preconditioned hearts, hence the additive protection of a pharmacological NHE inhibitor.

D.9 NHE Activation in Platelets and Neutrophils as Contributing Factors to Myocardial Reperfusion Injury

It is well-established that recruitment and activation of blood borne factors is an important mediator of the sequela of reperfusion injury in the heart in vivo.[75-77] While our discussion has thus far centered on the role of myocardial NHE in mediating such injury, there is extensive evidence that NHE in platelets and neutrophils is important for the activation of these cells. For example, platelet activation results in the release of cardiotoxic factors such as thromboxane A_2 which can produce coronary constriction and induce further platelet recruitment and aggregation.[78] Moreover, activated neutrophils synthesize and release a number of deleterious compounds including reactive oxygen species, which can directly damage the myocardium, as well as the potent chemotactic agent leukotriene B_4.[75]

There is compelling evidence that NHE regulates both platelet and neutrophil activation. With respect to the former, it has been demonstrated that the antiport mediates arachidonic acid mobilization and platelet activation and that inhibitors of the exchanger prevent calcium mobilization and stimulation of arachidonic acid release in response to various stimuli such as thrombin, collagen and ADP,[79-82] likely via pH dependent mechanisms.

The presence of NHE in neutrophils has been well-established and suggested to represent the primary mechanism for restoration of intracellular pH following an intracellular acid load.[83] Moreover, neutrophil activation with chemotactic agents results in concomitant intracellular, NHE-dependent alkalinization as well as chemotaxis; inhibition of the exchanger prevents both the alkalinization as well as chemotaxis.[84] It has also been demonstrated that the production of leukotriene B_4 by human neutrophils can be prevented by NHE inhibitors.[85] Recently, it has been shown that inhibition of cardiac NHE attenuates the deleterious effects of activated neutrophils in ischemic and reperfused isolated rat hearts.[65] Therefore, taken together, the above data are suggestive of an additional cardioprotective effect of NHE inhibition by virtue of attenuation of the activation of both platelets and neutrophils, thus further mitigating cardiac injury.

E. SUMMARY, CONCLUSIONS AND FUTURE DIRECTIONS

The NHE represents a major mechanism for restoring intracellular pH following ischemia-induced acidosis, and hence its activity is critical for normal cellular homeostasis. There is very compelling evidence, however, that activation of the antiport produces a paradoxical extension and acceleration of tissue damage particularly following reperfusion. The mechanism for this likely reflects the concomitant influx of Na^+ during NHE activation resulting in potentially deleterious consequences due to increases in intracellular Ca^{2+} concentrations. There is overwhelming evidence that inhibition of NHE affords substantial cardioprotection which is unmatched by any other approach; indeed there are virtually no studies which have failed to demonstrate salutary effects of NHE inhibition on the ischemic and reperfused myocardium. NHE inhibitors have a number of important advantages compared to other approaches for cardioprotection. For example, as discussed above, in addition to direct cardioprotection, these agents have the potential to attenuate both neutrophil as well as platelet activation, thereby providing added beneficial effects. Moreover, the potential for unwanted side effects from NHE inhibitors is, at least in theory, minimized based on the fact that under normal conditions the NHE operates at sub-maximum capacity or in fact not at all. While much more needs to be done, major advances in the past number of years have yielded important information concerning the molecular structure, regulation and pharmacological modulation of the NHE in the heart and how the activity of antiporter participates in cardiac pathology. These advances hold tremendous promise in the development of new strategies as adjunctive therapy for the protection of the myocardium during reperfusion protocols such as thrombolysis or angioplasty. NHE inhibitors would also likely be of substantial benefit as additions to cardioplegic solutions as well as in the treatments of cardiac disorders where attenuation of cell necrosis is a therapeutic goal.

ACKNOWLEDGMENTS

Studies from the author's laboratory were supported by the Medical Research Council of Canada. I am grateful to the many individuals who have contributed to the NHE-related research cited in this chapter including Jianmin Duan, James Haist, Josephine Ho, Nassirah Khandoudi, Sanjiv Mathur, Mary Lee Myers, Christopher Ward, Joanne Watson and especially (the late) Margaret P Moffat.

The author is a career investigator of the Heart and Stroke Foundation of Ontario.

REFERENCES

1. Fliegel L, Fröhlich O. The Na^+/H^+ exchanger: an update on structure, regulation and cardiac physiology. Biochem J 1993; 296:273-285.
2. Karmazyn M, Moffat MP. Role of Na^+/H^+ exchange in cardiac physiology and pathophysiology: mediation of myocardial reperfusion injury by the pH paradox. Cardiovasc Res 1993; 27:915-924.
3. Yun CH, Tse C-M, Nath S, et al. Structure/function studies of mammalian Na-H exchangers-an update. J Physiol 1995; 482:1S-6S.
4. Nöel J, Pouysségur J. Hormonal regulation, pharmacology, and membrane sorting of vertabrate Na^+/H^+ exchanger isoforms. Am J Physiol 1995; C283-C296.
5. Orchard CH, Kentish JC. Effects of changes of pH on the contractile function of cardiac muscle. Am J Physiol 1990; 258:C967-C981.
6. Lagadic-Gossman D, Vaughan-Jones RD. Coupling of dual acid extrusion in the guinea-pig isolated ventricular myocyte to α_1- and β-adrenoceptors. J Physiol 1993; 464:49-73.
7. Frelin C, Vigne P, Lazdunski M. The role of the Na^+/H^+ exchange system in cardiac cells in relation to the control of the internal Na^+ concentration. A molecular basis for the antagonistic effect of ouabain and amiloride on the heart. J Biol Chem 1984; 259:8880-8885.
8. Kim D, Smith TW. Effects of amiloride and ouabain on contractile state, Ca and Na fluxes, and Na content in cultured chick heart cells. Mol Pharmacol 1986; 29:363-371.
9. Finet M, Godfraind T. Selective inhibition by ethylisopropylamiloride of the positive inotropic effect evoked by low concentrations of ouabain in rat isolated ventricles. Br J Pharmacol 1986; 89:533-538.
10. Lazdunski M, Frelin C, Vigne P. The sodium/hydrogen exchange system in cardiac cells: its biochemical and pharmacological properties and its role in regulating internal concentrations of sodium and internal pH. J Mol Cell Cardiol 1985; 17:1029-1042.
11. Frelin C, Vigne P, Ladoux A, Lazdunski M. The regulation of the intracellular pH in cells from vertebrates. Eur J Biochem 1988; 174:3-14.
12. Piwnica-Worms D, Jacob R, Shigeto N, Horres CR, Lieberman M. Na/H exchange in cultured heart cells: Secondary stimulation of electrogenic transport during recovery from intracellular acidosis. J Mol Cell Cardiol 1986; 18:1109-1116.
13. Kim D, Cragoe EJ, Smith TW. Relations among sodium pump inhibition, NA-Ca and Na-H exchange activities and Ca-H interactions in cultured chick heart cells. Circ Res 1987; 60:185-193.
14. Siffert W, Akkerman JWN. Na^+/H^+ exchange and Ca^{2+} influx. FEBS Lett 1989; 259:1-4.
15. Piwnica-Worms D, Jacob R, Horres CR, et al. Na/H exchange in cultured chick heart cells. J Gen Physiol 1985; 85:43-64.
16. Duan J, Moffat MP. Contractile and electrophysiological effects of realkalization in cardiac tissue. Role of Na/H exchange and increased $[Ca]_i$. Adv Exp Med Biol 1992; 311:435-436.

17. Vigne P, Frelin C, Lazdunski M. The Na$^+$/H$^+$ antiport is activated by serum and phorbol esters in proliferating myoblasts but not in differentiated myotubules. Properties of the activation process. J Biol Chem 1985; 260:8008-8013.

18. Siffert W, Siffert G, Schied P. Activation of Na$^+$/H$^+$ exchange in human platelets by thrombin and phorbol ester. Biochem J 1987; 241:301-303.

19. Boscoboinik DO, Hensey CE, Azzi A. Effect of staurosporine on the phorbol ester-induced activation of the Na$^+$/H$^+$ antiporter in smooth muscle cells. Biochim Biophys Acta 1990; 1052:499-502.

20. Karmazyn M, Watson JE, Moffat MP. Mechanisms for cardiac depression induced by phorbol myristate acetate in working rat hearts. Br J Pharmacol 1990; 259:826-831.

21. Watson JE, Karmazyn M. Concentration-dependent effects of protein kinase C-activating and nonactivating phorbol esters on myocardial contractility, coronary resistance, energy metabolism, prostacyclin synthesis, and ultrastructure in isolated rat hearts. Effects of amiloride. Circ Res 1991; 69:1114-1131.

22. MacLeod KT, Harding SE. Effects of phorbol ester on contraction, intracellular pH and intracellular in isolated mammalian ventricular myocytes. J Physiol 1991; 444:481-498.

23. Ward CA, Moffat MP. Positive and negative inotropic effects of phorbol 12-myristate 13-acetate: relationship to PKC-dependence and $[Ca^{2+}]_i$. J Mol Cell Cardiol 1992; 24:937-948.

24. Karmazyn M. Ischemic and reperfusion injury in the heart. Cellular mechanisms and pharmacological interventions. Can J Physiol Pharmacol 1991; 69:719-730.

25. Kramer BK, Smith TW, Kelly RA. Endothelin and increased contractility in adult rat ventricular myocytes. Role of intracellular alkalosis induced by activation of the protein-kinase C-dependent Na$^+$/H$^+$ exchanger. Circ Res 1991; 68:269-279.

26. Puceat M, Clement-Chomienne O, Terzic A, et al. α_1-Adrenoceptor a purinoceptor agonists modulate Na-H antiport in single rat cardiac cells. Am J Physiol 1993; 264:H310-319.

27. Kabira M, Mitari S, Yano K et al. Involvement of Na$^+$/H$^+$ antiporter in regulation of L-type Ca^{2+} channel current by angiotensin II in rabbit ventricular myocytes. Circ Res 1994; 75:1121-1125.

28. Karmazyn, M. The role of endothelins in cardiac function in health and disease. In: Karmazyn M, ed. Myocardial Ischemia: Mechanisms, Reperfusion, Protection. Bikhauser Verlag. Basel, 1996; in press.

29. Khandoudi N, Ho J, Karmazyn M. Role of Na$^+$/H$^+$ exchange in mediating effects of endothelin-1 on normal and ischemic/reperfused hearts. Circ Res 1994; 75:369-378.

30. Yasutake M, Avkiran M. Exacerbation of reperfusion arrhythmias by α_1 adrenergic stimulation: a potential role for receptor mediated activation of sarcolemmal sodium-hydrogen exchange. Cardiovasc Res 1995; 29:222-230.

31. Karmazyn M. Amiloride enchances postischemic ventricular recovery: possible role of Na^+/H^+ exchange. Am J Physiol 1988; 255:H608-H615.
32. Duan J, Karmazyn M. Protective effects of amiloride on the ischemic reperfused rat heart. Relation to mitochondrial function. Eur J Pharmacol 1992; 210:149-157.
33. Tani M, Neely JR. Role of intracellular Na^+ in Ca^{2+} overload and depressed recovery of ventricular function of reperfused ischemic rat hearts. Possible involvement of H^+-Na^+ and Na^+-Ca^{2+} exchange. Circ Res 1989; 65:1045-1056.
34. Kitakaze M, Weisfeldt ML, Marban E. Acidosis during early reperfusion prevents myocardial stunning in perfused ferret hearts. J Clin Invest 1988; 82:920-927.
35. Duff HJ, Lester WM, Rahmberg M. Amiloride. Antiarrhythmic and electrophysiological activity in the dog. Circulation 1988; 78:1469-1477.
36. Duff HJ, Brown E, Cragoe EJ, et al. Antiarrhythmic activity of amiloride: mechanisms. J Cardiovasc Pharmacol 1991; 17:879-888.
37. Meng H-P, Pierce GN. Protective effects of 5-(N,N-dimethyl)amiloride on ischemia-reperfusion injury in hearts. Am J Physiol 1990; 258: H1615-H1619.
38. Khandoudi N, Bernard M, Cozzone P, et al. Intracellular pH and role of Na^+/H^+ exchange during ischaemia and reperfusion of normal and diabetic rat hearts. Cardiovasc Res 1990; 24:873-878.
39. Weiss RG, Lakatta EG, Gerstenblith G. Effects of amiloride on metabolism and contractility during reoxygenation in perfused rat hearts. Circ Res 1990; 66:1012-1022.
40. Dennis SC, Coetzee WA, Cragoe Jr EJ, et al. Effects of proton buffering and of amiloride derivatives on reperfusion arrhythmias in isolated rat hearts. Possible evidence for an arrhythmogenic role of Na^+/H^+ exchange. Circ Res 1990; 66:1156-1159.
41. Bond JM, Herman B, Lemasters JJ. Protection by acidotic pH against anoxia/reoxygenation injury to rat neonatal cardiac myocytes. Biochem Biphys Res Commun 1991; 179:798-803.
42. Murphy E, Perlman M, London RE, et al. Amiloride delays the ischemia-induced rise in cytosolic calcium. Circ Res 1991; 68: 1250-1258.
43. Hori M, Kitakaze M, Sato H, et al. Staged reperfusion attenuates myocardial stunning in dogs. Role of transient acidosis during early reperfusion. Circulation 1991; 84:2135-2145.
44. Meng H, Lonsberry BB, Pierce GN. Influence of perfusate pH of the postischemic recovery of cardiac contractile function: involvement of sodium-hydrogen exchange. J Pharmacol Exp Ther 1991; 258:772-777.
45. Matsuda N, Kuroda H, Mori T. Beneficial actions of acidotic initial reperfusate in stunned myocardium of rat hearts. Basic Res Cardiol 1991; 86:317-326.
46. Avkiran M, Ibuki C. Reperfusion-induced arrhythmias. A role for washout of extracellular protons? Circ Res 1992; 71:1429-1440.

47. Ibuki C, Hearse DJ, Avkiran M. Mechanisms of antifibrillatory effect of acidic reperfusion: role of perfusate bicarbonate concentration. Am J Physiol 1993; 264:H783-H790.
48. Bond JM, Chacon E, Herman B, et al. Intracellular pH and Ca^{2+} homeostasis in the pH paradox of reperfusion injury to neonatal rat cardiac myocytes. Am J Physiol 1993; 265:C129-C137.
49. Scholz W, Albus U, Lang HJ, et al. Hoe 694, a new Na^+/H^+ exchange inhibitor and its effects on cardiac ischaemia. Br J Pharmacol 1993; 109:562-568.
50. Mochizuki S, Seki S, Ejima M-A, et al. Na^+/H^+ exchanger and reperfusion-induced ventricular arrhythmias in isolated perfused heart: possible role of amiloride. Mol Cell Biochem 1993; 119:151-157.
51. du Toit EF, Opie LH. Role for the Na^+/H^+ exchanger in reperfusion stunning in isolated perfused rat heart. J Cardiovasc Pharmacol 1993; 22:877-883.
52. Moffat MP, Karmazyn M. Protective effects of the potent Na/H exchange inhibitor methyisobutyl amiloride against post-ischemic contractile dysfunction in rat and guinea-pig hearts. J Mol Cell Cardiol 1993; 25:959-971.
53. Meng H-P, Maddaford TG, Pierce GN. Effect of amiloride and selected analogues on postischemic recovery of cardiac contractile function. Am J Physiol 1993; H1831-H1835.
54. Karmazyn M, Ray M, Haist J. Comparative effects of Na^+/H^+ exchange inhibitors against cardiac injury produced by ischemia/reperfusion, hypoxia/reoxygenation and the Ca^{2+} paradox. J Cardiovasc Pharmacol 1993; 21:172-178.
55. Karmazyn M. Na^+/H^+ exchange inhibitors reverse lactate-induced depression in postischaemic ventricular recovery. Br J Pharmacol 1993; 108:50-56.
56. Sack S, Mohri M, Schwarz ER, et al. Effects of a new Na^+/H^+ antiporter inhibitor on postischemic reperfusion in pig heart. J Cardiovasc Pharmacol 1994; 23:72-78.
57. Hendrikx M, Mubagwa K, Verdonck F, et al. Ne Na^+/H^+ exchange inhibitor HOE 694 improves postischemic function and high-energy phosphate resynthesis and redices Ca^{2+} overload in isolated perfused rabbit heart. Circulation 1994; 89:2787-2798.
58. Hata K, Takasago T, Saeki A, et al. Stunned myocardium after rapid correction of acidosis. Increased oxygen cost of contractility and the role of the Na^+/H^+ exchange system. Circ Res 1994; 74:794-805.
59. Myers ML, Mathur S, Li G-H, et al. Sodium-hydrogen exchange inhibitors improve postischaemic recovery of function in the perfused rabbit heart. Cardiovasc Res 1995; 29:209-214.
60. Scholz W, Albus U, Counillon L, et al. Protective effects of HOE642, a selective sodium-hydrogen exchange subtype 1 inhibitor, on cardiac ischaemia and reperfusion. Cardiovasc Res 1995; 29:260-268.
61. Ward CA, Moffat MP. Modulation of sodium-hydrogen exchange activity in cardiac myocytes during acidosis and realkalinisation: effects on calcium, pH_i and cell shortening. Cardiovasc Res 1995; 29:247-253.

62. Bugge E, Ytrehus K. Inhibition of sodium-hydrogen exchange reduces infarct size in the isolated rat heart-a protective additive to ischaemic preconditioning. Cardiovasc Res 1995; 29:269-274.
63. Ladilov YV, Siegmund B, Piper HM, et al. Protection of reoxygenated cardiomyocytes against hypercontracture by inhibition of Na^+/H^+ exchange. Am J Physiol 1995; 268:H1531-H1539.
64. Klein HH, Pich S, Bohle RM, et al. Myocardial protection by Na^+/H^+ exchange inhibition in ischemic, reperfused porcine hearts. Circulation 1995; 92:912-917.
65. Faes FC, Sawa Y, Ichikawa H, et al. Inhibition of Na^+/H^+ exchanger attenuates neutrophil-mediated reperfusion injury. Ann Thorac Surg 1995; 60:377-81.
66. Kaplan SH, Yang H, Gilliam DE, et al. Hypercapnic acidosis and dimethylamiloride reduce reperfusion-induced cell death in ischaemic ventricular myocardium. Cardiovasc Res 1995; 29:231-238.
67. Anderson SE, Murphy E, Steenbergen C, et al. Na-H exchange in myocardium: effects of hypoxia and acidification on Na and Ca. Am J Physiol 1990; 259:C940-C948.
68. Imai S, Shi A-Y, Ishibashi T, et al. Na^+/H^+ exchange is not operative under low-flow ischemic conditions. J Mol Cell Cardiol 1991; 23:505-517.
69. Neely JR, Grotyohann LW. Role of glycolytic products in damage to ischemic myocardium. Dissociation of adenosine triphosphate levels and recovery of function of reperfused ischemic hearts. Circ Res 1984; 55:816-824.
70. Murry CE, Richard VJ, Reimer KA, et al. Ischemic preconditioning slows energy metabolism and delays ultrastructural damage during a sustained ischemic episode. Circ Res 1990; 66:913-931.
71. Asimakis GK, Inners-McBride K, Medellin G, et al. Ischemic preconditioning attenuates acidosis and postischemic dyfunction in isolated rat heart. Am J Physiol 1992; 263:H887-H894.
72. Steenbergen C, Perlman ME, London RE, et al. Mechanisms of preconditioning. Ionic alterations. Circ Res 1993; 72:112-135.
73. de Albuquerque CP, Gerstenblith G, Weiss RG. Importance of metabolic inhibition and cellular pH in mediating preconditioning contractile and metabolic effects in rat hearts. Circ Res 1994; 74:139-150.
74. Ramasamy R, Liu H, Anderson S, et al. Ischemic preconditioning stimulates sodium and proton transport in isolated rat hearts. J Clin Invest 1995; 96:1464-1472.
75. Mullane JM, Westlin W, Kraemer R. Activated neutrophils release mediators that may contribute to myocardial injury and dysfunction associated with ischemia and reperfusion. Ann NY Acad Sciences 1988; 524:103-121.
76. Smith EF, Egan JW, Bugelski PJ, et al. Temporal relation between neutrophil accumulation and myocardial reperfusion injury. Am J Physiol 1988; 255:H1060-H1068.

77. Westlin W, Mullane KM. Alleviation of myocardial stunning by leukocyte and platelet depletion. Circ Res 1989; 80:1828-1836.
78. Ogletree ML. Overview of physiological and pathophysiological effects of thromboxane A_2. Fed Proc 1987; 46:133-138.
79. Sweatt JD, Johnson SL, Cragoe EJ, et al. Inhibitors of Na^+/H^+ exchange block stimulus-provoked arachidonic acid release in human platelets. J Biol Chem 1985; 260:12910-12919.
80. Sweatt JD, Connolly TM, Cragoe EJ, et al. Evidence that Na^+/H^+ exchange regulates receptor-mediated phospholipase A_2 activation in human platelets. J Biol Chem 1986; 261:8667-8673.
81. Siffert W, Siffert G, Scheid P, et al. Activation of Na^+/H^+ exchange and Ca^{2+} mobilization start simultaneously in thrombin-stimulated platelets. Biochem J 1989; 258:521-527.
82. Siffert W, Siffert G, Scheid P, Akkerman JWN. Na^+/H^+ exchange modulates Ca^{2+} mobilization in human platelets stimulated by ADP and the thromboxane mimetic U 46619. J Biol Chem 1990; 264:719-725.
83. Simchovitz L, Roos A. Regulation of intracellular pH in human neutrophils. J Gen Physiol 1985; 85:443-470.
84. Simchowitz L, Cragoe EJ. Regulation of human neutrophil chemotaxis by intracellular pH. J Biol Chem 1986; 261:6492-6500.
85. Osaki M, Sumimoto H, Takeshige K, et al. Na^+/H^+ exchange modulates the production of leukotriene B_4 by human neutrophils. Biochem J 1989; 257:751-758.

CHAPTER 11

Expression and Activity of the Sodium-Hydrogen Exchanger in Cardiac Sarcolemma in Health and Disease

Grant N. Pierce, Tracy Slotin, Larry Fliegel,
James S.C. Gilchrist and Thane G. Maddaford

A. INTRACELLULAR HYDROGEN AND CELL FUNCTION

Hydrogen ions may be unique amongst the family of ionic molecules. Whereas many proteins and enzymes are functionally unaffected by, for example, Ca^{2+} or Na^+, there are remarkably few proteins that do not have their functions significantly altered by relatively small changes in cellular pH. These changes in protein function ultimately result in significant alterations in cell function. No cell type in the body appears to have immunity from the effects of H^+. Both extracellular and intracellular H^+ can have powerful effects on the performance of the cell, but it is probably the latter that has more lasting and wide ranging effects. Cellular metabolism, electrical behavior, gene transcription and translation, contractile performance, structural integrity and viability are all altered when cellular pH changes.

The functional significance of cellular H^+ concentrations extends beyond the minute by minute physiological existence of a healthy cell. It also plays important roles (both causative and secondary) in a variety of pathologies. For example, in the ischemic heart, increasing

---

The Na$^+$/H$^+$ Exchanger, edited by Larry Fliegel. © 1996 R.G. Landes Company.

evidence supports a direct involvement of acidosis in cellular contractile dysfunction, necrosis and death.[1-3] Alternatively, hypertrophic cardiomyopathy may represent a pathological condition in which the change in cellular H^+ regulatory systems may play a more supportive role in the disease process.

Because of this impressive range of effects in terms of cell type, molecules affected and functional implications, the mechanisms which directly regulate intracellular pH warrant intense scrutiny. Resting intracellular pH is ultimately a balance between H^+ production within the cell and those pathways which remove H^+ from the cell. Three or four pathways appear to be of importance in regulating intracellular pH in the heart (Fig. 11.1). The Na^+/H^+ exchanger[3-5] and the Na^+/HCO_3 cotransporter[6-8] are the primary acid extruding mechanisms in the heart. The Na^+/HCO_3 co-transporter is more important for regulating intracellular pH at rest than the Na^+/H^+ exchanger.[3,7,9,10] However, the Na^+/H^+ exchanger is more important for the restoration of pH during acid loads in the heart.[3] One, possibly two pathways may be important for regulating intracellular pH in the heart during an alkalinic load. The Na^+ independent Cl^-/HCO_3 exchanger is clearly important in reducing the pH from basic values to resting levels in the heart.[11] A Cl^-/OH^- exchanger has also been suggested to be present and active in the heart.

B. KINETICS OF Na^+/H^+ EXCHANGE IN CARDIAC SARCOLEMMA

The Na^+/H^+ exchange pathway was first identified in the heart using single, isolated cardiomyocytes.[4,5] This work was important not only because it provided the first evidence for the presence of the exchanger in cardiac tissue but also because it gave us our first information on the biochemical characteristics of the pathway and its relative importance to other Na^+ and H^+ transport pathways in cardiomyocytes. However, data obtained from preparations of whole cells is limited by many factors. First, it is difficult to employ incubation or assay conditions which are optimal for measuring transport activity. Many times these conditions are not conducive, for example, to cell survival. Secondly, it is often difficult if not impossible to isolate the transport pathway of interest from other ion transport pathways which are similar to the one to be studied. Following ^{22}Na as a radioactive tracer for the Na^+/H^+ exchanger may lead to erroneous conclusions if, for example, the Na^+ pump is also active. Third, there are technical difficulties encountered when measuring Na^+/H^+ exchange in whole cells.[12] Fourth, factors within the cell which alter its activity may be present and active in the whole cell but not in an isolated membrane preparation. For example, protein kinases or phosphatases in the cell are known to alter the phosphorylation status of the Na^+/H^+ exchanger[3] and significantly influence its activity. These proteins and/or their substrates

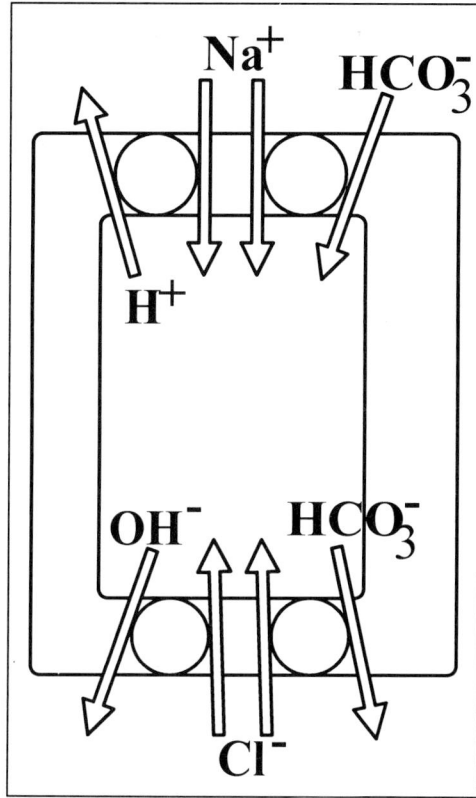

Fig. 11.1. The Na^+/H^+ exchanger, Na^+/HCO_3 cotransporter, Na^+ independent Cl^-/HCO_3 exchanger and Cl^-/OH^- exchanger are the primary pathways for the regulation of intracellular pH in the heart.

are effectively removed during the sarcolemmal membrane isolation procedure to provide a better estimate of the true kinetic profile of the exchanger. Finally, intracellular compartmentation of Na^+ or H^+ may complicate trans-sarcolemmal measurements. Intracellular compartmentation of Na^+ and H^+ have been documented.[13,14] Such intracellular pools would be expected to exert significant influence on trans-sarcolemmal Na^+/H^+ exchange kinetics. The use of purified sarcolemmal vesicles can obviate most of these problems and yield "cleaner" biochemical information on a transport protein. It is important to recognize, however, that the use of sarcolemmal vesicles is also fraught with significant limitations as well. A combination of information gleaned from both isolated, purified sarcolemmal vesicles and freshly isolated, whole cardiomyocytes is probably the best way to comprehensively characterize an exchanger.

It is somewhat surprising to find that there have been relatively few studies examining the biochemical characteristics of the Na^+/H^+

exchanger in sarcolemmal membranes isolated from the heart.[15-20] Na+/H+ exchange is primarily, if not selectively, located in the sarcolemmal membrane of the heart as opposed to other membrane fractions.[15] However, it is important to note that reports of Na+/H+ exchange in mitochondria have been published.[21] Activities for H+ dependent Na+ uptake at 1 mM extravesicular Na+ concentration ranges from ~20[16,18] to 60[15] nmol/mg/min in various labs. The variability is unlikely to be attributable to species differences. For example, one of us (G. Pierce) has reported values for H+ dependent Na+ uptake (at 0.05 mM Na+ for 5 seconds reaction time) of 0.15 nmol/mg in dog sarcolemma[15] as opposed to 0.18 nmol/mg in rat sarcolemma.[19] The H+ dependency for the reaction exhibits a very steep curve[15,16] consistent with its negligible activity at resting pH and rapid activation at an acidic intracellular pH.[3] Na+/H+ exchange is an electroneutral reaction.[16,17] The antiporter is blocked 80-100% by millimolar concentrations of amiloride.[15,16,20] Other drugs like quinacrine, quinidine and various amiloride analogs will block the exchanger effectively at much lower concentrations.[16] Based upon its sensitivity to dimethonium, an agent which screens membrane surface charge, Na+/H+ exchange appears sensitive to ionic charge distribution in the diffuse double layer region of the sarcolemmal membrane.[17] The first study[17] of ionic specificity of the exchanger described potent (~80%) inhibitory effects of Mg^{2+} and Li^+. Na^+, Ca^{2+}, Mg^{2+} and Li^+ were also potential confounding factors in studies of the exchanger when using sarcolemmal preparations because they could induce release of accumulated Na+ from the vesicles.[17] These findings were extended[18] to include K+ as an inhibitor of the exchanger and Na^+, K^+, Rb^+ and Li^+ as having the capacity to release intravesicular H+. All of these factors are important to consider when measuring Na+/H+ exchange in isolated vesicles.

C. RESPONSE OF THE NA+/H+ EXCHANGER TO DISEASE

C.1 ISCHEMIA AND HYPOXIA

There is a great deal of evidence presently available which supports a role for the Na+/H+ exchanger in both ischemic/reperfusion and hypoxic/reoxygenation injury to the heart.[1-3] The majority of the evidence is based upon the use of pharmacological blockers of Na+/H+ exchange which demonstrate cardioprotective capacities in both hypoxic and ischemic conditions. The adaptive response of the exchanger to ischemia/reperfusion is largely immediate in the majority of investigations which have examined this problem. Supranormal activation of the exchanger during ischemia and reperfusion occurs over a time span of several minutes or even seconds. It is not a chronic adaptive response but an immediate response induced by concentration differences of H+ and Na+ across the cardiac sarcolemma. Briefly, the results to date indicate that the intracellular accumulation of H^+ during ischemia (or hypoxia) stimulates the Na^+/H^+ exchanger[1] (Fig. 11.2). Whether this stimulation occurs during ischemia or during the initial minutes of reperfusion (or both) is unclear. This activation of the exchanger will remove intracellular H^+ through an influx of Na^+. Increased tissue [Na^+] have been reported by a number of labs.[1] This intracellular Na^+ will stimulate the Na^+/Ca^{2+} exchanger to bring Ca^{2+} into the myocardium. Excess Ca^{2+} entry has been directly associated with cellular dysfunction, damage and necrosis. Abnormally high [Ca^{2+}]$_i$ have been reported during reperfusion.[1] Drugs which inhibit the Na^+/H^+ exchanger protect the heart from ischemic/reperfusion[1] and hypoxic/reoxygenation injury.[22] A blocker of Na^+/Ca^{2+} exchange has also been shown to improve recovery of the heart from hypoxic/reoxygenation injury.[23]

These effects on the Na^+/H^+ exchanger, although important, are not adaptive responses to a disease which are long-lasting. Once the ion concentration gradients across the sarcolemma dissipate, the stimulation will disappear. It is possible, however, that in a condition where an acidic intracellular environment is more persistent, the exchanger may adapt more permanently. Chronic ischemia or hypoxia may be expected to occur whenever coronary atherosclerosis is severe enough to induce chest pain. Under these conditions, the exchanger may come under genetic control. Indeed, two recent studies demonstrated that chronic ischemia has the potential to induce changes in Na^+/H^+ exchange protein expression. In the first, relatively short ischemic times

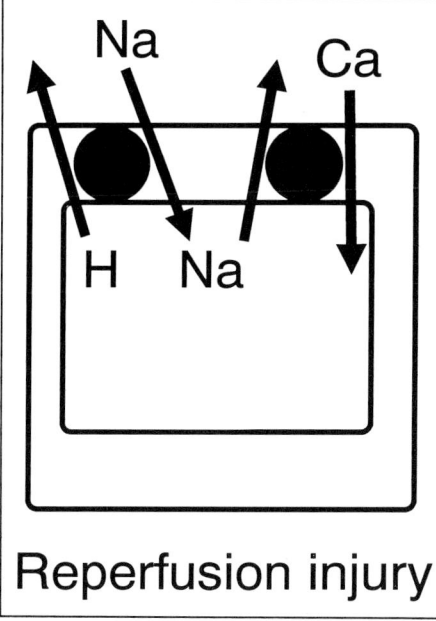

Fig. 11.2. Schematic depicting the ionic flux thought to occur in the heart during the initial phases of reperfusion after a period of ischemia. The stimulation of H^+ efflux from the acidic cardiomyocyte after ischemia occurs in exchange for extracellular Na^+ which in turn leaves the cell in exchange for extracellular Ca^{2+}.

(40-60 min.) did not change the 5 kb NHE1 Na$^+$/H$^+$ exchange message but did induce the expression of a smaller 3.8 kb message that was thought to be related to the amiloride sensitive Na$^+$/H$^+$ exchanger.[24] A second study[25] examined the response of the heart to more extended periods of low flow and no flow ischemia. Three hours of low flow ischemia (flow was reduced by approximately 70%) induced a significant increase in the expression of the NHE1 isoform of the Na$^+$/H$^+$ exchange message. More severe ischemic insults ($\geq$ 90% reduction in flow) prevented this increase. This was possibly due to a generation of damaged or necrotic tissue. The mechanism responsible for the stimulation in expression was thought to be chronic low external pH. Cultured myocytes increase their Na$^+$/H$^+$ exchange activity when exposed to low external pH for extended times.[25] Exactly how a decrease in external pH would induce a specific change in gene expression is unclear at present.

The clinical implications of an increased expression of the exchanger are also unclear. With more of the exchanger present, the heart would

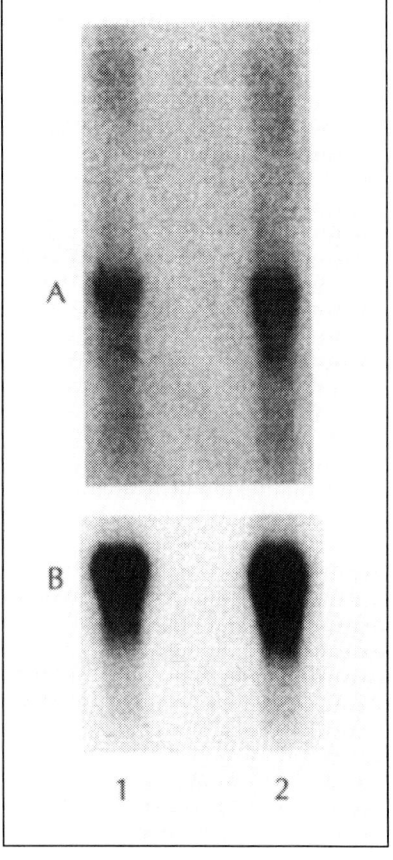

Fig. 11.3. Representative Northern blot analysis of NHE1 RNA. (A) message from control heart (2) or 8 week diabetic heart (1). The size of the message for the NHE1 is 5 kb. In B, the same blot has been reprobed for the 18S message for purposes of reference. Details of methodology are as described in reference 25.

certainly recover from an acidic insult much faster. However, because blocking Na^+/H^+ exchange is beneficial to the ischemic heart,[1] then it is reasonable to expect that the elevated exchange activity associated with the enhanced expression of the Na^+/H^+ exchanger will lead to more damage to the heart under ischemic conditions.

C.2 DIABETIC CARDIOMYOPATHY

The first direct demonstration of an adaptive change in Na^+/H^+ exchange in the heart to a chronic disease condition was demonstrated in sarcolemmal vesicles isolated from insulin-dependent diabetic rats.[19] Rats rendered chronically diabetic (eight weeks) after a single injection of streptozotocin exhibited ~60% decline in Na^+/H^+ exchange.[19] Because the sarcolemmal work does not rely on blockers of Na^+/H^+ exchange to prove its presence, curiously, this study remains as the sole direct determination of a change in Na^+/H^+ exchange in the heart during disease. It is, however, supported by several studies[26,27] which demonstrate that the diabetic hearts recover from an acidic load more slowly than control animals. These animals also appear to be more resistant to ischemic[28,29] and hypoxic[30] injury, observations which are again consistent with the causative role for Na^+/H^+ exchange in ischemic and hypoxic insults to the heart.

The mechanism responsible for the depression in cardiac Na^+/H^+ exchange during diabetes is not clear. From the work on the adaptive response of the exchanger to ischemia, a decrease in cell pH is unlikely to cause the depression in exchange activity. It is possible that a significant decrease in expression of the Na^+/H^+ exchanger may be responsible. However, we have examined mRNA message levels in hearts from diabetic rats 1 or 8 weeks after streptozotocin injection and found no significant change in comparison to control rats (Figs. 11.3 and 11.4). This rather surprising finding suggests that mechanisms other than genetic manipulation are responsible for the depressed Na^+/H^+ exchange activity. Two possibilities exist: (1) defects in translation of the exchanger; or (2) changes in membrane composition which then depress exchange function. Both of these alternative explanations are currently untested.

C3 NORMAL AND ABNORMAL CARDIAC GROWTH

The Na^+/H^+ exchanger has long been implicated as an important agent in inducing cell growth and proliferation.[31] The lines of evidence in support of such a concept are presented in Table 11.1. These results have led to the conclusion that although stimulation of Na^+/H^+ exchange is not essential for cell growth, it does have a permissive role.[31] The results in Table 11.1 have been obtained from a variety of cell types. Data on the role of the Na^+/H^+ exchanger in the growth of the heart are more scarce. Activity of the Na^+/H^+ exchanger in isolated cardiac sarcolemmal membrane fractions is significantly higher in newborn rabbits (one week of age) than in adult preparations.[32]

This would coincide temporally with the loss of proliferative potential in cardiomyocytes during early postnatal life. These data are corroborated by a study of Na^+/H^+ antiporter expression during cardiac development in the rabbit. Expression of the NHE1 mRNA levels decreased by approximately 40% from the fetal to the adult stage of development.[33]

There are more data available on the adaptive response of the exchanger during pathological cardiac hypertrophy, but it is still an area which requires additional attention. Indirect evidence suggests that during ischemia/reperfusion, Na^+/H^+ exchange may be stimulated to a greater extent in the hypertrophied heart than in the control. This may occur because hypertrophied hearts have an augmented glycolytic flux and an abnormal accumulation of metabolic by-products like H^+ during the ischemic period.[34] This may account for the increased susceptibility of hypertrophied hearts to injury from ischemic/reperfusion insult

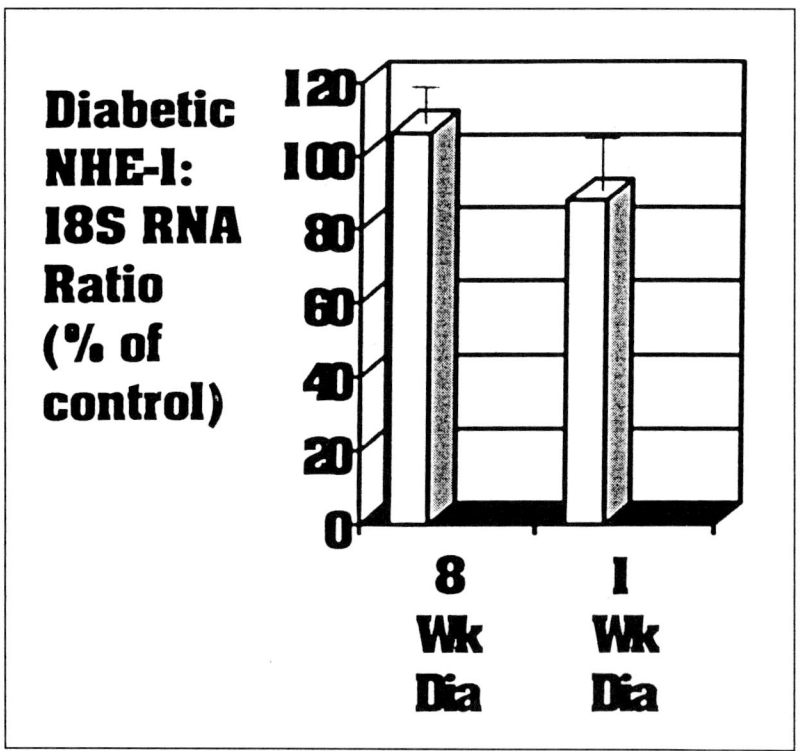

Fig. 11.4. Analysis of NHE1/18S RNA ratios of Northern blots prepared from control, 1 or 8 week diabetic rat hearts (n = 4 in each group) as described in Figure 11.3. Bands were quantitated with a scanning densitometer and the ratios to 18S used to compare the experimental levels of the NHE-1 to control values. Values are presented as a percent of control ratios. There were no significant differences detected.

Table 11.1. Evidence supporting a role for the Na^+/H^+ exchanger in cell growth

1. Inhibition of cell growth and proliferation in mutant cells devoid or depleted of the Na^+/H^+ exchange protein

2. Induction of cell proliferation and DNA synthesis by alkalinic cytosolic pH

3. Initiation of proliferation by extracellular sodium

4. Inhibition of cell growth by drugs which block the exchanger

5. Stimulation of Na^+/H^+ exchange by a great number of mitogens

6. Increased Na^+/H^+ exchange activity in cells exhibiting exponential growth

---

than control hearts.[35] Acidification in hypertrophied hearts is greater than normal in response to hypoxic challenge as well.[36]

More direct examinations of the status of the Na^+/H^+ exchanger during cardiac hypertrophy have yielded conflicting results which may be dependent upon the animal model of hypertrophy studied. For example, the viable myocardium in rats hypertrophied eight weeks after a myocardial infarction induced by a coronary ligature. However, Na^+/H^+ exchange did not change in the infarcted hearts in comparison to control hearts.[20] In contrast, in rabbit hearts hypertrophied by pressure overload via aortic constriction, the Na^+/H^+ exchange NHE1 mRNA message was significantly increased in comparison to sham-operated controls.[33] These workers suggested that Na^+/H^+ exchange may influence growth through an activation of MAP kinase activity, possibly via changes in intracellular H^+ or Ca^{2+}.[33] Clearly, more work on different models of hypertrophic cardiomyopathy are needed to understand the significance of the $Na^+/H+$ antiporter in this pathology and to recognize the therapeutic importance of blocking this ion transport pathway. Taking into account the clinical significance of hypertrophic cardiomyopathy as a cause of death and debilatative injury, this knowledge on the exchanger would be very valuable.

Acknowledgments

This work was supported by a grant from the Heart and Stroke Foundation of Manitoba. J.S.C Gilchrist is a scholar of the Heart and Stroke Foundation of Manitoba. L. Fliegel is a scholar of the Alberta Heritage Foundation for Medical Research. G.N. Pierce is a scientist of the Medical Research Council of Canada.

REFERENCES

1. Pierce GN, Czubryt MP. The contribution of ionic imbalance to ischemia/reperfusioin-induced injury. J Mol Cell Cardiol 1995; 23:53-63.
2. Karmazyn M, Moffat MP. Role of Na+/H$^+$ exchange in cardiac physiology and pathophysiology: mediation of myocardial reperfusion injury by the pH paradox. Cardiovasc Res 1993; 27:915-924.
3. Fliegel L, Fröhlich O. The Na$^+$/H$^+$ exchanger: an update on structure, regulation and cardiac physiology. Biochem J 1993; 296:273-285.
4. Piwnica-Worms D, Lieberman M. Microfluorometric monitoring of pH$_i$ in cultured heart cells: Na$^+$-H$^+$ exchange. Am J Physiol 1983; 244:C422-C428.
5. Frelin C, Vigne P, Lazdunski M. The role of the Na$^+$/H$^+$ exchange system in cardiac cells in relation to the control of the internal Na$^+$ concentration. J Biol Chem 1984; 259:8880-8885.
6. Grace A, Kirschenlohr HL, Metcalfe JC, et al. Regulation of intracellular pH in the perfused heart by external HCO$_3^-$ and Na$^+$-H$^+$ exchange. Am J Physiol 1993; 265:H289-H298.
7. Kusuoka H, Marban E, Cingolani HE. Control of steady-state pH in intact perfused ferret hearts. J Mol Cell Cardiol 1994; 26:821-829.
8. Camilion de Hurtado MC, Perez NG, Cingolani HE. An electrogenic sodium bicarbonate cotransport in the regulation of myocardial intracellular pH. J Mol Cell Cardiol 1995; 27:231-242.
9. Wallert MA, Frohlich O. Na$^+$-H$^+$ exchange in isolated myocytes from adult rat heart. Am J Physiol 1989; 257:C207-C213.
10. Pierce GN, Cole WC, Lui K, et al. Modulation of cardiac performance by amiloride and several selected derivatives of amiloride. J Pharmacol Exp Ther 1993; 265:1280-1291.
11. Xu P, Spitzer KW. Na-independent Cl$^-$HCO$_3^-$ exchange mediates recovery of pH$_i$ from alkalosis in guinea pig ventricular myocytes. Am J Physiol 1994; 267:H85-H91.
12. Grinstein S, Furuya W. Intracellular distribution of acridine derivatives in platelets and their suitability for cytoplasmic pH measurements. Biochim Biophys Acta 1984; 803:221-228.
13. Menard MR. The slowly exchanged pool of sodium in frog skeletal muscle is confined within a membranous organelle. Can J Physiol Pharmacol 1984; 62:49-52.
14. Sordahl LA, LaBelle EF, Rex KA. Amiloride and diltiazem inhibition of microsomal and mitochondrial Na$^+$ and Ca^{2+} transport. Am J Physiol 1984; 246:C172-C176.
15. Pierce GN, Philipson KD. Na$^+$-H$^+$ exchange in cardiac sarcolemmal vesicles. Biochem Biophys Acta 1985; 818:109-116.
16. Seiler SM, Cragoe Jr EJ, Jones LR. Demonstration of Na$^+$/H$^+$ exchange activity in purified canine cardiac sarcolemmal vesicles. J Biol Chem 1985; 260:4869-4876.
17. Pierce GN. Cationic interactions with Na$^+$-H$^+$ exchange and passive Na$^+$ flux in cardiac sarcolemmal vesicles. Mol Cell Biochem 1987; 78:89-94.

18. Periyasamy SM, Kakar SS, Garlid KD, et al. Ion specificity of cardiac sarcolemmal Na$^+$/H$^+$ antiporter. J Biol Chem 1990; 265:6035-6041.
19. Pierce GN, Ramjiawan B, Dhalla NS, et al. Na$^+$-H$^+$ exchange in cardiac sarcolemmal vesicles isolated from diabetic rats. Am J Physiol 1990; 258:H255-H261.
20. Docherty JC, Ramjiawan B, Afzal N, et al. Cardiac sarcolemmal Na$^+$/H$^+$ exchange after a myocardial infarction in the rat. In: Dhalla NS, Beamish RE, Takeda N, Singal PK, eds. Pathophysiology of Heart Failure. Kluwer Acad Publ, 1995:343-351.
21. Kakar SS, Mahdi F, Li XQ, et al. Reconstitution of the mitochondrial non-selective Na$^+$/H$^+$ (K$^+$/H$^+$) antiporter into proteoliposomes. J Biol Chem 1989; 264:5846-5851.
22. Anderson SE, Murphy E, Steenbergen C, et al. Na-H exchange in myocardium: effects of hypoxia and acidification on Na and Ca. Am J Physiol 1990; 259:C940-C948.
23. Ziegelstein RC, Zweier JL, Mellitz ED, et al. Dimethylthiourea, an oxygen radical scavenger, protects isolated cardiac myocytes from hypoxic injury by inhibitioin of Na$^+$-Ca^{2+} exchange and not by its antioxidant effects. Circ Res 1992; 70:804-811.
24. Dyck JRB, Lopaschuk GD, Fliegel L. Identification of a small Na$^+$/H$^+$ exchanger-like message in the rabbit myocardium. FEBS Lett 1992; 310:255-259.
25. Dyck JRB, Maddaford TG, Pierce GN, et al. Induction of expression of the sodium-hydrogen exchanger in rat myocardium. Cardiovasc Res 1995; 29:203-208.
26. Lagadic-Gossmann D, Chesnais JM, Feuvray D. Intracellular pH regulation in papillary muscle cells from streptozotocin diabetic rats: an ion-sensitive microelectrode study. Pflug Arch 1988; 412:613-617.
27. Khandoudi N, Bernard M, Cozzone P, et al. Intracellular pH and the role of Na$^+$/H$^+$ exchange during ischaemia and reperfusion of normal and diabetic rat. Cardiovasc Res 1990; 24:873-878.
28. Feuvray D, Idell-Wenger JA, Neely JR. Effects of ischemia on rat myocardial function and metabolism in diabetes. Circ Res 1979; 44:322-329.
29. Higuchi M, Ikema S, Toshihiro M, et al. Effects of norepinephrine on hypoperfusion-reperfusion injuries in hearts isolated from normal and diabetic rats. J Mol Cell Cardiol 1991; 23:137-148.
30. Savabi F, Kirsch A. Altered functional activity and anoxic tolerance in diabetic rat isolated atria. Arch Biochem Biophys 1990; 279:183-187.
31. Grinstein S, Rotin D, Mason MJ. Na$^+$/H$^+$ exchange and growth factor-induced cytosolic pH changes. Role in cellular proliferation. Biochem Biophys Acta 1989; 988:73-97.
32. Meno H, Jarmakani JM, Philipson KD. Developmental changes of sarcolemmal Na$^+$-H$^+$ exchange. J Mol Cell Cardiol 1989; 21:1179-1185.
33. Takewaki S, Kuro O-M, Hiroi Y. Activation of Na$^+$-H$^+$ antiporter (NHE-1) gene expression during growth, hypertrophy and proliferation of the rabbit cardiovascular system. J Mol Cell Cardiol 1995; 27:729-742.

34. Allard MF, Emanuel PG, Russell JA, et al. Preischemic glycogen reduction or glycolytic inhibition improves postischemic recovery of hypertrophied rat hearts. Am J Physiol 1994; 267:H66-H74.
35. Anderson PG, Allard MF, Thomas GD, et al. Increased ischemic injury but decreased hypoxic injury in hypertrophied rat heart. Circ Res 1990; 67:948-959.
36. Do E, Baudet S, Gow IF. Intracellular pH during hypoxia in normal and hypertrophied right ventricle of ferret heart. J Mol Cell Cardiol 1995; 27:927-939.

PART III

CHAPTER 12

MOLECULAR DISSECTION OF BACTERIAL Na^+/H^+ ANTIPORTERS

Shimon Schuldiner and Etana Padan

A. INTRODUCTION

The existence of cation/H^+ antiporters was first postulated by Mitchell[1] and demonstrated in mitochondria by Mitchell and Moyle.[2] In bacteria, antiporter activity was first reported in *Streptococcus faecalis*.[3] In *Escherichia coli* the antiporter activity has been studied using a wide variety of techniques at multiple levels from the intact cell to the pure protein reconstituted in proteoliposomes.[4-7] In *E. coli* cells there are two distinct systems that catalyze Na^+/H^+ exchange, namely NhaA and NhaB. NhaA was the first antiporter cloned, and NhaA and NhaB are the only antiporters which have been purified and reconstituted in a functional form.

This review will deal mostly with the biochemistry and molecular biology of these two antiporters and their bacterial relatives. We will only very briefly mention some aspects of the physiological role of the antiporters and regulation of expression of *nha*A by its positive activator NhaR. These topics are extensively discussed in recent reviews.[4-5]

A.1 THE ROLE OF THE Na^+/H^+ ANTIPORTERS IN HOMEOSTASIS OF Na^+

The physiological importance of the antiporters of *E. coli* has become apparent by construction of deletions mutations and ∆*nha*A, ∆*nha*B, ∆*nha*A∆*nha*B. Analysis of ∆*nha*A∆*nha*B showed that there are no specific Na^+/H^+ antiporters other than NhaA and NhaB in the membrane. Accordingly the ∆*nha*A∆*nha*B mutant is the most sensitive to Na^+ having an upper limit for growth of ≤ 0.05 M. It is also very sensitive to

The Na^+/H^+ Exchanger, edited by Larry Fliegel. © 1996 R.G. Landes Company.

Li⁺, yet in the absence of added Na⁺ (contaminating level of 10 mM) or Li⁺ it grows at the entire pH range, pH 6.5-8.4.[8] Hence the sensitivity of the mutant is specific to Na⁺ and Li⁺.

Deletion of each gene separately allowed identification of the relative role of the respective proteins in Na⁺ metabolism. Similar to $\Delta nhaA\Delta nhaB$, $\Delta nhaA$ also is Na⁺ sensitive. It cannot withstand sodium concentrations as high as the wild-type does (0.9 M NaCl at pH 7[9]), and it cannot challenge the toxic effects of Li⁺ ions (0.1 M). In two respects the $\Delta nhaA$ ion sensitivity differs from that of $\Delta nhaA\Delta nhaB$. It is less sensitive than the latter, and its Na⁺ sensitivity is pH-dependent, increasing at alkaline pH (from 0.4 M NaCl at pH 7 to 0.1 M NaCl at pH 8.5). It is thus concluded that *nhaA* is indispensable for adaptation to high salinity, for challenging Li⁺ toxicity, and for growth at alkaline pH (in the presence of Na⁺). In the absence of added Na⁺, $\Delta nhaA$ like $\Delta nhaA\Delta nhaB$, grows at the same rate as the wild-type strain. Hence the phenotype of both mutants exposes a specific sensitivity of the cytoplasm to Na⁺.

The $\Delta nhaB$ strain shows no impairment in its ability to adapt to high salt or alkaline pH nor in its resistance to Li⁺.[8] These findings suggest that NhaA alone can cope with the salt and pH stress, having adequate capacity for these functions. Nevertheless, in the absence of *nhaA*, *nhaB* confers a limited resistance to Na⁺ since $\Delta nhaA\Delta nhaB$ is more sensitive to Na⁺ than $\Delta nhaA$. The role of *nhaB* also becomes apparent in the wild-type at low pH and low Na⁺ when *nhaA* is not induced and the activity of NhaA is shut down.[8] Growth on Na⁺/symport substrates such as glutamate and proline does not occur in $\Delta nhaB$ under these conditions.

The phenotypes of the deletion mutations provide the most compelling evidence for the major role of Na⁺/H⁺ antiporter activity in the Na⁺ cycle of bacteria. Thus the level of the Na⁺/H⁺ antiporter activity in isolated membrane vesicle in each mutant accords the respective phenotype; $\Delta nhaA\Delta nhaB$ has no activity, $\Delta nhaA$ has partial activity (40-60%) and $\Delta nhaB$ has full activity.

A.2 REGULATION OF EXPRESSION OF *NHA*A

The NhaR protein, the product of the gene *nha*R, located downstream of *nha*A, regulates expression of *nha*A. Inactivation of chromosomal *nha*R yields cells sensitive to Li⁺ and Na⁺ in spite of having intact *nha*A. Furthermore, multicopy *nha*R increases expression of *nha*A, and this increase is completely Na⁺-dependent.[10] These results indicate that NhaR is a positive regulator of *nha*A, which functions in *trans*, and its effect is Na⁺-dependent. Accordingly, partially purified NhaR protein binds specifically to the promoter region of *nha*A and retards its mobility in a gel retardation system.[10]

The change in extracellular Na⁺ concentration must reach NhaR, since the effect of NhaR on *nha*A expression is Na⁺-dependent. It has been suggested that a change in intracellular Na⁺ which follows the change of the extracellular concentration of the ion, serves as the immediate signal. Since the induction of *nha*A by intracellular Na⁺ is dependent on NhaR, it is conceivable that a "Na⁺ sensor"-site exists on NhaR which can be identified by mutations affecting the Na⁺ sensitivity of the expression system. Should such a mutation increase the affinity to Na⁺ of the expression system, then at a given Na⁺ concentration, it may even increase the Na⁺/H⁺ antiporter activity in the membrane above the wild-type level. A previously isolated mutation, designated *ant*^{up} was found to increase the Na⁺/H⁺ antiporter activity, thereby conferring Li⁺ resistance to cells which are otherwise Li⁺ sensitive (Nha^{up} phenotype).[11-12] The Nha^{up} mutation resides in the C-terminus of NhaR and is the glu134 to ala134 substitution in the protein.[13] This mutation increases the affinity to Na⁺ of the NhaR mediated *nha*A transcription. Although Na⁺ may indirectly affect NhaR it is tempting to speculate that glu134 is part of the "Na⁺ sensor" of NhaR.

Based on the phenotypes of the mutants and on the fact that the expression of *nha*A is highly regulated and increases significantly under the conditions in which it is essential–high salt, alkaline pH (in the presence of Na⁺ ions), and the presence of toxic Li⁺ ions–we suggest that *nha*B is the housekeeping antiporter while *nha*A is the adaptive gene for Na⁺ tolerance.

B. MOLECULAR NATURE AND PROPERTIES OF THE Na⁺/H⁺ ANTIPORTERS

B.1 Na⁺/H⁺ Antiporters of *E. Coli*

NhaA and NhaB are two membrane proteins which share no significant homology but have a similar predicted secondary structure of 12 putative transmembrane segments (TMS) Rothman, Padan and Schuldiner, preliminary results[40] characteristic of transporters. Each protein was identified, overexpressed, purified and reconstituted, separately, in a functional form in proteoliposomes.[14-15] The turnover number of both purified NhaA and NhaB proteins is very similar to that in the native membrane. Hence, in both cases, a single polypeptide is enough to catalyze the exchange reaction. Because of the presence of two antiporters and many other ion leaks the purified functional proteins provided a system to study stoichiometry, sensitivity to inhibitors and pH sensitivity of the antiporters.

B.1.1 Stoichiometry of the Na⁺/H⁺ Antiporters

Although it was first suggested that Na⁺/H⁺ exchange is electroneutral,[1] further evidence has indicated an electrogenic activity, H⁺/Na⁺ > 1.[16] It has also been proposed that below a certain pH_{out} the antiporter is electroneutral, while above it, it is electrogenic.[17] This question was treated extensively in intact cells.[18-19] In these studies, the magnitudes of $\Delta\bar{\mu}_{H^+}$ and $\Delta\bar{\mu}_{Na^+}$ were measured at various external pH values using ^{23}Na$^+$ and ^{31}P-NMR spectroscopy. From these measurements, the apparent stoichiometry was found to change from 1.1 at pH$_{out}$ = 6.5 to 1.4 at pH$_{out}$ = 8.5. To explain this phenomenon the authors suggested a change in the relative contribution of two antiporters, one electroneutral and the other electrogenic with a stoichiometry of 2H$^+$/Na$^+$. While the rate of the first one is constant, the rate of the latter device is accelerated upon alkalinization of the medium. Indeed, two different specific Na$^+$/H$^+$ antiporters were found in *E.coli* namely NhaA and NhaB.[9] Measuring their stoichiometries therefore, was important both for the understanding of their mechanism as well as the way they participate in cell physiology.

Proteoliposomes reconstituted with purified functional antiporters have been proven to be the preferable system to measure stoichiometry of the ions. In this system passive and mediated leaks are minimal, and ion gradients can be imposed across the proteoliposome membrane and manipulated at will.

The first evidence that both NhaA and NhaB are electrogenic is that imposed $\Delta\psi$, negative inside, in proteoliposomes, enhances the rate of NhaA or NhaB-mediated Na$^+$ efflux.[14-20] A more direct demonstration of the rheogenic nature of the antiporters is that when a Na$^+$ gradient directed outward is imposed, both NhaA[20] and NhaB[15] generate a membrane potential positive inside. The membrane potential is enhanced when the ΔpH, formed by the antiporter activity is collapsed by nigericin.

Estimation of H$^+$ to Na$^+$ stoichiometry was achieved by a thermodynamic approach: the size of the membrane potential generated by the antiporters depends on the magnitude of the Na$^+$ gradient and its stoichiometry[20] according to ΔpNa = (n-1)$\Delta\psi$. From measurements of the size of the membrane potential as determined at steady-state at different Na$^+$ gradients, H$^+$/Na$^+$ ratio of 2 for NhaA[20] and 1.5 for NhaB[15] were calculated.

Another technique to measure stoichiometry was based on a kinetic approach; direct measurements of initial rates of H$^+$ and Na$^+$ yielded in the case of NhaA a value close to 2[20] but gave equivocal results in the case of NhaB due to difficulties in estimation of initial rates.[15] A modification of the kinetic approach was therefore undertaken with NhaB.[15] The activity of a rheogenic antiporter involves the electrophoretic movement of permeant ions to compensate for charge translocation. The ratio between the movement of the counter ion and the activity of the antiporter is therefore a measure of the number of net charges transferred in one catalytic cycle. K$^+$ and its analog Rb$^+$, in the presence of valinomycin was used as a counter ion in proteoliposomes reconstituted with NhaB. When conditions that drive Na$^+$ uptake were created, (ΔpH, acid inside), the initial rate of Na$^+$ uptake was stimulated 3- to 4-fold upon addition of 10 mM KCl and 1μM valinomycin.

This phenomenon was explained by the movement of K$^+$ to compensate for charge translocation which allowed for further uptake of ^{22}Na$^+$. Under these conditions, as expected, an amiloride (a specific inhibitor of NhaB, see II-A.2 and ref. 29) sensitive ^{86}Rb$^+$ uptake was observed. This uptake was dependent on the presence of valinomycin and on the Na$^+$ concentration in the assay medium. In these type of experiments initial rates of ^{22}Na$^+$ and ^{86}Rb$^+$ uptake were measured with and without 1 mM amiloride in the presence of 1 μM valinomycin and with different NaCl concentrations. The ratio of the fluxes of Na$^+$/Rb$^+$ was found to be very close to two. This is the ratio that is expected if one charge is translocated per 2 Na$^+$ ions. Such a case is consistent with a H$^+$/Na$^+$ stoichiometry of 3:2 for NhaB.[15]

B.1.2 Sensitivity of the Na$^+$/H$^+$ Antiporters to Inhibitors

The diuretic drug amiloride is a specific inhibitor of sodium transporting proteins in several cell types. It is a competitive inhibitor of plasma membrane Na$^+$/H$^+$ antiporter in animal cells as well as epithelial Na$^+$ channels.[21] In algae and plants it has been shown to inhibit tonoplast[22] and plasma membrane[23] Na$^+$/H$^+$ antiport activity. The Na$^+$/H$^+$ antiporter of methanogenic bacteria was also found to be inhibited by both amiloride and harmaline.[24]

There has been some controversy whether amiloride inhibits also the Na$^+$/H$^+$ antiport activity of *E. coli*.[6,25-26] Since they are weak amines, amiloride and some of its derivatives may inhibit due to their uncoupling activity, and therefore it has been hard to differentiate between specific and nonspecific effects of amiloride on ΔpH driven Na$^+$ transport.[26-27]

To avoid the possible uncoupling activity of the purified protein derivatives of amiloride it is preferable to test the effect of this drug on ΔpH independent reactions, namely passive Na$^+$ efflux and ^{22}Na$^+$:Na$^+$ exchange. No significant inhibition of down-hill sodium efflux was detected when catalyzed by purified NhaA in proteoliposomes.[14,28] Amiloride inhibited, the passive Na$^+$ efflux catalyzed by NhaB significantly (t1/2 > decreased from 25 sec to > 120 sec at 1 mM amiloride) as well as ^{22}Na$^+$:Na$^+$ exchange and ΔpH driven Na$^+$ uptake (competitive inhibition K$_i$ = 20 μM[29]).

NhaB differs in its sensitivity to amiloride derivatives from that of the mammalian exchanger isoforms. Rather, it resembles the amiloride sensitive sodium channel from epithelial cells or the alga *Dunaliella salina*,[23] in being more sensitive to amiloride than to its derivatives.[29]

It has been reported that amiloride inhibits growth of *E. coli* cells at an alkaline pH.[30-31] It was suggested that this phenomenon results from the inhibition of Na$^+$/H$^+$ exchange activity, which is responsible for pH regulation. This is probably not the case because of several reasons: (1) NhaA which is supposed to be very active at an alkaline pH is amiloride resistant, (2) an *E. coli* strain in which both *nha*A and *nha*B are deleted

($\Delta nhaA\Delta nhaB$) can grow at an alkaline pH;[8] and (3) growth of $\Delta nhaA\Delta nhaB$ is also inhibited by amiloride at alkaline pH, indicating that this growth inhibition is not due to inhibition of specific Na$^+$/H$^+$ antiport activity.

In a study aimed to characterize the amiloride binding site of the mammalian antiporter NHE1, Pouysségur and coworkers developed a procedure to isolate NHE1 mutants resistant to the amiloride analog 5-N-(methyl propyl) amiloride (MPA).[32] Two NHE1 mutants with decreased sensitivity to MPA were isolated, and both had a replacement of Leu-167 to Phe. This Leu is found in a sequence which is conserved among NHE isoforms, ^{164}VFFLFLLPPI^{173} which is located in the middle of the fourth putative transmembrane segment. Analysis of the sequence of NhaB reveals the pentamer ^{445}FLFLL^{450} which is identical to five amino acids of the putative amiloride binding site.[29] However it should be noted that in contrast to NHE1, this sequence resides in the middle of the eleventh putative transmembrane segment of NhaB, and the homology between NhaB and NHE1 is not extended beyond that in this region.

Harmaline and clonidine but not cimetidine, compounds known to inhibit the mammalian Na$^+$/H$^+$ antiporters, inhibit NhaB.[29]

Li$^+$ is a known substrate of both Na$^+$/H$^+$ antiporters of *E. coli*[9] and Mn^{++} is a substrate of mitochondrial Na$^+$/H$^+$ exchanger.[33] Both LiCl and MnCl$_2$ inhibit NhaB-dependent Na$^+$ uptake.[15,29] The $K_{0.5}$ values at a sodium concentration of 100 μM were found to be 1.2 and 5 mM for Mn^{++} and Li$^+$, respectively. These data indicate that, surprisingly, NhaB has a higher affinity for Mn^{++} than to Li$^+$, a situation similar to that observed for the mitochondrial Na$^+$/H$^+$ exchanger.[33]

B.1.3 NhaA Functions Simultaneously as a pH Sensor and a Titrator

NhaA is highly sensitive to pH, changing its rate more than 1000-fold between pH 7-8, thus functioning simultaneously as a pH "sensor" and a "titrator."[14,16] This distinct pH dependence of the Na$^+$/H$^+$ antiporter of *E. coli*, was first demonstrated in right side out membrane vesicles measuring ^{22}Na efflux driven by imposed artificial ΔpH or $\Delta\psi$[16] and then in purified protein functionally reconstituted in proteoliposome, measuring passive ^{22}Na efflux.[14]

Since flux of cations via the cation/H$^+$ antiporters affects the ΔpH across the membrane, cation induced changes in the fluorescence of acridine orange, and similar probes measuring ΔpH, have proven a fast and reproducible way to monitor the activity of H$^+$/cation antiporters in isolated membrane vesicles.[17,34] Calculation of the kinetic parameters of an antiporter with this technique is complicated due to the indirect nature of the measurement. However, when measurements are conducted at cation concentrations above the Km of the antiporter, it most probably reflects the V_{max} of the system and is thus most suitable for comparison of antiporters activity.[34]

Indeed with this technique a pronounced pH dependence of NhaA activity was found, similar to that found by direct flux measurements of ^{22}Na.[35] Furthermore, the experimental system used was everted membrane vesicles isolated from $\Delta nhaA\Delta nhaB$ mutant transformed with plasmids bearing *nha*A. This system has proven most suitable to identify (by site directed mutagenesis) residues in NhaA involved in the pH sensitive domain of the protein. A mutation can be easily introduced into the plasmid containing NhaA and its effect tested both in vitro, in everted isolated membrane vesicles, and in vivo, in both cases with no background of chromosomal encoded either NhaA or NhaB Na$^+$/H$^+$ antiporters.

Using this approach it was found that none of the eight histidines of NhaA are essential for Na$^+$/H$^+$ exchange activity of the protein.[35] However, the replacement of His-226 by Arg (H226R) markedly changed the pH dependence of the antiporter. A strain deleted of both antiporters genes, *nha*A and *nha*B, transformed with multicopy plasmid bearing wild-type *nha*A, exhibited both Na+ and Li$^+$ resistance throughout the pH range of 6-8.5. In marked contrast, transformants of plasmid bearing H226R-*nha*A are resistant to Li$^+$ and to Na$^+$ at neutral pH, but became sensitive to Na$^+$ above pH 7.5. Analysis of the Na$^+$/H$^+$ antiporter activity of membrane vesicles derived from H226R cells showed that the mutated protein was activated by pH to the same extent as the wild-type. However, whereas the activation of the wild-type NhaA occurred between pH 7 and pH 8, that of H226R antiporter occurred between pH 6.5 and pH 7.5. In addition, while the wild-type antiporter remains almost fully active, at least up to pH 8.5, H226R is reversibly inactivated above pH 7.5, retaining only 10-20% of the maximal activity at pH 8.5.[35]

Histidine and arginine share certain properties: both most probably bear a positive charge and both are polar residues capable of hydrogen bonding. The question as to which of these properties is important for the pH sensitivity of NhaA was then tested.

The fact that H226R is inactive at alkaline pH has provided a very powerful system to isolate revertants or suppressors to H226R.[35A] First site suppressors were obtained in which cysteine and serine replace H226.

In solution both cysteine and histidine have a pK in the physiological range,[36] whereas serine does not ionize under these conditions. Both histidine and cysteine have been shown to be ionized in proteins with pK not very different from that observed in solution.[36-37] Nevertheless it is impossible to predict the pK of an amino acid in a protein neither from its ionization constant in solution, nor from its pK in another protein. It has been suggested that serine is also capable of ionization in a protein of the bacterial photosynthetic reaction center, but only when its electrostatic environment is drastically changed by electron transport,[38] an unlikely situation in the case of NhaA.

Therefore, we suggest that ionization of residue 226 is not important for pH regulation of the antiporter, rather it is the polarity and/or capacity to form hydrogen bonds, properties shared by histidine, serine and cysteine, which is essential.

To test the effect of negative charge, aspartate was introduced at position 226 (H226D). As opposed to Arg which shifted the pH profile towards acidic pH, Asp shifted the pH profile towards basic pH. When Ala (H226A) was introduced, which is neither polar nor charged, the antiporter showed no activity. In conclusion, a polar group is essential; charge shifts the pH profile, positive, towards acidic pH, negative towards basic pH.

As yet it is not known how these residues exert their effect. Amino acid residues are known to cause micro changes within proteins reflected in the electrostatic microenvironment, stability of acidic or basic residues, density of local protons and even binding of water molecules.[36-38] It is assumed that at a particular pH, a certain proton density, at a given site, is crucial for the pH sensitivity of NhaA. According to this reasoning at a pH below this critical value, a positive charge, at position 226, would be beneficial to reduce the excess proton concentration. At the alkaline range negative charge would be useful to attract the scarce protons. It should be emphasized however, that as long as the structure of the protein is unsolved, short-range steric effects or long-range conformational effects, not directly related to the "pH sensor," cannot be excluded.

It is of interest that the difference between the mutants is not expressed in growth phenotype at the acidic pH range. H226D with the lowest activity at this pH range grows like the wild-type. On the other hand, at alkaline pH growth is detected only in the mutants which are substantially activated by pH, thus H226A and H226R stop growing beyond pH 8.4. These results emphasize the physiological importance of NhaA at alkaline pH in the presence of Na$^+$.

B.1.4 pH Dependence of the NhaB Protein

In contrast to NhaA, whose activity is extremely pH dependent (Section B.1.3), NhaB activity was considered to be pH independent.[8-9] Indeed, when ammonium gradient driven ^{22}Na$^+$ uptake was measured in NhaB-proteoliposomes at external pH's of 7.2, 7.6 and 8.5, the V$_{max}$ found was 107, 67.5 and 87.8 μmol/min/mg, respectively.[15] However, a 10-fold increase in the K$_m$ of NhaB to Na$^+$ was detected (from 1.5 to 16.6 mM) upon decrease of the pH in these experiments.

A similar effect of pH on the K$_m$ of Na$^+$/H$^+$ antiporter activity was reported by Leblanc and collaborators.[16] These studies were performed in membrane vesicles isolated from cells containing both *nha*A and *nha*B genes. However, the cells were grown under conditions (low Na$^+$, neutral pH) in which the level of *nha*A expression is known to be low (Section A). These results reflect therefore the changes in affinity of NhaB. It has been suggested that the effect of pH on the K_m of NhaB is due to competition of both ions H^+ and Na^+, on a common site.[16]

B.2 Na^+/H^+ Antiporters of Other Bacteria

A comprehensive review of the properties of bacterial Na^+/H^+ antiporters, based on their activities in isolated membrane vesicles and intact cells, has recently been published.[4] Here we will focus only on the few antiporters whose genes have recently been cloned. As mentioned above only the *E. coli* NhaA and NhaB proteins have been purified, reconstituted in proteoliposomes in a functional form and properties of the pure proteins studied in detail.

Two strategies for cloning of Na^+/H^+ antiporter genes have been advanced. Each based on one of the two substrates, common to all Na^+/H^+ antiporters, Li^+ and Na^+. On a concentration basis, Li^+ is 10 times more toxic than Na^+, both to wild-type *E. coli*[11] and fission yeast.[39] Therefore Li^+ provides a screen for cells capable of maintaining low internal Na^+ or Li^+ levels without selecting for osmotolerance. Another advantage of Li^+ selection over that of Na^+ is that it can be applied directly to wild-type cells. Realizing this advantage of Li^+ selection, *nha*A was cloned using a DNA library containing sequences overlapping the *nha*A locus.[11] The *sod*2 gene has been cloned from *Schizosaccharomyces pombe* using a similar approach.[39]

Although wild type *E. coli* cells transformed with multicopy plasmids bearing *nha*A become Li^+ resistant as compared to the wild-type, their tolerance to Na^+ is unchanged. This implies that other factors, such as adaptation to increased osmolarity determines the upper level of resistance to Na^+. In contrast to wild-type cells, mutants, Δ*nha*A or Δ*nha*AΔ*nha*B, which are Na^+ sensitive due to the lack of the antiporters are most suitable to apply Na^+ selection and clone by complementation DNA inserts encoding Na^+/H^+ antiporter genes. With this approach and the *E. coli* mutants various antiporter genes have been cloned from very different bacteria including *nha*B[40] and *cha*A[41] from *E coli* and *nha*C from an alkaliphile *Bacillus firmus* OF4.[42] A similar approach applied to *Enterococcus hirae* yielded napA antiporter.[43] Utilizing this approach several new Na^+/H^+ antiporters have recently been cloned from bacteria (Section B.2.1).

B.2.1 Na^+/H^+ Antiporters of *Vibrio alginolyticus* and *Vibrio parahaemolyticus*

In contrast to *E. coli*, whose growth is independent of Na^+, the marine bacterium *Vibrio alginolyticus* requires 0.5 M NaCl for optimal growth.[44] In addition it has a primary Na^+ extrusion system, which is an electrogenic Na^+-translocating NADH-quinone reductase in the respiratory chain.[45-46] Like *E. coli* it has a Na^+/H^+ antiporter which is driven by a protonmotive force[47] and suggested to play a primary role in Na⁺ circulation at neutral pH, when the respiratory linked Na⁺ pump is not operating. Many transport systems for carbohydrates and amino acids of *V. alginolyticus* are Na⁺-substrate symport systems that utilize $\Delta\bar{\mu}_{Na^+}$ as a driving force.[45] Also, flagellar rotation is driven by an influx of Na⁺.[48] Thus, Na⁺ circulation across the cell membrane is very important for energy transduction in this organism.

A gene has been cloned from a DNA library from the marine bacterium *Vibrio alginolyticus* that functionally complements the Δ*nha*A mutant strain of *E. coli* conferring resistance to Na⁺ (0.5 M NaCl at pH 7.5) and concomitantly increasing Na⁺/H⁺ antiport activity measured in isolated everted membrane vesicles.[44] The *V. alginolyticus* protein also allowed Na⁺ preloaded cells of *E. coli* NM81 (Δ*nha*A) and RS1 (Δ*nha*A, *cha*A⁻),[49] to extrude Na⁺ at alkaline pH.[50] The extrusion of Na⁺ occurred against its chemical gradient in the presence of membrane-permeable amine and Δψ. Thus, the protein is functional as an electrogenic Na⁺/H⁺ antiporter in *E. coli* cells, and in this respect it is functionally similar to *E. coli* NhaA.

The nucleotide sequence of the cloned fragment revealed an open reading frame, which encodes a protein with a predicted 383 amino acid sequence and molecular mass of 40,400 D. The hydropathy profile is characteristic of a membrane protein with 11 membrane spanning regions. The deduced protein shows no significant sequence similarity with any cation/H⁺ antiporter except for the *E. coli* NhaA. Since it is 58% identical to the *E. coli* NhaA it was designated *nha*Av.[43] We however, suggest to designate it Va-*nha*A to follow a more general convention used also with eukaryotic genes.

Vibrio parahaemolyticus is also a slightly halophilic marine bacterium, which requires at least 30 mM Na⁺ for growth[51] and is very similar in its Na⁺ metabolism to *V. alginolyticus*.[51] Also in this bacterium, a Na⁺/H⁺ antiporter is very important for Na⁺ circulation, especially at neutral pH when the primary respiratory linked pump is inactive.[52]

Recently, the Na⁺/H⁺ antiporter activity of *V. parahaemolyticus* has been characterized.[51] It seems that there are two Na⁺/H⁺ antiporters in the cell membrane of *V. parahaemolyticus*, resembling NhaA and NhaB of *E. coli*. One system is pH-dependent (activity increases as pH increases between 7.0 and 9.0) and the other is pH-independent and less active. The K_m values and V_{max} values of the *V. parahaemolyticus* Na⁺/H⁺ antiporters are considerably larger than those of *E. coli*.[51] Since *V. parahaemolyticus* lives in sea water and large quantities of Na⁺ enter *V. parahaemolyticus* cells, a low-affinity and high-capacity Na⁺ extruding system would be necessary for *V. parahaemolyticus* cells.

A gene encoding a Na⁺/H⁺ antiporter was cloned from the chromosomal DNA of *Vibrio parahaemolyticus* using a Na⁺ and Li⁺ sensitive Δ*nha*AΔ*nha*B *E. coli* mutant.[51] As expected, membrane vesicles prepared from the original *E. coli* mutant did not show any detectable

Na⁺/H⁺ (and Li⁺/H⁺) antiport activity, while a high Na⁺/H⁺ (and Li⁺/H⁺) antiport activity was observed in membrane vesicles prepared from the transformed cells. The predicted protein encoded by the cloned gene consists of 383 amino acid residues putatively forming 12 TMS and showing high homology (59% identity and 87% similarity) with the NhaA Na⁺/H⁺ antiporter of *E. coli* and 97.4% identify to that of *V. alginolyticus* Va-NhaA. These results imply the presence of NhaA type antiporter in marine bacteria. We suggest designating the *V. parahaemolyticus* gene Vp-*nha*A. The activity of Vp-NhaA increased dramatically when the pH of the assay medium was raised from 7.0 to 8.5, similar to its *E. coli* homolog. Interestingly, however, in contrast to the amiloride-insensitive *E. coli* NhaA, the activity of the pH-dependent Na⁺/H⁺ antiporter of *V. parahaemolyticus* is inhibited by the drug.[51]

It should be noted that the activity conferred by the *V. parahaemolyticus* gene expressed in *E. coli*, a heterologous host, differs from the native activity in untransformed *V. parahaemolyticus*. Drastic decrease in K_m values for Li⁺ and Na⁺ were observed in the heterologous expression system.

B.2.2 Search for Common Denominators Among the Na⁺/H⁺ Antiporters

Although there is quite a high degree of sequence conservation among the mammalian Na⁺/H⁺ antiporters,[53-54] there is no significant homology between them and NhaA, and not even between NhaA and NhaB.[4,55] These findings would suggest that there are no universal Na⁺ and H⁺ recognition and exchange sites and that, instead, different sequences accomplish similar functions. However, it is possible that very few conserved residues are adequate to carry out these functions, even when they are dispersed throughout the protein. Identification of the domains necessary for activity regulation, and H⁺ and Na⁺ sensing in the various proteins is certainly one of the future challenges.

NhaB displays a weak but distinct homology to a membrane protein with unknown function in *Mycobacterium leprae*.[29] There is also a limited homology to other transporters found in bacteria and animal cells: the rat sodium-dependent sulfate transporter, the product of the mouse and human pink-eyed dilution gene, and the ArsB protein.[56] A domain of high homology is present twice in the pink-eyed dilution gene (Fig. 12.1), and a signature characteristic of the family can be easily detected:

$$(S,T,A)X\ (L,I)GGXXTXXGX(P,S)XN(L,I,V)(I,V).$$

Interestingly, NhaB but not NhaA of *E. coli* is inhibited by amiloride.[29] The Vp-NhaA Na⁺/H⁺ antiporter is also inhibited by amiloride. A search for homologous domains in NhaB revealed one sequence (⁴⁴⁵FLFLL⁴⁵⁰) that was very similar to the amiloride binding

```
NhaB         r s LMM HAG VGTAL GG VM TMV GE PQ N LIIA KAA-
H-NhaB       r s LMM HAG VGTAL GG VM TVV GE PQ N LIIA EQA-
M97900-1     r q v LI AEV IFTNI GG AA TA I GDPP N VII -----
M97900-2     -- LMY ALA LGACL GG NG TL I GASTN VVCA GIA-
M99564-1     r q v LI AEV IFTNI GG AA TA I GDPP N VII -----
M99564-2     -- LMY ALA FGACL GG NG TL I GASAN VVCA GIA-
L10660       f -- LM AEV FASNI GG AR TL V GDPP N IIIa sra-
L19102       k l MCL CIA YSSTI GG LT TI T GTSTN LIFS ehfn
```

Fig. 12.1. Multiple alignment of NhaB homologs. Blocks of similarity were first identified using Pileup and Gap (Wisconsin Package, 1994) and then MACAW.[79] In addition to the segments from E. coli and H. influenza NhaB (starting at amino acid 190), the other sequences are listed by accession number: M97900 and M99564 are the mouse and human pink-eyed dilution gene, respectively; in both cases 1 and 2 denote two segments of the protein starting at amino acids 460 and 769, respectively. L19102 is the accession number of the rat renal Na^+/SO_4 cotransporter (segment compared starts at amino acid 246). L10660 is the accession number of a membrane protein from Mycobacterium leprae with unknown function (the segment compared starts at amino acid 123).

domain in human NHE (^{164}VFFLFLLPPI^{173}).[29] A VFFLLI sequence was found in TMS2 of Vp-NhaA.[51] Since amiloride appears to compete with Na^+, this region may be involved in Na^+ binding. A similar motif was found in the Na^+/H^+ antiporter of *Enterococcus hirae* (NapA)[43] and other Na^+ driven transporters.[51] It is not known, however, whether the activity of these transporters is inhibited by amiloride or not.

The sequence of the entire genome of *Haemophilus influenzae* Rd has recently been published.[57] It includes genes encoding proteins homologous to NhaB, NhaC and NhaA (87%, 62%, 75%, similarity respectively). This newcomer joins the growing family of NhaA which includes the Na^+/H^+ antiporter from *Salmonella enteriditis*,[58] *V. alginolyticus*[44] and *V. parahaemolyticus*[51] and shows substantial homology with the *E. coli* protein (Fig. 12.2).

Pinner et al[40] reported that in one domain of the *E. coli* NhaB (starting at amino acid position 295), the homology with NhaA (starting at 300) at practically the equivalent position in the protein was quite significant (43% identity and 65% similarity). A diffuse and restricted homology with other Na^+-transporting proteins was found in this area (GXXLXXXXA). Between the Gly and Ala, no charged residues were found in any of the sequences. Also, this stretch of the homology overlaps with part of the "sodium consensus box" previously identified in other Na^+-translocating systems.[59] However, replacement in NhaA of the conserved Gly with Cys did not affect activity (Dibrov P, Padan E and Schuldiner S, unpublished results). Similar sequences were found in Vp-NhaA.[51] At present, the functional significance of these domains is not clear.

In the *E. coli* NhaA, His-226 has been assigned a role in the pH sensitivity of the protein.[35] Most interestingly, this histidine is conserved in all the members of the NhaA family, suggesting that the "pH sensor" domain is similar in all of them.

There are 9 aspartic acids and 13 glutamic acids in Va-NhaA protein. Of 22 negatively charged residues, only 4 amino acids are predicted to be in TMS and all of them are aspartic acid.[44,50] These 4 amino acids are conserved in NhaA of *E. coli*, *Salmonella entiritidis* and Vp-NhaA. Furthermore, D-111, which is predicted in a loop region between the TMS III and IV is also conserved in NhaA. Each of these conserved aspartic acids was replaced by asparagine. *E. coli* NM81 cells containing a plasmid harboring the Va-*nha*A gene mutated at D-125, -155 or -156 could neither grow in a high NaCl medium nor extrude Na^+ at analkaline pH against its chemical gradient.[50] It is reasonable to assume that negatively charged amino acid residues localized in putative TMS are important for binding or translocation of cations like H^+ and Na^+. Interestingly, replacement of a conserved membrane Glu in NHE1 (E262I) renders the mammalian antiporter inactive.[32] Nevertheless it should be stressed that intramembrane charges can also be involved in the assembly and folding of the protein.[60-61] Hence the suggestion that these aspartates form the Na^+ binding site,[50] although appealing, may be premature.

Eight consecutive amino acid residues starting at position 45 in *E. coli* NhaA are missing in Va-*nha*A.[44] Since these two NhaA antiporters possess similar properties, these eight residues probably have no significant role in the antiporter protein.

The expression of *nha*A in *E. coli* has been shown to be regulated by Na^+ and by a regulatory protein encoded by *nha*R located just downstream from *nha*A (Section A). The structural gene corresponding to *nha*R, however, was not found in the upstream and downstream regions of either Va-*nha*A or Vp-*nha*A. Since these marine bacteria require Na^+ for optimal growth, it has been suggested that the Na^+/H^+ antiporter is constitutively expressed for the extrusion of cellular Na^+ in marine bacteria.[43]

B.2.3 The Chromosomal Tetracycline—Resistance Locus tetB(L) of *Bacillus subtilis* Encodes a Protein Which Confers Na^+/H^+ Antiporter Activity

Tetracycline-resistant determinants that function *via* proteins catalyzing efflux of the antibiotic have long been recognized in numerous plasmids and transposons from both gram-negative and gram-positive bacteria.[62] In addition, an efflux type of *tet* gene exists in single copy near the origin of replication of some *B. subtilis* strains that are nonetheless tetracycline-sensitive. This *tet*B gene, which is referred to as *tet*B(L), has remained an unexplained, presumably cryptic oddity.[63]

```
NhaA       mk------------- HLH RFF SSDAS GGI ILIIA AIL AM I        27
H-NhaA     vn fllcifkgvyvik LIQ RFF KLESA GGI LLLFS AVV AM L       40
S-NhaA     vk------------- HLH RFF SSDAS GGI ILIIA AAL AM L        27
Va-NhaA    mnd------------ VIR DFF KMESA GGI LLVIA AAI AM T        28
Vp-NhaA    mnd------------ VIR DFF KMESA GGI LLVIA AAI AM T        28

NhaA       MAN Sga tsgwyhd fletpvql rVGSL EINKNM LLWIND AL          67
H-NhaA     LAN Spl snqyndf lnlpvslq -IGSF SINKTL IHWIND GF          79
S-NhaA     MAN Mga tsgwyhd fletpvql rVGAL EINKNM LLWIND AL          67
Va-NhaA    IAN Spl getyqsv --------- LHTY VFGMSV SHWIND GL          59
Vp-NhaA    IAN Spl getyqsl --------- LHTY VFGMSV SHWIND GL          59

NhaA       MAVFF LLVGLEVK RELMQ GSLASLRQ AAFPV IAA IGGM IV          107
H-NhaA     MAVFF VLVGMEVK KELFE GALSTYQQ AIFPA IAA IGGM VI          119
S-NhaA     MAVFF LLIGLEVK RELMQ GSLASLRQ AAFPV IAA IGGM IV          107
Va-NhaA    MAVFF LLIGLEVK RELLE GALKSKETA IFPAI AA VGGM LA           99
Vp-NhaA    MAVFF LLIGLEVK RELLE GALKSKETA IFPAI AA VGGM LA           99

NhaA       PALL YLAFN YA DPITR EGWAI PAATDIAFALG VLALL GSR          147
H-NhaA     PAVV YWFIA KQ DPSLA NGWAI PMATDIAFALG IMALL SKQ          159
S-NhaA     PALL YLAFN YS DPVTR EGWAI PAATDIAFALG VLALL GSR          147
Va-NhaA    PALI YVAFN AN DPEAI SGWAI PAATDIAFALG IIALL GKR          139
Vp-NhaA    PALI YVAFN AN DPEAI SGWAI PAATDIAFALG IMALL GKR          139

NhaA       VPLALK IFL MALAIIDDLG AIII IALFY TNDLS MASL GVA          187
H-NhaA     VPLPLK IFL LALAIIDDLG AIVV IALFF SHGLS VQAL IFS          199
S-NhaA     VPLALK IFL MALAIIDDLG AIVI IALFY TSDLS IVSL GVA          187
Va-NhaA    VPVSLK VFL LALAIIDDLG VVVI IALFY TGDLS TMAL LVG          179
Vp-NhaA    VPVSLK VFL LALAIIDDLG VVVI IALFY TGDLS SMAL LVG          179

NhaA       AVAI AVLAVLN LCG ARRTGV YIL VGVV LWTA VLKSGV HAT         227
H-NhaA     AVAI IVLILLN RFR VSALCA YMV VGAI LWAS VLKSGV HAT         239
S-NhaA     AFAI AVLALLN LCG VRRTGV YIL VGAV LWTA VLKSGV HAT         227
Va-NhaA    FIMT GVLFMLN AKE VTKLTP YMI VGAI LWFA VLKSGV HAT         219
Vp-NhaA    FVMT GVLFMLN AKE VTKLTP YMI VGAI LWFA VLKSGV HAT         219

NhaA       LAGV IVGFF IPLK EKHG-RS PAK RLEHVLH PWVA YLIL PL         266
H-NhaA     LAGV IIGFS IPLK GKKG-ER PLD DFEHILA SWSS FVIL PL         278
S-NhaA     LAGV IVGFF IPLK EKHG-RS PAK RLEHVLH PWVA YLIL PL         266
Va-NhaA    LAGV VIGFA IPLK GKQGEHS PLK HMEHALH PYVA FGIL PL         259
Vp-NhaA    LAGV VIGFA IPLK GKQGEHS PLK HMEHALH PYVA FGIL PL         259
```

Fig. 12.2. (part 1) Multiple alignment of NhaA Na$^+$/H$^+$ antiporters. The alignment was first performed using Pileup (Wisconsin Package, 1994) and then MACAW.[79]

```
NhaA     FAFANAG VSLQGVTLDGLT SILPLGIIAGLLIGKPLGISL   306
H-NhaA   FAFANAG VSFAGIDVNMIS SPLLLAIASGLIIGKPVGIFG   318
S-NhaA   FAFANAG VSLQGVTIDGLT SMLPLGIIAGLLIGKPLGISL   306
Va-NhaA  FAFANAG ISLEGVSMSGLT SMLPLGIALGLLVGKPLGIFT   299
Vp-NhaA  FAFANAG ISLEGVSMSGLT SMLPLGIALGLLIGKPLGIFS   299

NhaA     FCWLALRLKL AHLPEGTTYQQIMVVGILCGIGFTMS IFIA   346
H-NhaA   FSYISVKLGL AKLPDGINFKQIFAVAVLCGIGFTMS MFLA   358
S-NhaA   FCWLALRFKL AHLPQGTTYQQIMAVGILCGIGFTMS IFIA   346
Va-NhaA  FSWAAVKMGV AKLPEGVNFKHIFAVSVLCGIGFTMS IFIS   339
Vp-NhaA  FSWAAVKLGV AKLPEGINFKHIFAVSVLCGIGFTMS IFIS   339

NhaA     SLAF-GSVDPEL INWAK LGIL VGSISSAVI GYSWLRvrlr   385
H-NhaA   SLAFdANAGESV NTLSR LGIL LGSTVSAIL GYLFLKqttk   398
S-NhaA   SLAF-GNVDPEL INWAK LGIL IGSLLSAVV GYSWLRarln   385
Va-NhaA  SLAF-GNVSPEF DTYAR LGIL MGSTTAALL GYALLHfslp   378
Vp-NhaA  SLAF-GNVSPEF DTYAR LGIL MGSTTAVL GYALLHfslp   378

NhaA     psv--       388
H-NhaA   ln---       400
S-NhaA   apa--       388
Va-NhaA  kkaga       383
Vp-NhaA  kkagd       383
```

Fig. 12.2. (part 2)

Transpositional disruption of the chromosomal *tet*B(L) locus of *B. subtilis* led to impaired growth at alkaline pH in the presence of Na$^+$ and reduced rates of electrogenic Na$^+$ efflux at alkaline pH.[64] The mutant phenotype was reversed by transformation with a plasmid expressing the *tet*B(L) gene, suggesting a physiological role for this locus in Na$^+$-resistance and Na$^+$-dependent pH homeostasis. TetB(L) was also inferred to have a modest capacity for K$^+$ efflux.[64] Energy-dependent tetracycline efflux rates in the wild-type were larger than in the transposition mutant, but were not sufficient to confer resistance to the antibiotic. The plasmid *tet*B(L) conferred only a limited Tet resistance.

To further explore its function, the *tet*B(L) gene was expressed in *E. coli* Δ*nha*A (NM81) which is sensitive to Na$^+$. Δ*nha*A is also completely inhibited by tetracycline concentrations as low as 2 μg/ml. Both Na$^+$ resistance up to 0.6 M and tetracycline resistance (4 μg/ml) was conferred upon transformation of Δ*nha*A with the *tet*B(L)-bearing plasmid. Furthermore, everted membrane vesicles isolated from the transformants catalyze Na$^+$/H$^+$ antiport even more actively than tetracycline/H$^+$ antiport.[64]

These results for the first time raise the possibility of the activity of Tet proteins in Na$^+$ efflux, and of physiological roles for *Tet* gene products that are unrelated to antibiotic resistance. TetB(L) is suggested to function as a Na$^+$/H$^+$ antiporter at neutral pH as well as at pH 8.5. Nevertheless, this interesting suggestion will be experimentally proven only when the TetB(L) protein is purified in a functional form.

B.2.4 Na$^+$/H$^+$ Antiporter of Extreme Alkaliphiles

The pioneering studies of Krulwich and colleagues on the physiology of extreme alkaliphiles, demonstrated that Na$^+$/H$^+$ antiporters play a crucial role in pH homeostasis in these organisms.[65-66] Intracellular pH regulation in these bacteria, which at an alkaline pH involves acidification of the cytoplasm, requires Na$^+$, and a mutant lacking the antiporter activity is no longer alkaliphilic. *nha*C was cloned from *B. firmus* OF4[42] using *E. coli* Δ*nha*A as a host and its Na$^+$ sensitivity for functional complementation. *nha*C confers Na$^+$ resistance to the *E. coli* mutant and increases Na$^+$/H$^+$ activity in isolated everted membrane vesicles, suggesting that it encodes a Na$^+$/H$^+$ antiporter. Further evaluation of the role of *nha*C in alkaliphiles awaits the development of an appropriate genetic systems in the extreme alkaliphiles.

Recently, a series of alkali-sensitive mutants of alkaliphilic *Bacillus* sp. C-125 were isolated.[67] One of the mutants, 38154 could not grow above pH 9.5 and was unable to sustain a low internal pH in an alkaline environment. A DNA fragment from the wild-type parent was cloned by functional complementation of the mutant restoring alkaline pH regulation.[68] Direct sequencing of the mutant's DNA corresponding region revealed that the mutation resulted in an amino acid substitution from Gly-393 to Arg of the putative ORF1 product. ORF1 is predicted to encode an 804-amino-acid polypeptide with a molecular weight of 89,070 and hydrophobicity characteristics of a membrane protein. The N-terminal part of the putative ORF1 product showed amino acid similar to those of the chain-5 products of eukaryotic NADH quinone oxidoreductases. Membrane vesicles prepared from 38154 did not show membrane potential (Δψ)-driven Na$^+$/H$^+$ antiporter activity. Antiporter activity was regained by introducing the cloned gene. These results indicate that the mutation in 38154 affects, whether directly or not, the electrogenic Na$^+$/H$^+$ antiporter activity. This is the first report which directly shows that a gene encoding a putative Na$^+$/H$^+$ antiporter is important in the pH regulation of alkaliphilic microorganisms at high external pH.

B.2.5 Expression of Homologous and Heterologous Antiporter Genes in *E. Coli* Devoid of Its Own Na$^+$/H$^+$ Antiporter Genes

As previously mentioned, heterologous expression of antiporter genes in *E. coli* Δ*nha*AΔ*nha*B provided a very powerful tool for cloning of antiporter genes. This expression system was also used for the study of the activity of antiporter proteins.[41,44,50,51,64,69] However, it should be noted that the affinity of the *V. parahaemolyticus* Vp-NhaA antiporter for Li$^+$ and Na$^+$ is significantly increased when the protein is expressed heterologously in *E. coli* cells.[51] Since the pH activity relationship was similar in both *E. coli* and *V. parahaemolyticus*, it seems that no significant change in the affinity for H⁺ had occurred. Although we do not know the reason(s) for the changes in affinities, several possibilities can be considered. (1) A difference in membrane lipids in *V. parahaemolyticus* and *E. coli* cells may affect the activity of the Na⁺/H⁺ antiporter. The composition of lipids in the membrane have been shown to change the properties of transporters.[70] (2) The Vp-*nha*A gene or its protein may be specifically modified in some way in *E. coli* or in *V. parahaemolyticus* cells to account for the differences. (3) Na⁺/H⁺ antiporter activities in the two organisms are regulated by different mechanisms.

C. Na⁺ EXCRETION MACHINERY OTHER THAN THE Na⁺/H⁺ ANTIPORTERS IN *E. COLI*

The high Na⁺/H⁺ antiporter activity in the cytoplasmic membrane due to the Na⁺/H⁺ antiporters implies that as long as they are in the membrane it is very hard, if not impossible, to reach a conclusion about the existence of additional Na⁺ fluxes. Hence, a strain deleted of both antiporters, Δ*nha*AΔ*nha*B due to its Na⁺ sensitivity, is a most suitable system to apply selection for Na⁺ resistance. This technique represents a most powerful tool to clone and express genes conferring Na⁺ resistance. These already include homologous or heterologous antiporters genes (Section B.1), and even *cha*A a Ca²⁺/H⁺ antiporter with apparently low affinity to Na⁺.[41] Most interestingly, since Δ*nha*AΔ*nha*B bears deletion mutations that cannot revert, it also allows the search for suppression mutations that restore resistance and thus unravel novel systems that are not necessarily antiporters.[71]

Suppression mutations in Δ*nha*AΔ*nha*B were found to be very frequent. Several were isolated, and it was found that they confer Na⁺ but not Li⁺ resistance. All map in the same locus which we call MH1.[71]

MH1 does not encode for a Na⁺/H⁺ antiporter since such activity was not detected in its membrane. The mutation also affects neither the K⁺/H⁺ nor the Ca²⁺/H⁺ activity. Remarkably, however, MH1 maintains a Na⁺ gradient of 5-8 (directed inward) in the presence of 50 mM [Na⁺]$_{out}$, as does the wild-type. Furthermore, up to 350 mM [Na⁺]$_{out}$, the gradient of MH1 is only slightly lower than that of the wild-type and only beyond it, decreases. Most interestingly, at 400 mM [Na⁺]$_{out}$ when [Na⁺]$_{in}$ concentration of MH1 reaches 90 mM, as compared with 15 mM in the wild-type, a difference in growth rate between MH1 and wild-type becomes apparent. These results imply that the mutation MH1 exposes a Na⁺ export machinery which at least up to 350 mM [Na⁺]$_{out}$ is similar in its capacity to that of the wild-type.[71]

Δ*unc* strains cannot interconvert phosphate bond energy with the electrochemical proton gradient, allowing conditions to be established in which the sources of energy available for transport are both phosphate bond energy and an electrochemical gradient (during metabolism of glucose), only phosphate bond energy (glucose metabolism in the presence of uncouplers or respiratory inhibitors) or only an electrochemical gradient of protons or redox (during respiration of substrates of the electron transport chain). As previously shown[72] the driving force for Na$^+$ extrusion in wild-type *E. coli* (in our case TA15) is $\Delta\tilde{\mu}_{H^+}$. Thus TA15Δunc maintains a Na$^+$ gradient like TA15 and the uncoupler CCCP collapses the gradient in TA15Δunc.[71] In contrast to TA15Δunc, MH1Δunc maintains the Na$^+$ gradient even in the presence of an uncoupler.[71] It could be argued that MH1Δunc is insensitive to uncouplers. This is highly unlikely since, as described above, the isogenic strain TA15Δunc is uncoupler sensitive. Furthermore, the collapse of the Na$^+$ gradient in MH1 by anaerobiosis is accelerated in the presence of uncouplers. Hence Na$^+$ extrusion in MH1Δunc is not driven by $\Delta\tilde{\mu}_{H^+}$. Since only anaerobiosis in the presence or absence of uncouplers collapsed the gradient, we have suggested that the Na$^+$ extrusion in MH1 can be directly coupled to electron transport, demonstrating the presence of a respiration dependent, $\Delta\tilde{\mu}_{H^+}$ independent, Na$^+$ extrusion mechanism in *E. coli*. This mechanism became evident in MH1, a mutant devoid of both antiporters and carrying a mutation. Hence the triple mutant, MH1, allowed us to describe a mechanism which in the wild-type, if present, is masked by the activity of two antiporters.

MH1 can be a mutation in a Na$^+$ pump itself increasing its activity and/or expression. On the other hand, it is also possible that MH1 affects another system which is needed for expression or which limits the activity of the pump. In this respect, $\Delta nhaA\Delta nhaB$ (EP432) in the presence of high K$^+$, shows a limited Na$^+$ extrusion capacity.[71] It is possible that this low activity can be increased by induction, under conditions unfavorable to the H$^+$ cycle (in the presence of protonophores or at alkaline pH) as suggested before[73-78] or by MH1-like mutations.[71]

ACKNOWLEDGMENTS

The research in the authors' laboratory was supported by grants from the Israel Academy of Sciences to E. Padan and by the Moshe Shilo Center for Biogeochemistry.

REFERENCES

1. Mitchell P. Coupling of phosphorylation to electron and hydrogen transfer by a chemiosmotic type of mechanism. Nature (London) 1961; 191:144-146.
2. Mitchell P, Moyle J. Respiration-driven proton translocation in rat liver mitochondria. Biochem J 1967; 105:1147-1162.
3. Harold F, Papineau D. Cation transport and electrogenesis by *Streptococcus faecalis*. J Membr Biol 1972; 8:45-62.
4. Padan E, Schuldiner S. Molecular physiology of Na$^+$/H$^+$ antiporters, molecular devices that couple the Na$^+$ and H$^+$ circulation in cells. Biochim Biophys Acta 1994; 1185:129-151.

5. Padan E, Schuldiner S. Na$^+$/H$^+$ antiporters, molecular devices that couple the Na$^+$ and H$^+$ circulation in cells. J Bioenerg Biomemb 1993; 25:647-669.
6. Schuldiner S, Padan E. Na$^+$/H$^+$ antiporters in *E. coli*. In: Bakker E, ed. Alkali Cation Transport Systems in Prokaryotes. Boca Raton, Fl: CRC Press, 1992:3-24.
7. Schuldiner S, Padan E. Molecular analysis of the role of Na$^+$/H$^+$ antiporters in bacterial cell physiology. In: Friedlander M, Mueckler M, eds. Molecular Biology of Receptors and Transporters. San Diego: Academic Press, Internation Review of Cytology Edition, 1993; 137C:229-266.
8. Pinner E, Kotler Y, Padan E, et al. Physiological role of NhaB, a specific Na$^+$/H$^+$ antiporter in *Escherichia coli*. J Biol Chem 1993; 268:1729-1734.
9. Padan E, Maisler N, Taglicht D, et al. Deletion of ant in *E. coli* reveals its function in adaptation to high salinity and an alternative Na$^+$/H$^+$ antiporter system(s). J Biol Chem 1989; 264:20297-20302.
10. Rahav-Manor O, Carmel O, Karpel R, et al. NhaR, a protein homologous to a family of bacterial regulatory proteins (LysR) regulates *nha*A, the sodium proton antiporter gene in *Escherichia coli*. J Biol Chem 1992; 267:10433-10438.
11. Goldberg B, Arbel T, Chen J, et al. Characterization of Na$^+$/H$^+$ antiporter gene of E. coli. Proc Natl Acad Sci USA 1987; 84:2615-2619.
12. Niiya S, Yamasaki K, Wilson T, et al. Altered cation coupling to melibiose transport in mutants of *Escherichia coli*. J Biol Chem 1982; 257:8902-8906.
13. Carmel O, Dover N, Rahav-Manor O, et al. A single amino acid substitution (Glu134 > Ala) in NhaR1 increases the inducibility by Na$^+$ of the product of *nha*A, a Na$^+$/H$^+$ antiporter gene in *Escherichia coli*. EMBO J 1994; 13:1981-1989.
14. Taglicht D, Padan E, Schuldiner S. Overproduction and purification of a functional Na$^+$/H$^+$ antiporter coded by *nha*A (*ant*) from *Escherichia coli*. J Biol Chem 1991; 266:11289-11294.
15. Pinner E, Padan E, Schuldiner S. Kinetic properties of NhaB, a Na$^+$/H$^+$ antiporter from *E. coli*. J Biol Chem 1994; 269:26274-26479.
16. Bassilana M, Damiano E, Leblanc G. Kinetic properties of Na$^+$ antiport in *Escherichia coli* membrane vesicles. Effects of imposed electrical potential, proton gradient and internal pH. Biochemistry 1984a; 23:5288-5294.
17. Schuldiner S, Fishkes H. Sodium-proton antiport in isolated membrane vesicles of *Escherichia coli*. Biochemistry 1978; 17:706-710.
18. Castle A, Macnab R, Schulman R. Measurements of intracellular sodium concentration and sodium transport in *E. coli* by 23Na$^+$ NMR. J Biol Chem 1986a; 261: 3288-3294.
19. Pan J, Macnab R. Steady-state measurements of *Escherichia coli* sodium and proton potentials at alkaline pH support the hypothesis of electrogenic antiport. J Biol Chem 1990; 265:9247-9250.
20. Taglicht D, Padan E, Schuldiner S. Proton-sodium stoichiometry of NhaA, an electrogenic antiporter from *Escherichia coli*. J Biol Chem 1993; 268:5382-5387.
21. Kleyman T, Cragoe Jr E. Amiloride and its analogs as tools in the study of ion transport. J Membrane Biol 1988; 105:1-21.

22. Barkla B, Blumwald E. Identification of a 170-kDa protein associated with the vacuolar Na$^+$/H$^+$ antiport of *Beta vulgaris*. Proc Natl Acad Sci USA 1991; 88:11177-11181.
23. Katz A, Kleyman T, Pick U. Utilization of amiloride analogs for characterization and labeling of the plasma membrane Na$^+$/H$^+$ antiporter from *Dunaliella salina*. Biochemistry 1994; 33:2389-2393.
24. Schonheit P, Beimborn D. Presence of a Na$^+$/H$^+$ antiporter in *Methanobacterium thermoautotrophicumat* and its role in Na$^+$ dependent methanogenesis. Arch Microbiol 1985; 142:354-360.
25. Mochizuki-Oda N, Oosawa F. Amiloride sensitive Na$^+$/H$^+$ antiporter in *Escherichia coli*. J Bacteriol 1985; 163:395-397.
26. Leblanc G, Bassilana M, Damiano E. Na$^+$/H$^+$ exchange in bacteria and organelles. In: Grinstein S, ed. CRC Press 1988.
27. Davies K, Solioz M. Assessment of uncoupling amiloride analogs. Biochemistry 1992; 31:8055-8058.
28. Taglicht D. PhD Thesis. Hebrew University 1992.
29. Pinner E, Padan E, Schuldiner S. Amiloride and harmaline are potent inhibitors of NhaB, a Na$^+$/H$^+$ antiporter from *Escherichia coli*. FEBS Letters 1995; 365:18-22.
30. McMorrow I, Shuman H, Sze D, et al. Sodium/proton antiport is required for growth of *Escherichia coli* at alkaline pH. Biochim Biophys Acta 1989; 981: 21-26.
31. Onoda T, Oshima A, Fukunaga N, et al. Effect of Ca^{2+} and K$^+$ on the intracellular pH of an *Escherichia coli* L-form. J Gen Microbiol 1992; 138:1265-1270.
32. Counillon L, Franchi A, Pouysségur J. A point mutation of the Na$^+$/H$^+$ exchanger gene (NHE1) and amplification of the mutated allele confer amiloride resistance upon chronic acidosis. Proc Natl Acad Sci USA 1993; 90:4508-4512.
33. Garlid K, Shariat-Madar Z, Nath S, et al. Reconstitution and partial purification of the Na$^+$-selective Na$^+$/H$^+$ antiporter of beef heart mitochondria. J Biol Chem 1991; 226:6518-6523.
34. Rosen B. Ion extrusion systems in *Escherichia coli*. Methods Enzymol 1986; 125:328-336.
35. Gerchman Y, Olami Y, Rimon A, et al. Histidine 226 is part of the pH sensor of NhaA, a Na$^+$/H$^+$ antiporter in *Escherichia coli*. Proc Natl Acad Sci USA 1993; 90:1212-1216.
35 Rimon A, Gerchman Y, Olami Y, et al. Replacements of histidine 226 (H226) of NhaA-Na$^+$/H$^+$ antiporter of *Escherichia coli*: cysteine (H226C) or serine (H226S) retain both normal activity and pH sensitivity, aspartate (H226D) shifts the pH profile toward basic pH, and alanine (H226A) inactivates the carrier at all pH values. J Biol Chem 1995; in press.
36. Fersht A. In: Fersht A, ed. New York: WH Freeman Co, 1985.
37. Antosiewitcz J, McCammon J, Gilson M. Prediction of pH-dependent properties of proteins. J Mol Biol 1994; 238:415-436.

38. Okamura M, Feher G. Proton transfer in reaction centers from photosynthetic bacteria. Annu Rev Biochem 1992; 61:861-896.
39. Jia Z, McCullough N, Martel R et al. Gene amplification at a locus encoding a putative Na+/H+ antiporter confers sodium and lithium tolerance in fission yeast. Embo J 1992; 11:1631-1640.
40. Pinner E, Padan E, Schuldiner S. Cloning, sequencing and expression of *nha*B gene encoding a Na+/H+ antiporter in *Escherichia coli*. J Biol Chem 1992; 267:11064-11068.
41. Ivey D, Guffanti A, Zemsky J, et al. Cloning and characterization of a putative CA++/H+ antiporter gene from *Escherichia coli* upon functional complementation of Na+/H+ antiporter deficient strains by the over-expressed gene. J Biol Chem 1993; 268:11296-11303.
42. Ivey D, Guffanti A, Bossewitch J, et al. Molecular cloning and sequencing of a gene from alkaliphilic *Bacillus firmus* OF4 that functionally complements an *Escherichia coli* strain carrying a deletion in the *nha*A Na+/H+ antiporter gene. J Biol Chem 1991; 266:23483-23489.
43. Waser M, Hess-Bienz D, Davies K, et al. Cloning and disruption of a putative Na/H antiporter gene of *Enterococcus hirae*. J Biol Chem 1992; 267:5396-5400.
44. Nakamura T, Komano Y, Itaya E, et al. Cloning and sequencing of an Na+/H+ antiporter gene from the marine bacterium *Vibrio alginolyticus*. Biochim Biophys Acta 1994; 1190:465-468.
45. Tokuda H, Unemoto T. Characterization of the respiration-dependent Na+ pump in the marine bacterium *Vibrio alginolyticus*. J Biol Chem 1982; 257:10007-10014.
46. Unemoto T, Hayashi M. Na(+) translocating NADH-quinone reductase of marine and halophilic bacteria. J Bioenerg Biomembr 1993; 25:385-391.
47. Nakamura T, Kawasaki S, Unemoto T. Roles of K+ and Na+ in pH homeostasis and growth of the marine bacterium *Vibrio alginolyticus*. J Gen Microbiol 1992; 138:1271-1276.
48. Atsumi T, McCarter L, Imae Y. Polar and lateral flagellar motors of marine *Vibrio* are driven by different ion-motive forces. Nature 1992; 355:182-184.
49. Ohyama T, Imaizumi R, Igarashi K, et al. *Escherichia coli* is able to grow with negligible sodium ion extrusion activity at alkaline pH. J Bacteriol 1992; 174:7743-7749.
50. Nakamura T, Komano Y, Unemoto T. Three aspartyl residues in membrane-spanning regions of Na+/H+ antiporter from *Vibrio alinolyticus* are essential for electrogenic Na+ extrusion. Biochim Biophys Acta 1995; in press.
51. Kuroda T, Shimamoto T, Inaba K, et al. Properties and sequence of the NhaA Na+/H+ antiporter of *Vibrio parahaemolyticus*. J Biochem 1994; 116:1030-1038.
52. Tsuchiya T, Shinoda S. Respiration-driven Na+ pump and Na+ circulation in *Vibrio parahaemolyticus*. J Bacteriol 1985; 162:794-798.
53. Tse M, Levine S, Yun C, et al. Structure/function studies of the epithelial isoforms of the mammalian Na+/H+ exchanger gene family. J Membr Biol 1993; 135:93-108.

54. Tse M, Levine S, Yun C, et al. The mammalian Na$^+$/H$^+$ exchanger gene family – initial structure function studies. Am Soc Nephrol 1993; 4:969-975.
55. Padan E, Schuldiner S. Na$^+$ transport systems in prokaryotes. In: Bakker E, ed. Alkali Cation Transport Systems in Prokaryotes. Boca Raton, Fl: CRC Press, 1992: 3-24.
56. Lee S-T, Nicholls R, Jong M, et al. Organization and sequence of the human P gene and identification of a new family of transport proteins. Genomics 1995; 28:354-363.
57. Fleischmann R, Adams M, White O, et al. Whole-genome random sequencing and assembly of *Haemophilus influenzae* Rd. Science 1995; 269:496-512.
58. Pinner E, Carmel O, Bercovier H, et al. Cloning, sequencing and expression of the *nha*A and *nha*R genes from *Salmonella entiritidis*. Arch Microbiol 1992; 157: 323-328.
59. Deguchi Y, Yamato I, Anraku Y. Nucleotide sequence of gltS, the Na$^+$/Glutamate symport carrier gene of *Escherichia coli*. J Biol Chem 1990; 265: 21704-21708.
60. King S, Hansen C, Wilson T. The interaction between aspartic acid 237 and lysine 358 in the lactose carrier of *Escherichia coli*. Biochim Biophys Acta 1991; 1062:177-186.
61. Kaback H, Jung K, Jung H, et al. What's new with lactose permease. J Bioenerg Biomemb 1993; 25:627-636.
62. Schwarz S, Cardoso M, Wegener H. Nucleotide sequence and phylogeny of the tet(L) tetracycline resistance determinant encoded by plasmid pSTE1 from *Staphylococcus hyicus*. Antimicrob. Agents Chemother 1992; 36:580-588.
63. Salyers A, Speer B, Shoemaker N. New perspectives in tetracycline resistance. Molec Microbiol 1990; 4:151-156.
64. Cheng J, Guffanti A, Krulwich T. The chromosomal tetracycline-resistance locus of *Bacillus subtilis* encodes a Na$^+$/H$^+$ antiporter that is physiologically important at elevated pH. J Biol Chem 1994; 269:27365-27371.
65. Krulwich T. Na$^+$/H$^+$ antiporters. Biochim Biophys Acta 1983; 726: 245-264.
66. Krulwich T, Guffanti A. The Na$^+$ cycle of extreme alkalophiles: a secondary Na$^+$/H$^+$ antiporter and Na$^+$/solute symporters. J Bioenerg Biomemb 1989; 21:663-677.
67. Kudo T, Hino M, Kitada M, et al. DNA sequences required for the alkalophily of *Bacillus* sp strain C-125 are located close together on its chromosomal DNA. J Bacteriol 1990; 172:7282-7283.
68. Hamamoto T, Hashimoto M, Hino M, et al. Characterization of a gene responsible for the Na$^+$/H$^+$ antiporter system of alkalophilic *Bacillus* species strain C-125. Molec Microbiol 1994; 14:939-946.
69. Strausak D, Waser M, Solioz M. Functional expression of the *Enterococcus hirae* Na/H-antiporter in *Escherichia coli*. J Biol Chem 1993; 268:26334-26337.

70. Van de Vossenberg J, Ublink-Kok T, et al. Ion permeability of the cytoplasmic membrane limits the maximum growth temperature of bacteria and archaea. Molec Microbiol 1995; in press.
71. Harel-Bronstein M, Dibrov P, Olami Y, et al. MH1, a second-site revertant of an *E. coli* mutant, lacking Na$^+$/H$^+$ antiporters (ΔnhaAΔnhaB) regains Na$^+$ resistance and a capacity to excrete Na$^+$ in a $\Delta\bar{\mu}_{H^+}$ independent fashion. J Biol Chem 1995; 270:3816-3822.
72. Borbolla M, Rosen B. Energetics of sodium efflux from *Escherichia coli*. Arch Biochem Biophys 1984; 229:98-103.
73. Avetisyan A, Dibrov P, Skulachev V, et al. The Na$^+$ motive respiration in *Escherichia coli*. FEBS Lett 1989; 254:17-21.
74. Avetisyan A, Bogachev A, Murtasina R, et al. Involvement of a d-type oxidase in the Na($^+$) motive respiratory chain of *Escherichia coli* growing under low delta mu H$^+$ conditions. FEBS Lett 1992; 306:199-202.
75. Avetisyan A, Bogachev A, Murtasina R, et al. ATP-driven Na$^+$ transport and Na$^+$ dependent ATP synthesis in *Escherichia coli* grown at low $\Delta\bar{\mu}_{H^+}$. FEBS Lett 1993; 3:267-270.
76. Skulachev V. Sodium bioenergetics. Trends Biol Sci 1984; 9:483-485.
77. Skulachev V. In: Skulachev V, ed. The sodium world. Berlin: Springer-Verlag, 1988.
78. Dibrov P. The role of sodium ion transport in *Escherichia coli* energetics. Biochim Biophys Acta 1991; 1056:209-224.
79. Schuler GD, Altschul SF, Lipman DJ. A workbench for multiple alignment construction and analysis. PROTEINS: Structure, Function and Genetics 1991; 9:180-190.

CHAPTER 13

SODIUM TOLERANCE AND EXPORT FROM YEAST CELLS

Karen M. Hahnenberger, Zhengping Jia, Larry Fliegel,
Sean Hemmingsen and Paul G. Young

A. INTRODUCTION

Most cells maintain high internal potassium and low internal sodium concentrations, and in most species high internal sodium is toxic. In the face of ubiquitous sodium in the environment, sodium enters cells on various channels and transporters and must be exported. Thus, low internal sodium is achieved directly in animal cells by the Na^+/K^+-ATPase which generates a primary Na^+ gradient ($Na_{internal}$ low, $Na_{external}$ high) satisfying the need to maintain low internal Na^+. This also provides an internally directed sodium gradient necessary for various secondary transport systems. In plant and fungal cells, the H^+-ATPase establishes a primary proton gradient across the plasma membrane ($H_{internal}$ low, $H_{external}$ high) which is then utilized by secondary transport systems. In this case, low internal sodium concentrations must be maintained by transporters specialized for this purpose. The mechanism for sodium export from plant cells is not yet known, however they are adapted for accumulation and sequestration of Na^+ internally in the vacuole.

In the yeasts, two quite distinct sodium export mechanisms have evolved: in fission yeast, *Schizosaccharomyces pombe*[1] and in *Zygosaccharomyces rouxii*[2] a sodium/proton antiporter utilizes the proton gradient to export sodium, and in budding yeast, *Saccharomyces cerevisiae*, a family of P-type Na^+-ATPases actively pump sodium from the cell.[3] At the present time it is not known whether both mechanisms exist in all types of yeast or if these are unique in the alternative systems. This short review addresses the characteristics of these two mechanisms with an emphasis, given the context, on the antiports. We also review the literature showing that the budding yeast Na$^+$-ATPase can be functionally expressed in fission yeast[4] and that the fission yeast Na$^+$/H$^+$ antiport can be functionally expressed in budding yeast.[5] In both cases the heterologous expression conveys Na$^+$ export capacity and sodium tolerance.

B. THE FUNGAL SODIUM/PROTON ANTIPORTER

The survival of a yeast cell in high external sodium concentrations requires two distinct physiological mechanisms: the activation of pathways to raise the internal concentration of various osmolytes, mostly glycerol, through activation of the HOG1 MAP kinase pathway by osmosensing membrane proteins[6-8] and the export of sodium from the cell through antiports or P-type ATPases. In fission yeast, the ability of the cell to grow in high concentrations of external sodium or lithium is strongly pH dependent, i.e., cells can tolerate higher ion concentrations at low pH than at pHs near neutrality.[1,9] This is not true for potassium where the tolerance is similar at all pHs and for which any growth retardation is largely due to the osmotic effect at high concentrations.

The similarity in pH dependency of the response for Na$^+$ and Li$^+$ suggested that the two ions are probably exported on the same carrier. In addition lithium is toxic at much lower concentrations with 50% growth rate inhibition occurring at about 400 mM NaCl and at about 1 mM LiCl at pH 6.0.[1,9] Therefore in order to genetically target sodium export mechanisms in fission yeast Jia et al[9] utilized Li$^+$ as the selection since the high toxicity would allow for selection under conditions which did not necessitate an osmotic adjustment by the cell. This extreme sensitivity to Li$^+$ is not seen in *S. cerevisiae* unless the cells are grown in the absence of glucose suggesting that at least some of the internal targets for lithium toxicity are associated with glucose repressible pathways (Jia and Young, unpublished).

A number of lithium tolerant fission yeast strains were selected. These strains were strongly tolerant to lithium (Fig. 13.1) and to sodium in a pH dependent manner. All of them were at the same locus, named *sod2*, were dominant, and displayed a degree of genetic instability. In addition,^{22}Na uptake and export experiments showed that net uptake was reduced somewhat in the mutants; however, export was greatly enhanced in the *sod2* strains. These characteristics were consistent with a gene amplification event leading to overproduction of the export pump. The gene was subsequently cloned by simple overexpression of a wild-type gene bank in a wild-type background and selecting for Li$^+$ tolerance. The single gene obtained mapped to the *sod2* locus and encoded a highly hydrophobic membrane protein with 12 putative transmembrane domains[9] displaying substantial sequence similarity to the hydrophobic domain of mammalian sodium/proton antiporters especially within the transmembrane domains (Fig. 13.2). A number of motifs are strongly conserved within domains V (DPV), VI (NDG) and XI.

Sodium Tolerance and Export from Yeast Cells

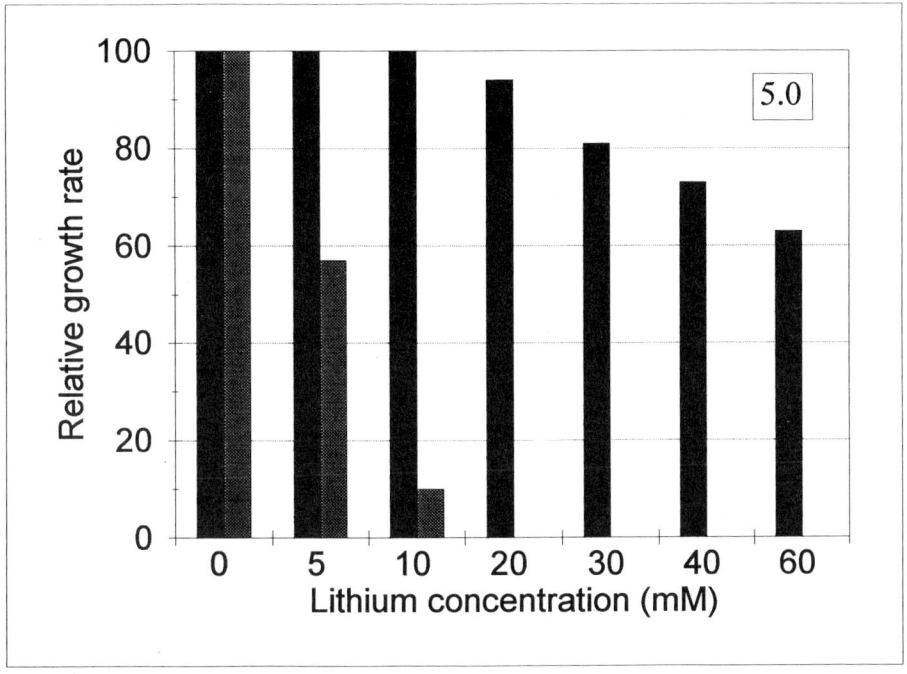

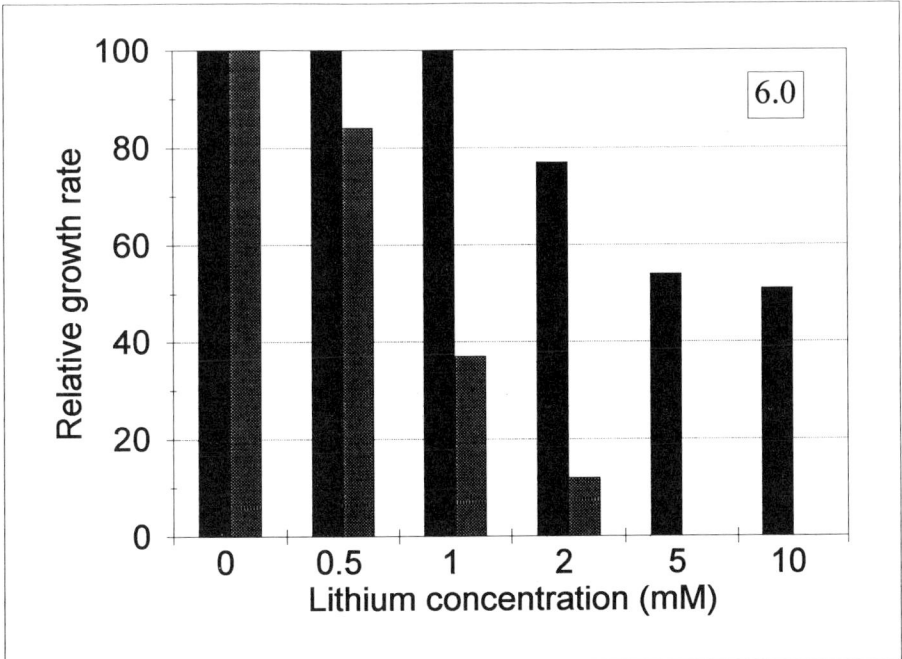

Fig. 13.1. Effect of pH and lithium on the growth of wild-type and sod2-1 fission yeast. Wild-type (black) and sod2-1 (gray) cells were grown on plates at pH 5.0 (top panel), pH 6.0 (middle panel) and pH 6.8 (bottom panel) in the presence of LiCl as indicated. Growth was assessed as colony diameter.[9] (For bottom panel see next page.)

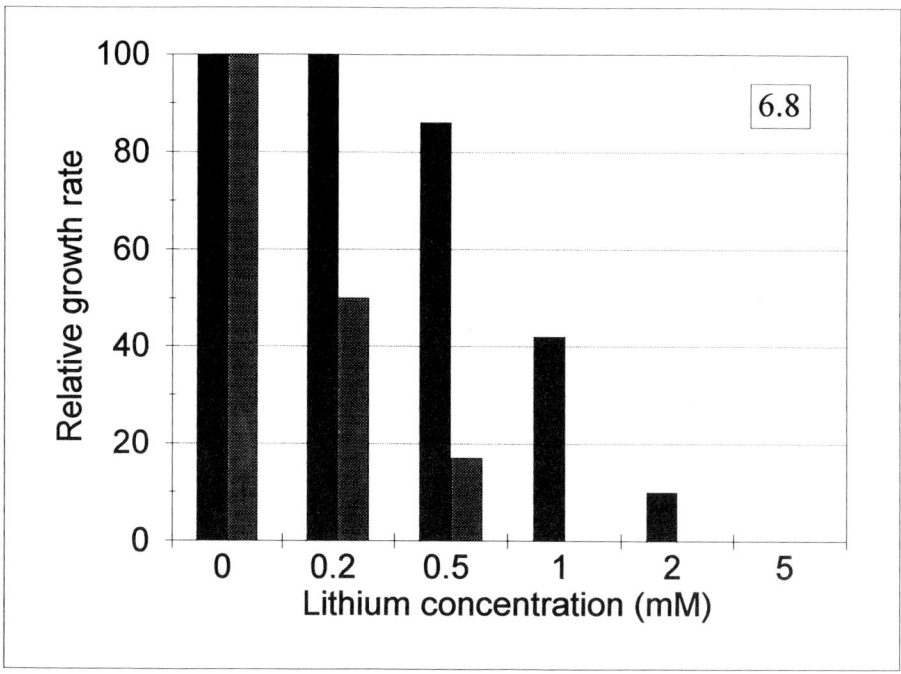

Fig. 13.1 (bottom panel)

The putative amiloride binding site of human antiports (here found in domain III) is poorly if at all conserved in the yeasts.[10]

Analysis showed that fission yeast normally has only one copy of the *sod2* gene but that resistant strains had undergone gene amplification of the *sod2* locus and that this simple amplification event was sufficient to upregulate sodium export and provide sodium and lithium tolerance.[9] Isolation of a cDNA clone from the tolerant yeast showed that the amplified gene was wild-type in sequence. Disruption of the gene (*sod2::ura4*) resulted in a strain that was hypersensitive to Na^+ and Li^+ and which was completely unable to export $^{22}Na^+$ from the cell demonstrating that *sod2* was the only operative sodium export mechanism under the conditions tested. The disruption did not affect osmotolerance. Transport experiments also showed that as sodium was exported from sodium-loaded cells, protons were taken up, demonstrating the antiport function, although the stoichiometry of exchange was not determined. No sensitivity to amiloride could be detected, however carbonyl cyanide *m*-chlorophenylhydrazone (CCCP) or dinitrophenyl (DNP) could block export. The gene was constitutively expressed and was not transcriptionally regulated by external sodium. The *sod2* gene behaved as an electroneutral antiport.

Sodium uptake, as measured by net $^{22}Na^+$ influx in 6 mM external sodium, was not affected in *sod2::ura4* cells, however in the sodium

```
                            I                         II
PSOD2   1   MGWRQLDIDKVHLALIVAGGFITFFCYFSEVFRKKLL--VGEAVLGSITG    46
ZSOD2   1   MVWRQLEVTKAHVAYSCLGIFSSIFSLVSLFVKERLY--IGESMVASVFG    46
NHE4   67   VQIPYEVTLWILLASLAKIGFHLYHRLPHLMPESCLLIIVGALVGSIIFG   116
NHE1  104   VRTPFEISLWILLACLMKIGFHVIPTISSIVPESCLLIVVGLLVGGLIKG   152

                                                III
PSOD2       LIF-GPHAAKLVDPF----SWGDHGDYLTVEICRIVLDVRVFASAIELPG    93
ZSOD2       LIV-GPHCLNWFNPL----SWGN-TDSITLEISRILLCLQVFAVSVELPR    92
NHE4        THHKSPPVMD------------------------SSIYFLYLLPPI      138
NHE1        VGET-PPFL-------------------------QSDVFFLFLLPPI     173

                           IV
PSOD2       AYFQ----------HNFRSIIVMLL-PVMAYGWLVTAGFAYAL-------   125
ZSOD2       KYMQ----------KHWLSVT-MLLVPVMTSGWLVIALFVWIL-------   124
NHE4        VLESGYFMPTRPFFENIGSILWWAGLGALINAFGIGLSLYFICQIKAFG-   187
NHE1        ILDAGYFLPLRQFTENLGTILIFAVVGTLWNAFFLGGLMYAVCLVGGEQ-   222

                       V
PSOD2       FPQINFLGSLLIAGCITSTDPVLSALIVGEGPLAKKTPERIRSLLIAESG   175
ZSOD2       VPGLNFPASLLMGACITATDPVLAQSVVS-GTFAQKVPGHLRNLLSCESG   173
NHE4        LGDINLLQNLLFGSLISAVDPVAVLAVFEEA----RVNEQLYMMIFGEAL   233
NHE1        INNIGLLDNLLFGSIISAVDPVAVLAVFEEI----HINELLHILVFGESL   268

                       VI                       VII
PSOD2       CNDGMAVPFFYFAIKLLTVKPSRNAG-RDWVLLVVLYECAF--GIFFGCV   222
ZSOD2       CNDGLAFPFVFLSIDLLLYPGRGGEIVKDWICVTILWECIF--GSILGCI   221
NHE4        LNDGISVVLYNILIAFTKMHKFEDIEAVDILAGCARFVIVGCGGVFFGII   283
NHE1        LNDAVTVVLYHLFEEFAN---YEHVGIVDIFLGFLSFFVVALGGVLVGVV   315

PSOD2       IGYLLSFILKHAQKYRLIDAI----SYYSLPLAIPLL-CSGIGTIIGVDD   267
ZSOD2       IGYCGRKAIRFAEGKRIIDRE----SFLAFYLILALT-CAGFGSMLGVDD   266
NHE4        FGFISAFITRFTQNISAIEPLIVFMFSYLSYLAAETLYLSGILAITACAV   333
NHE1        YGVIAAFTSRFTSHIRVIEPLFVFLYSYMAYLSAELFHLSGIMALIASGV   365

                    VIII                      IX
PSOD2       LLMSFFAGILFNWNDLFSKNISACSVPAFIDQTFSL---LFFTYYGTIIP   314
ZSOD2       LLVSFFAGTAFAWDGWFATKTHESNVSNVIDVLLNY---AYFVVLGSILP   313
NHE4        TMKKYVEENVSQTSYTTIKY--------FMKMLSSVSETLIFIFMGVSTV   375
NHE1        VMRPYVEANISHKSHTTIKY--------FLKMWSSVSETLIFIFLGVSTV   407

                                           X
PSOD2       WNNFNWSVEGLPVWRLIVFSILTLVCRRLPVVFSVKPLVPDIKTW----K   360
ZSOD2       WKDFNNADIGLDVWRLIILSLVVIFLRRIPAVLLLKPLIPDIKSW----R   359
NHE4        GKNHEWN------WAFVCFTLAFCQIWRAISVFTLFYVSNQFRTFPFSIK   419
NHE1        AGSHHWN------WTFVISTLLFCLIARVLGVLGLTWFINKFRIVKLTPK   451

                             XI
PSOD2       EALFVGHFGPIGVCAVYMAFLAKLLLSPDEIEKSI---YESTTVFSTLNE   407
ZSOD2       EAMFIGHFGPIGVGAVYAAIMSKSQLESHLTDEETPLNYTPGKGSKHWQA   409
NHE4        DQLIIFYSGVRGAGSFSLAFLLPLTLFPRKKLF----------------   452
NHE1        DQFIIAYGGLRGAIAFSLGYLLDKKHFPMCDLF----------------   484

                      XII
PSOD2       --IIWPIISFVILSSIIVHGFSIHVLVIWGKLKSLYLNRKVTKSDSDLEL   455
ZSOD2       MACLWPITCFSIITSVIVHGSSVAVIMLGRYLNTVTLTAAPTSRTASTST   459
NHE4        VTATLVVTYFTVFFQGITIGPLVRYLDVRKTNK-KESINEELHIRLMDHL   501
NHE1        LTAIITVIFFTVFVQGMTIRPLVDLLAVKKKQETKRSINEELHTQFLDHL   534

PSOD2       QVIGVDKSQEDYV                                        468
ZSOD2       KNSWLQSLPPFDKSGRPFSLQRLDKETSPTPGQIDVRTSGMIAAPALGMR   509
NHE4        KAGIEDVCGQWSHYQVRDKFKKFDHRYLRKILIRRNQPKSSIVSLYKKLE   551
NHE1        LTGIEDICGHYGHHHWKDKLNRFNKKYVKKCLIAGERSKEPQLIAFYHKM   583
```

Fig. 13.2. Alignment of antiporter genes from yeast and rat. The sequence of the Schizosaccharomyces pombe sod2 protein[9] with its transmembrane domains indicated by roman numerals and shading is aligned with the sequence of the Zygosaccharomyces rouxii zsod2 protein,[2] the rat NHE1 protein[20] and the rat NHE4 protein.[20] Identity of three or four amino acids at a single position are indicated in bold. The amino termini of NHE1 and NHE4 and the carboxy termini of zsod2, NHE1 and NHE4 are not shown.

resistant strain, *sod2-1,* net uptake was sharply reduced.[9] The rate of efflux by the antiporter in *sod2-1* cells was sufficient to export $^{22}Na^+$ almost as fast as it could leak in by other routes resulting in very low net uptake. When the sod2 antiporter was inhibited with CCCP in the *sod2-1* strain, the net $^{22}Na^+$ influx increased to close to the level seen in wild type. These results demonstrated that sodium uptake into the fission yeast cell is via carriers or channels other than the antiport and that loss of antiport function did not materially affect the process.[9] They also demonstrated just how effectively the net sodium content of the cell could be controlled by this gene.

C. THE *ZSOD2* SODIUM/PROTON ANTIPORTER FROM *ZYGOSACCHAROMYCES ROUXII*

A close homolog of *sod2, zsod2,* has very recently been cloned from the very salt tolerant yeast, *Zygosaccharomyces rouxii.*[2] Rather surprisingly for such distantly related yeasts, the *zsod2* gene was cloned by probing genomic libraries with the *sod2* gene. The *zsod2* predicted protein shows many regions of sequence identity with that of *sod2* (Fig. 13.2). One major difference between zsod2 and sod2 is that the *Zygosaccharomyces* gene includes a carboxy-terminal hydrophilic tail similar to, though not showing sequence identity with, the NHE type mammalian antiporters (COOH terminal regions not included in Fig. 13.2). The functional significance of the carboxy terminus in this case is not known but based on the mammalian case would be presumed to be regulatory.[11] The gene disruption strain is sensitive to sodium and could not grow in the extreme three molar sodium concentrations which *Z. rouxii* is able to tolerate. It could however still tolerate 2 M NaCl, a remarkable achievement for any cell. This suggests either that other mechanisms operate (such as Na^+-ATPases) or that other copies of the *zsod2* gene exist in the cell. To this end the authors note a second genomic hybridization signal to a *sod2* probe, although the nature of it is unknown. Given the results in fission yeast, duplication (or amplification) of this locus would seem to be something which would be selected for in such an extreme halophilic yeast. As in fission yeast, the gene is constitutively expressed and not transcriptionally regulated by external sodium. It isn't yet known whether *zsod2* can functionally complement *sod2;* the high degree of conservation makes it very likely that it would.

D. USE OF THE CYTOSENSOR MICROPHYSIOMETER TO DEMONSTRATE BIDIRECTIONAL EXCHANGE ON *SOD2*

In the initial transport experiments with fission yeast, we were not able to demonstrate bidirectional sodium/proton exchange through the antiporter. Very recently using more sensitive techniques we have shown that the yeast antiporter is capable of exchanging sodium in both the inward and outward direction.[5] The Cytosensor microphysiometer (Molecular Devices Corp., Sunnyvale, CA) allows one to immobilize living cells in a three microliter flow chamber. The instrument then measures extracellular acidification rates using a pH sensitive silicon sensor.[12,13] The cells are maintained under constant osmotic potential and the buffer switched to include external NaCl or not (using tetramethyammonium as the alternative cation). Wild type and *sod2-1* cells respond to the switch from 0 to 150 mM external Na$^+$ by extruding protons into the medium (Fig. 13.3A). As the movement of Na$^+$ into the cells slows, the internal concentration rises, the acidification rate decreases until after a few minutes an equilibrium is restored and there is no net proton flux. This transient response shows the direct coupling of proton flux to that of sodium. When the NaCl containing buffer is switched to solution without Na$^+$ the response is in the opposite direction; the cells take up protons as the previously accumulated Na$^+$ is exported from the cell via the antiporter. These responses are abolished in the *sod2::ura4* gene disruption strain and can be reconstituted by expressing the *sod2* gene on a plasmid in the *sod2::ura4* gene disruption strain. When tested with amiloride and several more potent derivatives of the drug this response was unaffected. This is in keeping with earlier work showing that the yeast antiporter is not sensitive to amiloride,[9,5] and it is perhaps not surprising since the putative amiloride binding site is not well conserved (Fig. 13.2).[10]

E. *PMR2/ENA1* P-TYPE SODIUM EXPORT PUMPS

In *Saccharomyces cerevisiae*, a clustered family of genes encoding P-type Na$^+$-ATPases are responsible for sodium and lithium efflux (*PMR2/ENA1*[3,14-17]). These ATPases directly export Na$^+$ and Li$^+$ from the cell. Disruption of the sodium pumps in *S. cerevisiae* results in a phenotype very similar to disruption of *sod2* in *S. pombe*; the cells become hypersensitive to Na$^+$ and Li$^+$ and cannot export these ions. Overexpression of the genes causes an increase in Na$^+$ and Li$^+$ export and tolerance. There is evidence of functional differentiation amongst the members of the gene family with the PMR2A gene being a more effective Na$^+$ exporter and the PMR2B gene being more effective exporting Li$^+$.[16] This is presumably reflected in the different binding affinities of these ions for the two proteins.

In contrast to the fission yeast antiport system there is evidence that the PMR2 export pumps are transcriptionally regulated by Na$^+$ or high pH and that this induction depends upon the calcineurin phosphatase.[18,19] Calcineurin gene disruption strains are very sodium and lithium sensitive. Very recently, results utilizing both calmodulin and calcineurin mutants have been presented showing that calmodulin can affect sodium tolerance by mechanisms which are calcineurin independent and which appear to operate post-transcriptionally.[16]

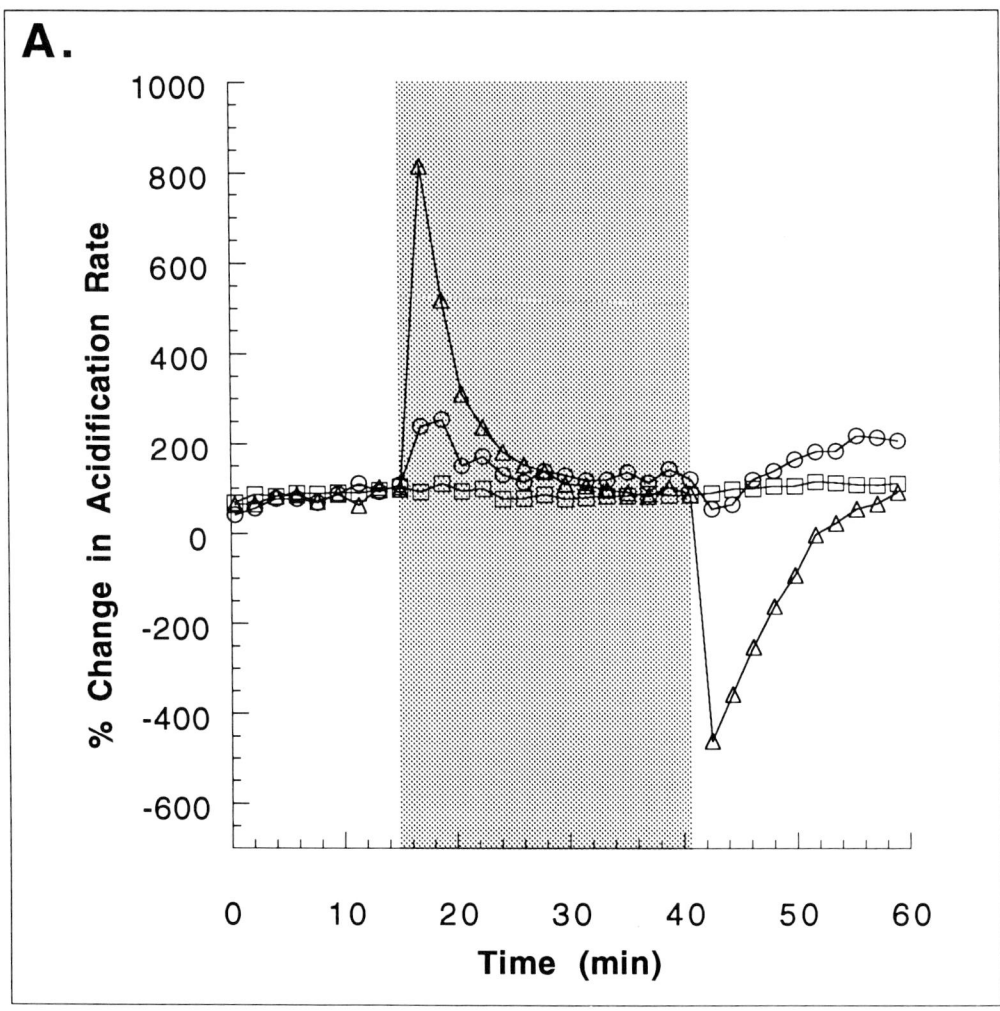

Fig. 13.3A. Cytosensor microphysiometer analysis of antiport activity. Various mutants of Schizosaccharomyces pombe were analyzed for proton extrusion or uptake following shifts from media containing 150 mM tetramethylammonium (nontransportable cation) to 150 mM NaCl (shaded time period) and back to tetramethylammonium. The y-axis indicates the net rate of proton extrusion (positive values) or uptake (negative values). wild-type (circles), sod2::ura4 (squares), sod2-1 (triangles).

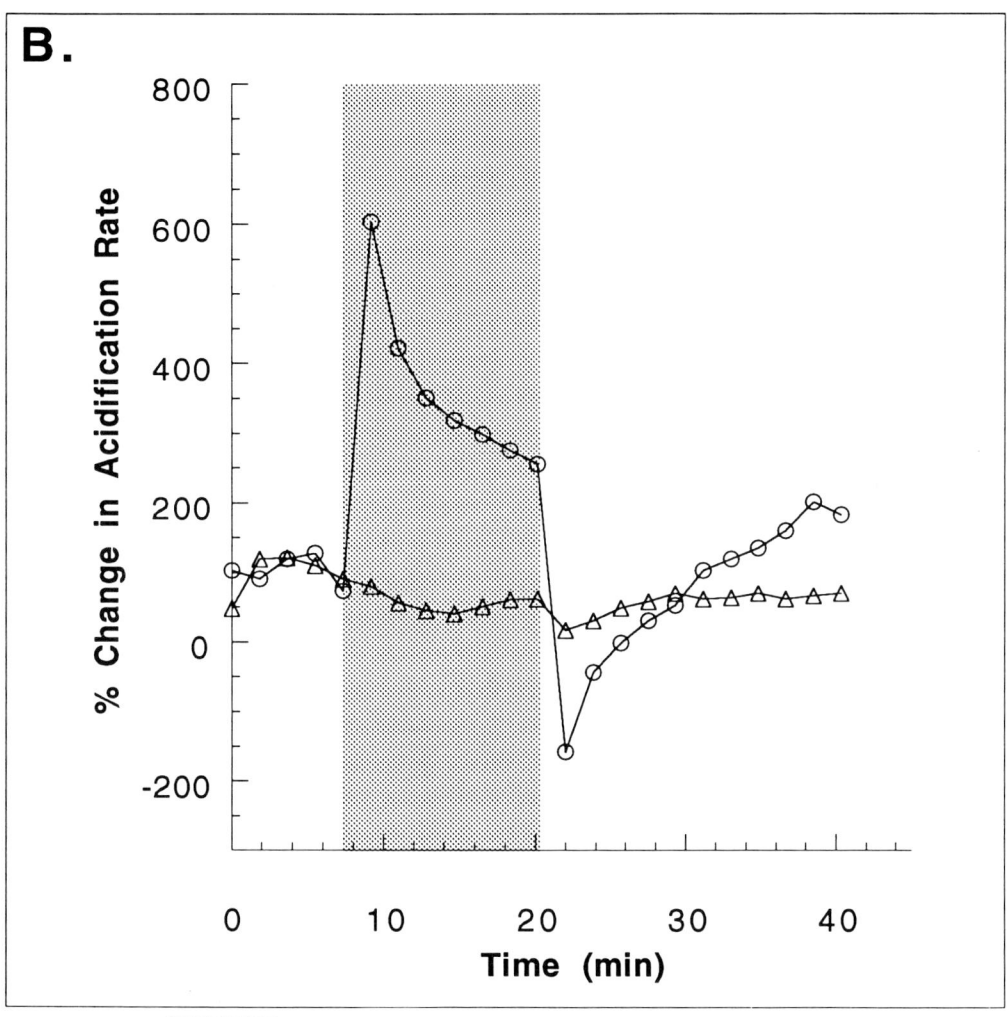

Fig. 13.3B. Cytosensor microphysiometer analysis of antiport activity. Saccharomyces cerevisiae RH16.6 (ena1::leu2::ena4) cells with a deletion of the PMR2/ENA1 gene cluster carrying either an empty plasmid (triangles) or a plasmid expressing the S. pombe sod2 sodium/proton antiporter gene behind a PGK1 promoter (circles) were subjected to a shift from 300 mM tetramethylammonium containing buffer to 300 mM NaCl (shaded time period) and back to tetramethylammonium.[5]

F. HETEROLOGOUS EXPRESSION OF THE *SOD2* SODIUM/PROTON ANTIPORTER

Given the two different systems operative in these two groups of yeasts the question arises of whether expression of a P-type ATPase in fission yeast or the expression of a sodium/proton antiport in budding yeast is capable of rescuing the sodium/lithium sensitivity seen in the various disruption strains. This has now been done in both directions, and the answer is yes.

Banuelos et al[4] successfully expressed the *PMR2 (ENA1)* ATPase behind the PGK1 promoter in the *sod2::ura4* strain of *S. pombe*. This strain recovered sodium and lithium export capacity and tolerance in a pH independent fashion and even allowed the transformed fission yeast cells to grow at pHs slightly above 7.0, a pH not tolerated by wild type fission yeast.

The reciprocal experiment, expression of the *sod2* antiport gene in a *PMR2* gene deletion strain of *S. cerevisiae* has also been done.[5] Expression was also on the PGK1 promoter. The functional expression of the antiport was demonstrated directly by activity measurements using microphysiometry. Upon exposure to 300 mM NaCl, *S. cerevisiae ena1::leu2::ura4* cells expressing the *sod2* gene displayed the characteristic increase in the rate of proton export while withdrawal of the sodium was accompanied by a transient alkalinization of the external environment as seen for antiport function in fission yeast cells (Fig. 13.3B). As a control, cells transformed with the empty vector showed no change in proton flux in response to the external concentration of sodium. As would be expected, the presence of the functional antiporter conferred growth tolerance to high external sodium and lithium to the *ena1::leu2::ena4* strain which lacked the PMR2/ENA1 gene cluster. No effects on osmotolerance were found.

G. CONCLUDING REMARKS

Within the fungi, evolution has clearly provided two routes for dealing with the problem of internal sodium or lithium. Both the P-type ATPases found in the budding yeast and the antiporter found in fission yeast are sufficient to export these ions from the cell. The unanswered question at the moment is whether *S. cerevisiae* harbors a cryptic antiporter and whether fission yeast has cryptic P-type ATPase genes for this purpose. Recently an apparent sod2 homolog has been identified within the budding yeast genome sequencing project (Genpept SCIIRA-18). No functional characterization is available. It is already clear however that even if such genes are present their regulation is such that they cannot normally be expressed/function in such a way that they can compensate for the loss of function of the primary Na^+ export system in either case.

ACKNOWLEDGMENTS

This work was supported by grants from ARPA, U.S.A. to K. Hahnenberger and in Canada from NSERC to L. Fliegel, from NSERC to S. Hemmingsen and from NSERC and NCIC to P. G. Young.

REFERENCES

1. Jia Z. Molecular genetic analysis of sodium transport in fission yeast, *Schizosaccharomyces pombe*. Ph.D. Thesis, Queen's University, Canada 1991.
2. Watanabe Y, Miwa S, Tamai Y. Characterization of Na^+/H^+-antiporter gene closely related to salt-tolerance of the yeast *Zygosaccharomyces rouxii*. Yeast. 1995; 11:829-838.
3. Haro R, Garciadeblas B, Rodriquez-Navarro A. A novel P-type ATPase from yeast involved in sodium transport. FEBS Lett. 1991; 291:189-191.
4. Banuelos MA, Quintero FJ, Rodriquez-Navarro J. Functional expression of the ENA1(PMR2)-ATPase of *Saccharomyces cerevisiae* in *Schizosaccharomyces pombe*. Biochim Biophys Acta 1995; 1229:233-238.
5. Hahnenberger KM, Jia Z, Young PG. Functional expression of the *Schizosaccharomyces pombe* Na^+/H^+ antiporter gene, *sod2*, in *Saccharomyces cerevisiae*. Proc Natl Acad Sci USA 1996; in press.
6. Brewster JL, de Valoir T, Dwyer ND, et al. An osmosensing signal transduction pathway in yeast. Science 1993; 259:1760-1763.
7. Millar JBA, Buck V, Wilkinson MG. Pyp1 and Pyp2 PTPases dephosphorylate an osmosensing MAP kinase controlling cell size division in fission yeast. Genes Dev 1995; 9:2117-2130.
8. Maeda T, Takekawa M, Saito H. Activation of yeast PBS2 MAPKK by MAPKKKs or by binding of an SH3-containing osmosensor. Science 1995; 269:554-558.
9. Jia Z, McCullough N, Martel R, et al. Gene amplification at a locus encoding a putative Na+/H+ antiporter confers sodium and lithium tolerance in fission yeast, EMBO Journal 1992; 11:1631-1640.
10. Counillon L, Franchi A, Pouysségur J. A point mutation of the Na^+/H^+ exchanger gene (NHE1) and amplification of the mutated allele confer amiloride resistance upon chronic acidosis. Proc Natl Acad Sci USA 1993; 90:4508-4512.
11. Sardet C, Counillon L, Franchi A, et al. Growth factors induce phosphorylation of Na^+/H^+-antiporter, a glycoprotein of 110 Kda. Science 1990; 247:723-726.
12. Hafeman DG, Parce JW, McConnell HM. Light-addressable potentiometric sensor for biochemical systems. Science 1988; 240:1182-1185.
13. McConnell HM, Owicki JC, Parce JW, et al. The cytosensor microphysiometer: biological applications of silicon technology. Science 1992; 257:1906-1912.
14. Rudolf HK, Antebi A, Fink GR, et al. The yeast secretory pathway is perturbed by mutations in PMR1, a member of a Ca^{2+} ATPase family. Cell 1989; 58:133-145.

15. Garciadeblas B, Rubio F, Quintero FJ, et al. Differential expression of two genes encoding isoforms of the ATPase involved in sodium efflux in *Saccharomyces cerevisiae.* Mol Gen Genet 1993; 236:363-368.
16. Wieland J, Nitsche AM, Strayle J, et al. The PMR2 gene cluster encodes functionally distinct isoforms of a putative Na+ pump in the yeast plasma membrane. EMBO J. 1995; 14:3870-3882.
17. Rodriquez-Navarro A, Quintero FJ, Garciadeblas B. Na^+-ATPase and Na^+/H^+ antiporters in fungi. Biochim Biophys Acta 1994; 1187:203-205.
18. Nakamura T, Liu Y, Hirata D, et al. Protein phosphatase type 2B (calcineurin-mediated, FK506-sensitive) regulation of intracellular ions in yeast is an important determinant for adaptation to high salt stress conditions. EMBO J 1993; 12:4063-4071.
19. Mendoza I, Rubio F, Rodriguez-Navarro A, et al. The protein phosphatase calcineurin is essential for NaCl tolerance of *Saccharomyces cerevisiae.* J Biol Chem 1994; 269:8792-8796.
20. Orlowski J, Kandasamy RA, Shull GE. Molecular cloning of putative members of the Na/H exchanger gene family. cDNA cloning, deduced amino acid sequence, and mRNA tissue expression of the rat Na/H exchanger NHE1 and two structurally related proteins. J Biol Chem 1992; 267:9331-9339.

PART IV

CHAPTER 14

THERAPEUTIC POTENTIAL OF INHIBITORS OF Na^+/H^+ EXCHANGE ACTIVITY IN TUMOR SELECTIVE THERAPY

Motoyuki Yamagata and Ian F. Tannock

A. INTRODUCTION

Since Na^+/H^+ exchange plays an important role in regulation of intracellular pH (pH_i) and is essential for homeostasis of cells, inhibitors of the exchanger might influence the viability and/or proliferation of both normal and tumor cells. Many solid tumors have a microenvironment that is acidic compared to that of most normal tissues. Therefore the role of the Na^+/H^+ antiport in maintaining pH_i in tumor cells might be more critical than in normal tissue. We describe here the possible use of inhibitors of Na^+/H^+ exchange as therapeutic agents in tumor-selective therapy.

B. THE ACIDIC MICROENVIRONMENT OF TUMORS

The development of effective therapy for malignant disease has been inhibited by the lack of consistent biological or biochemical differences between tumors and normal tissues. Thus, it has been difficult to develop therapeutic strategies which have major toxic effects against tumors, without causing damage to normal cells. One major difference between many solid tumors and surrounding normal tissues is the nutritional and metabolic environment.[1] The functional vasculature of tumors if often inadequate to supply the nutritional needs of the expanding population of tumor cells, leading to deficiency of oxygen and many other nutrients.[2] Under these conditions, acidic products

The Na^+/H^+ Exchanger, edited by Larry Fliegel. © 1996 R.G. Landes Company.

of metabolism, such as lactic acid produced by aerobic and anaerobic glycolysis,[3,4] carbonic acid and others[5] accumulate because of poor clearance from tumors. These acids contribute to the acidic microenvironment which has been found in many types of solid tumors.

Most estimates of pH in tissue have been obtained by insertion of pH electrodes.[3,6] Measurements made by electrodes are presumed to reflect predominantly the extracellular pH (pH_e). Microelectrode measurements of pH_e in human and animal tumors reviewed by Wike-Hooley et al[3] demonstrated that, in general, tumors are more acidic than normal tissues with median pH_e values of about 7.0 in tumors and about 7.5 in normal tissues. There is considerable heterogeneity in the distribution of pH_e, both within and between tumors, but recent work suggests that lower values of pHe are found distant from tumor blood vessels.[7]

^{31}P-nuclear magnetic resonance (NMR) spectroscopy has been adapted to measure tissue pH. The method is based on the pH-dependent shift of the resonance frequency of phosphate.[8] Since phosphate is largely intracellular, the method leads to an estimate of pH_i. Although pH_e in solid tumors tends to be acidic, pH_i measured by ^{31}P-NMR-spectroscopy is usually found to have similar values in solid tumors and normal tissues with the possible exception of brain tumors which appear to be more alkaline than normal brain tissue.[4,9] These results indicate that tumor cells are exposed frequently to an acidic environment and that the cells have active mechanisms which regulate their pH_i to physiological levels.

The deficiency of nutrients and the presence of an acidic microenvironment might contribute to cell death and necrosis within tumors. The microenvironment in tumors may lead also to resistance of cells to radiation and to conventional chemotherapy because of hypoxia (hypoxic cells are resistant to radiation), because of effects to decrease the rate of cell proliferation (most anticancer drugs are more active against proliferating cells) and because of limited drug access. Therefore new therapies which are active under acidic and hypoxic conditions might have considerable potential to improve tumor therapy. The difference in pH_e between tumors and normal tissues provides an opportunity for tumor-selective therapy through the development of drugs which have increased cell toxicity at low pH_e.[1] In the present chapter we review the evidence that inhibitors of Na$^+$/H$^+$ exchanger might contribute to tumor-selective therapy.

C. REGULATION OF INTRACELLULAR pH (pH_i)

Major components to regulation of pH_i in mammalian cells under acidic conditions include buffering capacity and two membrane-based ion exchangers: the Na$^+$/H$^+$ exchanger and the Na$^+$-dependent HCO$_3^-$/Cl$^-$ exchanger (Fig. 14.1).[10,11] The Na$^+$/H$^+$ exchanger is a plasma membrane-associated transporter found in most animal cells which catalyzes

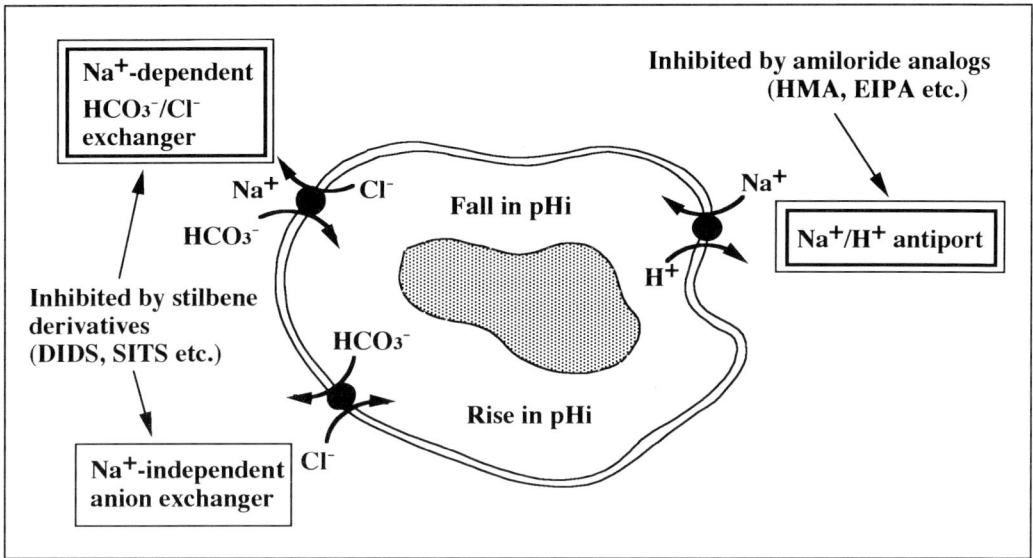

Fig. 14.1. Schematic diagram illustrating the ion exchange mechanisms that are involved in regulation of intracellular pH (pH_i).

electroneutral transport of Na^+ and H^+ across the cell membrane with 1:1 stoichiometry.[10,12,13] The exchanger is inhibited by amiloride and its substituted analogs such as 5-(N-ethyl-N-isopropyl) amiloride (EIPA) or 5-(N,N-hexamethylene) amiloride (HMA).[14] Recent genetic studies have revealed that there are at least four isoforms of the Na^+/H^+ exchanger, named NHE1 to NHE4, which have been cloned and characterized.[15-19] The NHE1 transcript is expressed ubiquitously, but relative tissue specificity exists among other isoforms. NHE3 and 4 are expressed preferentially in the gastrointestinal tract, whereas NHE2 is expressed in ileum and kidney.[16]

There are two known bicarbonate-based anion exchangers in the membranes of most mammalian cells. The Na^+-dependent HCO_3^-/Cl^- exchanger is active under acidic conditions; the Na^+ gradient from outside to inside the cell is believed to drive HCO_3^- into the cell, where it can buffer H^+, while Cl^- leaves the cell against its concentration gradient.[11] The stoichiometry of the exchanger is not known with certainty, and its gene(s) has not been characterized. The Na^+-independent HCO_3^-/Cl^- exchanger operates under alkaline conditions, and is probably not important in tumors. Both exchangers are inhibited by stilbene derivatives such as 4,4-diisothiocyanstilbene 2,2-disulfonic acid (DIDS).

A method to study the regulation of pHi by these exchangers involves intracellular acidification by artificial means followed by the observation of mechanisms which attempt to restore pHi to the physiological range. The preferred method for measuring pHi uses a pH-sensitive fluorescent dye such as 2',7'-bis-(2-carboxyethyl)-5-(and

6)-carboxyfluorescein acetoxymethyl ester (BCECF-AM).[20] The uncharged BCECF-AM diffuses into the cell where nonspecific esterases cleave the ester groups, leaving the fluorescent charged BCECF molecule. There is a linear relationship between the fluorescence intensity of BCECF and pH_i within the range of 6.5-7.5 at excitation and emission wavelengths of 495 nm and 525 nm, respectively. Fluorescence intensity at the excitation wavelength of 440 nm and the same emission wavelength is independent of pH_i, and depends only on the amount of BCECF present.[21] The ratio of fluorescence at pH-dependent and independent wavelengths provides therefore an estimate of pHi that is independent of the amount of BCECF or the number of cells present. Detection of fluorescence by fluorometry or flow cytometry allows estimation of pH_i with a very rapid response time.

To assess the activity of exchangers cells are exposed to BCECF-AM in medium not containing Na^+ or HCO_3^- (e.g., N-methyl glucamine). Cells are acidified by using nigericin which allows entry of protons in exchange for K^+ which leaves the cell down its chemical gradient, or by the ammonium prepulse technique.[22,23] In the latter method, cells are exposed to NH_4Cl which causes initial cellular alkalinization. Removal of external NH_4Cl then leads to rapid escape of NH_3 from the cytoplasm which causes a decrease of pH_i due to the reaction $NH_4^+ \rightarrow NH_3 + H^+$. Fluorescence traces from cells loaded with BCECF-AM are shown in Figure 14.2. The activity of the Na^+/H^+ exchanger and the Na^+-dependent HCO_3^-/Cl^- exchanger can be estimated from the rate of change of pHi following addition of NaCl or $NaHCO_3$, respectively, in the presence or absence of their inhibitors (Fig. 14.2).

Although both the Na^+/H^+ exchanger and the Na^+-dependent HCO_3^-/Cl^- exchanger contribute to the regulation of pH_i of tumor cells, the activity of the exchangers might be influenced by microenvironmental conditions which exist within tumors. Studies from our laboratory have quantitated the relative importance of the exchangers in human bladder carcinoma cells (MGH-U1) and experimental murine tumor cells (EMT-6) under different conditions.[24] The activity of the exchangers depends on pH_e and on the pH gradient (i.e., pH_e-pH_i) across the cell membrane. The relative activity of the exchangers as a function of pH gradient, when the pH_e is maintained at 7.4, is shown in Figure 14.3. Moreover our group has demonstrated that the Na^+/H^+ exchanger is the major mechanism for regulation of pH_i at values of pH_e much lower than the physiological level, while the importance of the Na^+-dependent HCO_3^-/Cl^- exchanger is greater at values of pH_e closer to 7.4.

Boyer et al[25] have shown also that cultured cells exposed for several hours to low pH_e develop increased activity of both Na^+/H^+ and Na^+-dependent HCO_3^-/Cl^- exchangers. They also studied exchanger activity in multicellular tumor spheroids and in murine tumors by using a fluorescent probe (Hoechst 33342) to separate cells in different microenvironments. Nutrient-deprived (presumably acidic) cells from the

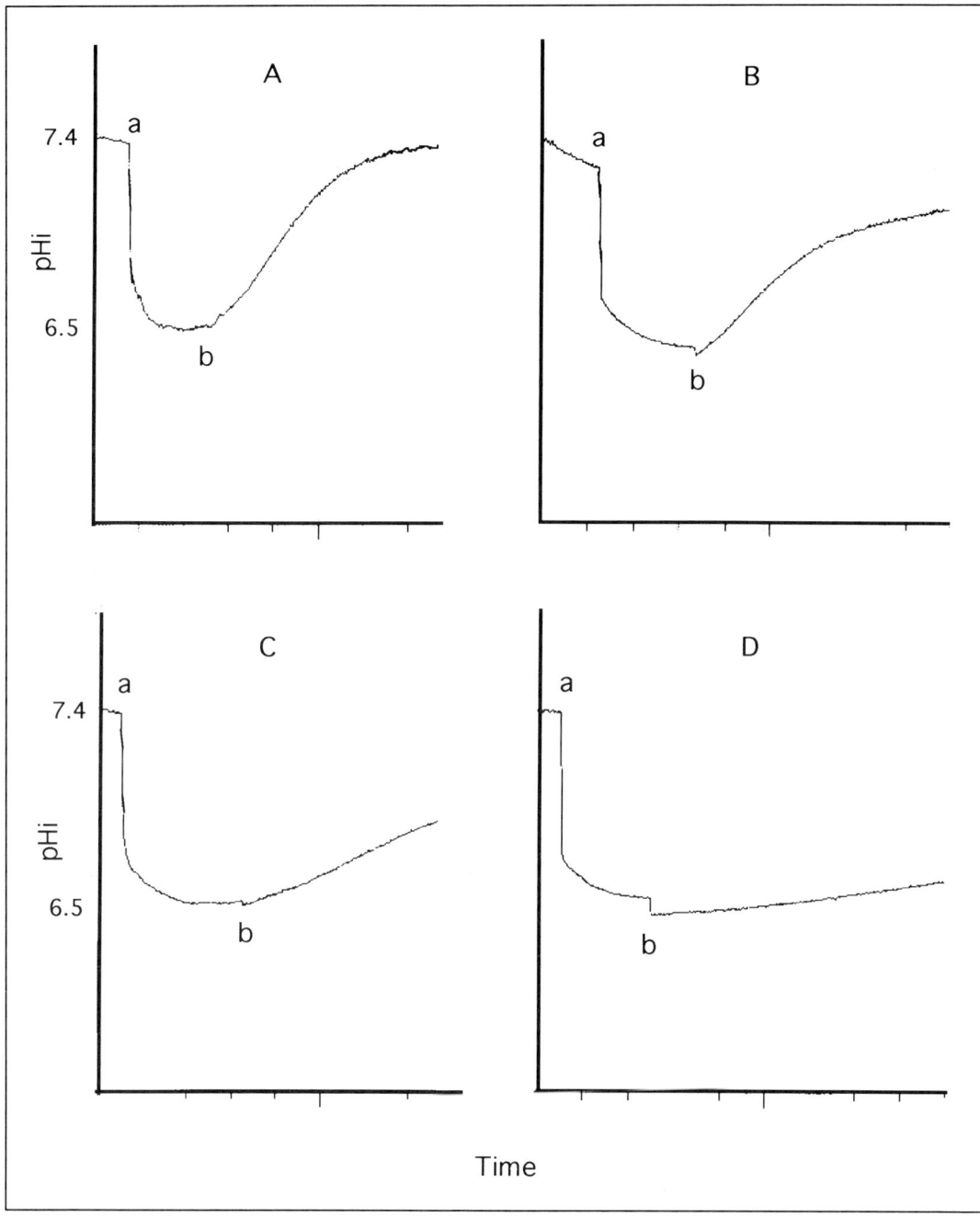

Fig. 14.2. Fluorescence traces from cells loaded with BCECF-AM after acidification by the ammonium prepulse technique (detail is described in the text). At point **a**, NH_4Cl solution was replaced by N-methyl glucamine solution. **A**: $NaHCO_3$ is added at point **b** and both exchangers contribute to the recovery of pH_i. **B**: NaCl is added at point **b** and only the Na^+/H^+ exchanger contributes to the recovery of pH_i. **C**: $NaHCO_3$ is added at point **b** with EIPA, an inhibitor of the Na^+/H^+ exchanger. Only the Na^+-dependent HCO_3^-/Cl^- exchanger contributes to the recovery of pHi. **D**: $NaHCO_3$ is added at point **b** together with EIPA and DIDS. Neither exchanger can contribute to the recovery of pH_i.

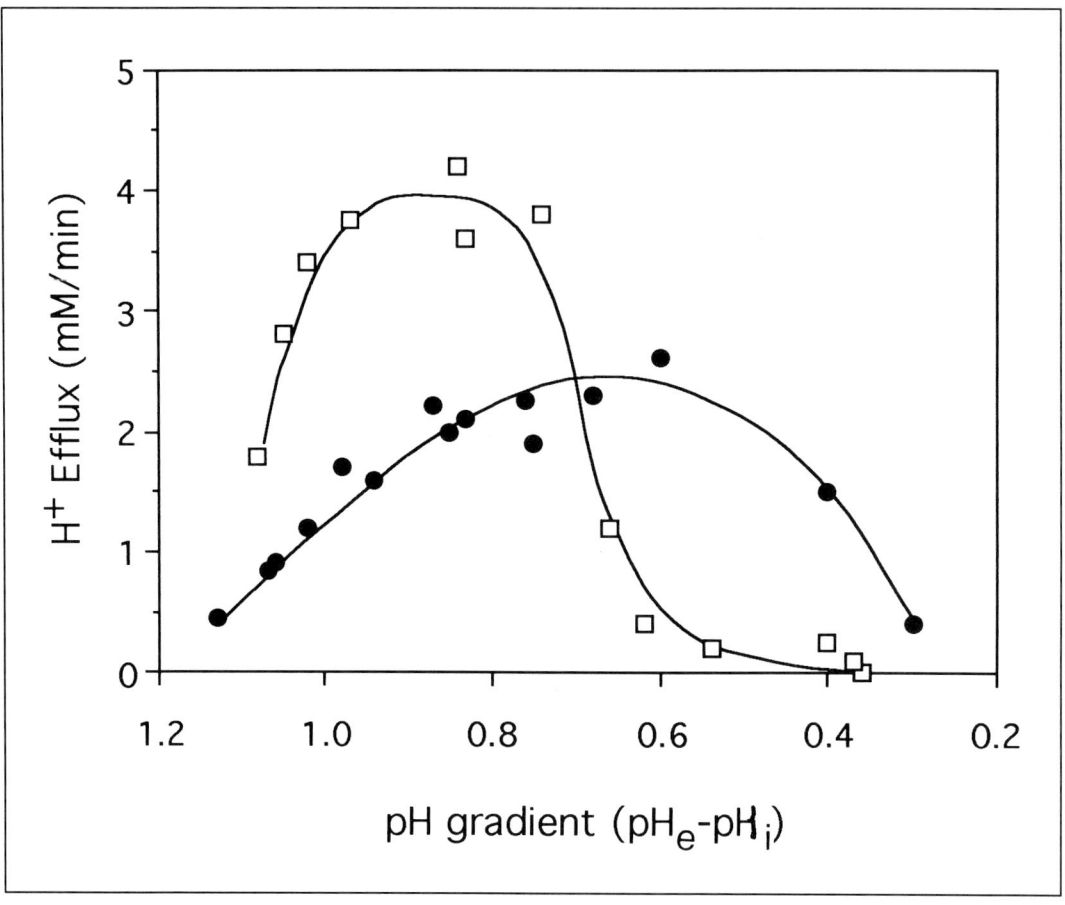

Fig. 14.3. Rate of H⁺ efflux due to the activity of the exchangers as a function of pH gradient in MGH-U1 cells. pH_e was maintained at 7.4. ☐: Na⁺/H⁺ exchanger, ●: Na⁺-dependent HCO_3^-/Cl^- exchanger, operating independently. (Reproduced with permission from Boyer, Tannock, Cancer Res 1992; 52:4441-4447.)

centers of EMT-6 and MGH-U1 spheroids which were grown in medium at pHe 6.6 had increased activity of the Na⁺/H⁺ exchanger as compared to well nourished cells from the periphery.[25] This observation is consistent with the hypothesis that chronic exposure to acidic conditions leads to upregulation of the exchangers which contribute to regulation of pH_i. However, we were unable to demonstrate significant differences in the activity of the Na⁺/H⁺ exchanger in cells from different subpopulations of EMT-6 tumors or from MGH-U1 xenografts in nude mice.

D. ROLE OF THE Na⁺/H⁺ EXCHANGER IN PROLIFERATION AND SURVIVAL OF TUMOR CELLS

Numerous studies have presented evidence that activation of the Na⁺/H⁺ exchanger and/or the consequent cytoplasmic alkalinization can be important precursors of cell proliferation. Grinstein et al[10] reviewed these studies and summarized that: (1) Many mitogens and comitogens, including growth factors, and the product of the *ras* oncogene, cause rapid activation of the Na^+/H^+ exchanger in the absence of bicarbonate in the medium, resulting in an elevation of pH_i; (2) Amiloride and its analogs inhibit growth factor-induced DNA synthesis in several mammalian cell types; and (3) DNA synthesis and cell proliferation can be induced by cytoplasmic alkalinization in the absence of mitogens in some cells. In contrast to the above evidence, many studies have shown that activation of the exchanger and elevation of pH_i are probably not sufficient and may not even be necessary for proliferation of many types of cell. Thus, Ganz et al[26] have shown that arginine vasopressin stimulates both Na^+-dependent and -independent HCO_3^-/Cl^- exchangers in the presence of bicarbonate, as well as the Na^+/H^+ exchanger, leading to a net decrease of pHi in growth-stimulated renal mesangial cells. Also variant cells lacking Na^+/H^+ exchange activity are able to proliferate if the pH_e is within a permissive range.[27,28] It appears that the Na^+/H^+ exchanger may be essential for cell proliferation only under restricted conditions such as an acidic pHe where bicarbonate levels are low and/or the Na^+-dependent HCO_3^-/Cl^- exchanger is not active.

Studies from our laboratory have provided evidence that the Na^+/H^+ exchanger is important for growth of some types of tumor.[29] Na^+/H^+ exchange-deficient variant cells (HSPD) were established from MGH-U1 cells by using the protein suicide selection technique developed by Pouysségur et al.[27] Subcutaneous inoculation of HSPD cells into immunodeficient mice led to either complete absence or marked retardation of tumor growth as compared to parental MGH-U1 cells. Loss of tumorigenicity was not due to pretreatment with a mutagen but to the loss of activity of the Na^+/H^+ exchanger, because (1) cells which survived the same selection procedure but which preserved Na^+/H^+ exchange activity retained their tumorigenic capacity; and (2) revertant cells which regained Na^+/H^+ exchange function also regained the ability to generate tumors. Also tumors that grew slowly from variant cells contained revertant cells.

The above evidence suggests that the Na^+/H^+ exchanger may play an important role in cell proliferation and tumor growth at least under restricted conditions. The requirement for Na^+/H^+ exchange activity might be due to mediation of the activity or some growth factors, or (more likely) because of dependence of regulation of pH_i by tumor cells within the acidic microenvironment that develops in many solid tumors. Inhibitors of the membrane-based exchangers which regulate pHi might have potential to inhibit cell proliferation and/or tumor growth, and hence to contribute to anticancer therapy.

E. POTENTIAL THERAPEUTIC APPLICATION OF INHIBITORS OF THE Na^+/H^+ EXCHANGER

Amiloride, a potassium-sparing diuretic, is known to inhibit the activity of the Na^+/H^+ exchanger by binding to the antiport at a site on the extracellular side of the membrane.[10] Kleyman and Cragoe[14] showed that analogs of amiloride such as EIPA or HMA were more potent and specific inhibitors of Na$^+$/H$^+$ exchange activity than amiloride. In our laboratory, Maidorn et al[30] have compared the ability of amiloride and three analogs to inhibit Na$^+$/H$^+$ exchange in tumor cells (Fig. 14.4). Each of the analogs was more potent than amiloride, and EIPA was found to be a potent and almost complete inhibitor of Na$^+$/H$^+$ exchange activity.

E.1 SHORT-TERM EFFECTS OF THE INHIBITORS IN VITRO

We have shown in multiple experiments that short-term exposure (1-6 hr) of cultured cells to amiloride or its analogs leads to little or no toxicity within the range of pH$_e$ 6.0-7.5.[30,31] Increased toxicity was observed however when amiloride or its analogs were combined with agents that cause intracellular acidification.[30-34] The ionophores nigericin

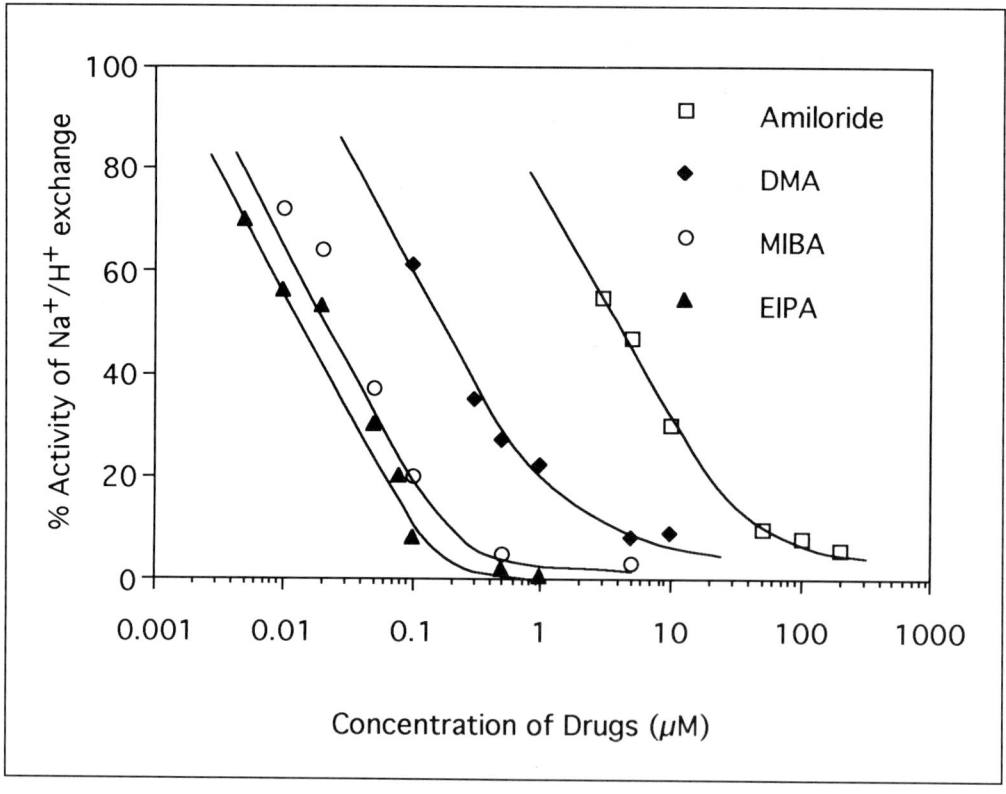

Fig. 14.4. Per cent activity of the Na$^+$/H$^+$ exchanger in MGH-U1 cells in the presence of different concentration of amiloride, DMA, MIBA or EIPA. Per cent activity was measured as the rate of recovery of pH$_i$ relative to control (no inhibitor) after acidification with nigericin. (Reproduced with permission from Maidorn, Cragoe; Br J Cancer 1993; 67: 297-303.)

and carbonylcyanide-3-chlorophenylhydrazone (CCCP) lower pH_i in an acidic microenvironment, and are toxic to cells at low but not at physiological pH_e. Rotin et al[31] have shown that the survival of Chinese hamster ovary (CHO) cells exposed to nigericin for up to six hours declines rapidly as pH_e is reduced below pH_e 6.5. Similar toxicity of CCCP has been shown for EMT-6 murine tumor cells grown in suspension at pHe 6.5 or below, or as solid nodules in culture (spheroids) that allow for penetration of tissue.[35] By inhibiting the regulation of pH_i, amiloride or its analogs lead to lower values of pH_i when cells are exposed to nigericin or CCCP under acidic conditions, and increase the amount of cell killing.

The amiloride analogs, EIPA and HMA, which are more potent and complete inhibitors of Na^+/H^+ exchanger activity, have been shown to enhance the toxicity of nigericin to cultured cells at low pH_e to a greater extent than amiloride even if used at one tenth of the concentration.[30,33] For EMT-6 murine tumor cells, a relative survival of 10^{-4}-10^{-3} as compared to control (no drugs used) could be achieved for a six hour exposure to nigericin and EIPA or HMA at pH_e 6.5 (Fig. 14.5). Moreover, cell killing was observed in the range of pH_e 6.5-7.0 when EIPA or HMA was added to nigericin, but only at pH_e < 6.5 when nigericin was used alone. This result has important implications for tumor therapy, because values of pH_e in solid tumors are often in the range of 6.5-7.0, but rarely below pH_e 6.5. We also obtained evidence for a cause and effect relationship between inhibition of Na^+/H^+ exchange activity and enhancement of cell killing by nigericin, since there was no added effect of EIPA against variant cells that lacked Na^+/H^+ exchange activity.[30]

The regulation of pHi depends on Na^+-dependent HCO_3^-/Cl^- exchange activity as well as on the Na^+/H^+ exchanger. DIDS, an inhibitor of the Na^+-dependent HCO_3^-/Cl^- exchanger, has been found not to be toxic to CHO cells when used alone for up to six hour exposure at pHe in the range 6.0-7.5 but also increases the cytotoxicity of nigericin at low pH_e.[31] Moreover, the combination of DIDS and amiloride (or its analogs) with nigericin is more toxic at low pHe than any combination of two of these agents. This result is consistent with the hypothesis that cell killing is due to inhibition of regulation of pH_i when cells are acidified with nigericin.

E.2 Acute Administration of the Inhibitors In Vivo

Stimulated by the above observations in tissue culture, in vivo experiments using combinations of agents which inhibit the regulation of pH_i have been performed. Using ^{31}P-NMR, Newell et al[32] showed that nigericin could lead to a fall in mean pH_i in an experimental murine tumor. Studies of nigericin used with amiloride, HMA or EIPA given by single injection, or three injections at two hour intervals led however to minimal killing of tumor cells with surviving fraction per

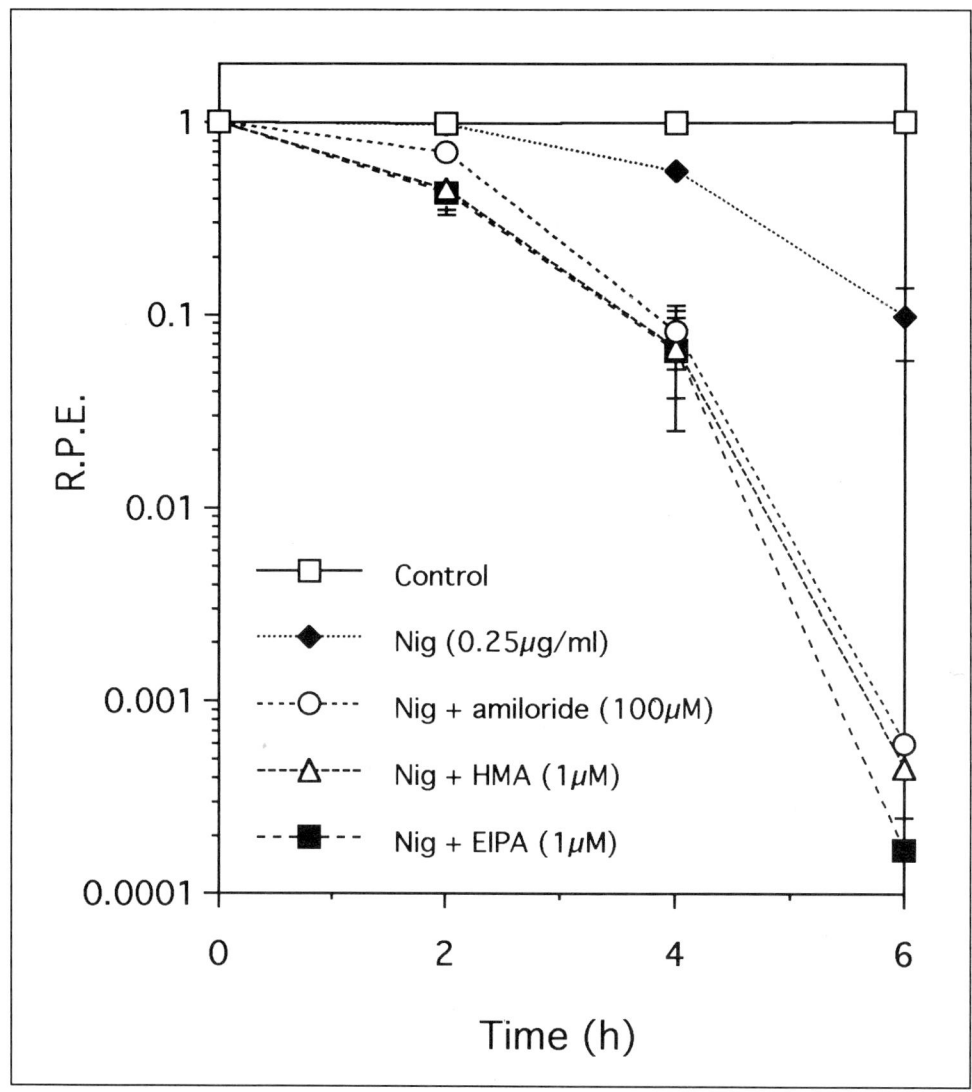

Fig. 14.5. Relative plating efficiency (RPE) of EMT-6 cells treated for varying times with a given amiloride, EIPA or HMA in the presence of 0.25 μg/ml nigericin at pHe 6.5. □ control, ◆: nigericin alone, ○: nigericin + amiloride (100 μM), Δ: nigericin + HMA (1 μM), ■: nigericin + EIPA (1 μM). (Reproduced with permission from Luo, Tannock; Br J Cancer 1994; 70:617-624).

tumor (assessed by a colony-forming assay) greater than 10%.[33,34] Failure to show greater in vivo cytotoxicity was most likely due to: (1) a higher level of pHe in tumors as compared to the medium in which the agents showed major effects to kill cells in vitro; and (2) other toxicities of these agents which led to insufficient concentrations of the drugs in tumors at doses that were tolerated by mice.

In attempts to increase cell killing by the combination of nigericin and amiloride analogs, we investigated strategies to lower pH_e in tumors. Numerous studies have shown that infusion of glucose may lead to tumor acidification.[32,36-41] Also, hydralazine has been shown previously to produce a dose-dependent reduction in tumor blood flow,[42,43] which may also lead to a decrease in tumor pH_e.[32] Experiments in our laboratory have shown that I.P. injections of glucose and/or of hydralazine prior to injection of nigericin and EIPA caused a fall in mean tumor pH_e and led to increased cell killing (Fig. 14.6).[34] The lowest mean values of pH_e that could be achieved were about pHe 6.6, and the surviving fraction of KHT fibrosarcoma tumors was reduced to a minimum of about 5×10^{-3}, when a combination of glucose, hydralazine, nigericin and EIPA was used. This result establishes the potential for pH-dependent tumor-selective cell killing in vivo, but represents a relatively small step towards elimination of a tumor that may contain more than 10^9 cells. Also, our unpublished data demonstrate that DIDS increases cell killing by nigericin + EIPA + hydralazine suggesting the importance of inhibition of both the Na^+/H^+ antiport and the Na^+-dependent HCO_3^-/Cl^- exchanger in maximizing this approach to tumor therapy.

In order to optimize the effects of these drugs, it is necessary to study pharmacokinetics and to provide a schedule of administration that ensures maintenance of an adequate concentration of drugs to ensure inhibition of exchanger activity. It is difficult to detect nigericin in vivo since it does not have appropriate spectral properties. However a bolus injection of 2.5 mg/kg body weight into mice suppresses pH_i for about two hours.[32] Amiloride and its analogs can be detected by their absorption at a wavelength of 365 μm, and the pharmacokinetics of amiloride have been studied in animals and humans using high-performance liquid chromatography (HPLC).[44-48] We have adapted these methods to study the pharmacokinetics of EIPA and HMA in murine plasma and tumors after I.P. injection (Fig. 14.7).[33] A solid-phase extraction method was used to prepare the samples, and an isocratic method was used for their HPLC analysis.[47,48] Our results demonstrate that peak concentrations of EIPA and HMA in mouse plasma were in the range of 10-40 μM after injection of a tolerated dose of 10 μg/g body weight, and then decayed with a half-life of about 30 minutes. In tumors, the peak concentration of EIPA was 3 mM with a slow rate of decay. Since these agents are time-dependent in their cytotoxic effects as well as dose-dependent, the use of a continuous infusion may be a more appropriate strategy to maintain an effective concentration that will inhibit Na^+/H^+ exchange activity without major toxicity to host animals.

E.3 PROLONGED ADMINISTRATION OF THE INHIBITORS

Time-dependent effects of the agents which are involved in regulation of pH_i to kill KHT cells in culture are shown in Figure 14.8.

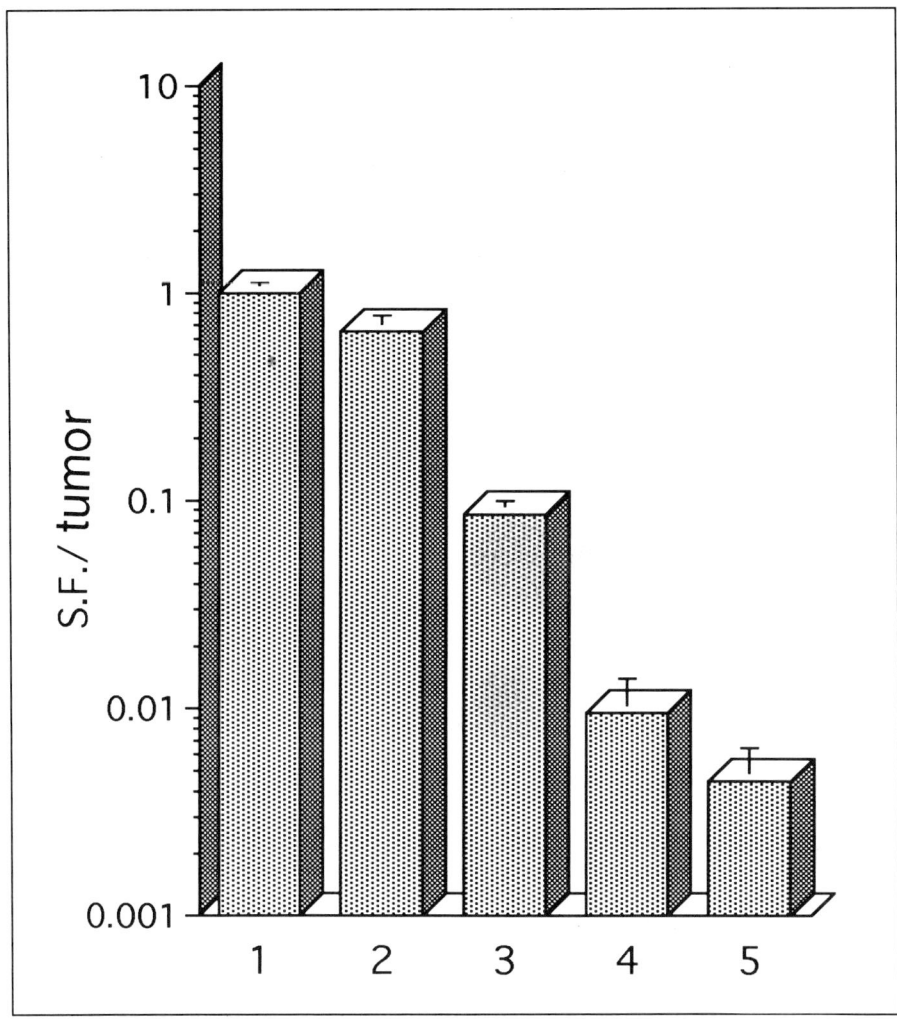

Fig. 14.6. Effect of treatments which acidify tumor pHe on survival of KHT tumors treated with nigericin (1.25 mg/kg) and EIPA (10 mg/kg). (1) control; (2) nigericin + EIPA; (3) nigericin + EIPA + glucose; (4) nigericin + EIPA + hydralazine (5) nigericin + EIPA + glucose + hydralazine. Glucose (6 g/kg) was given 60 minutes prior to nigericin and EIPA. Hydralazine (10 mg/kg) was given simultaneously with nigericin and EIPA.

In these studies, pH_e of medium was adjusted to pH_e 6.8 or 7.2, which may be representative of the pH_e in some tumors and normal tissues, respectively. The concentrations of drugs were about one tenth of the minimal effective doses when used for short-term exposure up to six hours. The results show that nigericin causes time-dependent cell killing. Only minor effects to enhance cell killing induced by nigericin were observed for EIPA at concentrations up to 5 µM. However, exposure to DIDS led to a greater dose-dependent enhancement of cell

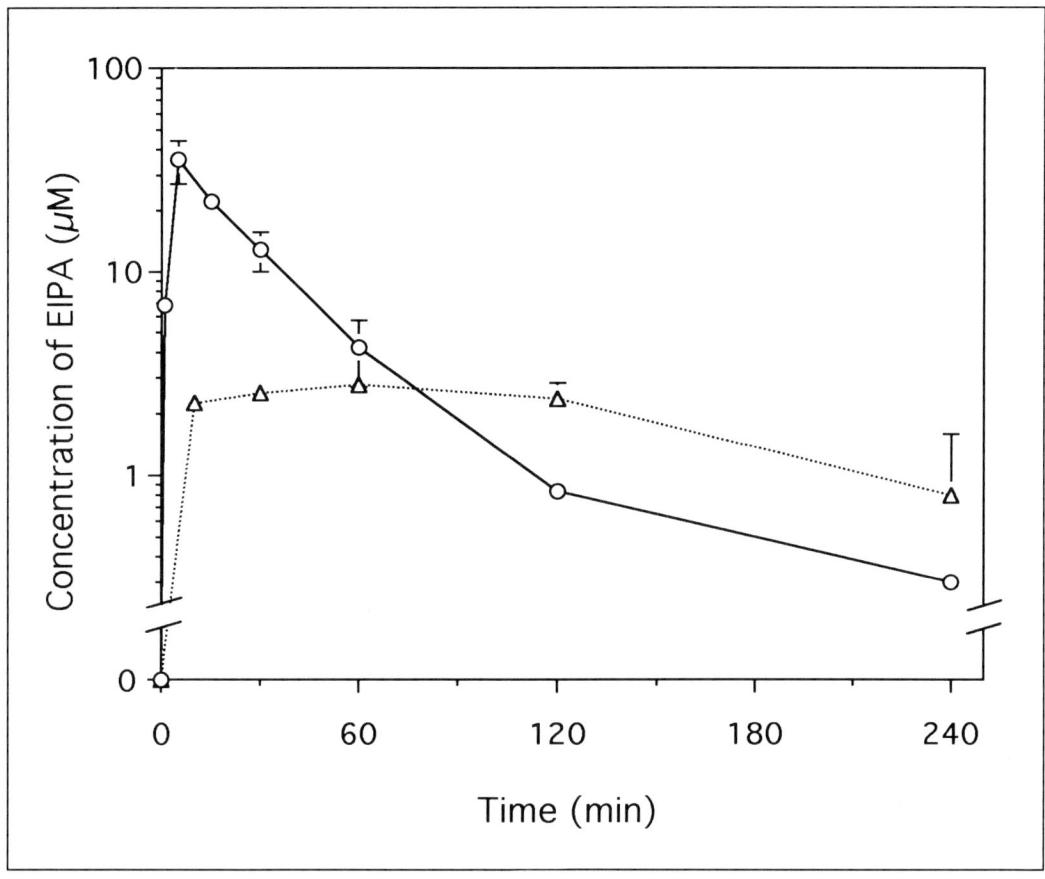

Fig. 14.7. Pharmacokinetics of EIPA in plasma (○) and in KHT tumors (△) following an I.P. injection of 10 μg/g EIPA to C3H/HeJ mice. The limit of detection is about 0.1 μM in plasma, and ~1 μM in tumors (where other tissue elements contribute to spectral noise).

killing by nigericin. These results indicate that the Na^+-dependent HCO_3^-/Cl^- exchanger is probably dominant in regulation of pH_i under the conditions in which cells are exposed under slightly acidic conditions such as pH_e 6.8, which is consistent with results reported previously by Boyer et al[24] as shown in Figure 14.3.

Figure 14.9 shows the pH-dependent effect of a 72 hour exposure to nigericin with or without DIDS + EIPA on KHT and EMT-6 cells in vitro. There is strong dependence of cell killing on pHe within the narrow range of 6.7-6.9, indicating that even a minor decrease in pHe around pH 6.8 may result in great enhancement of cytotoxicity induced by agents which inhibit the regulation of pH_i.

Sparks et al[49] have demonstrated suppression of growth of the H6 hepatoma and DMA/J mammary carcinoma in mice treated repeatedly

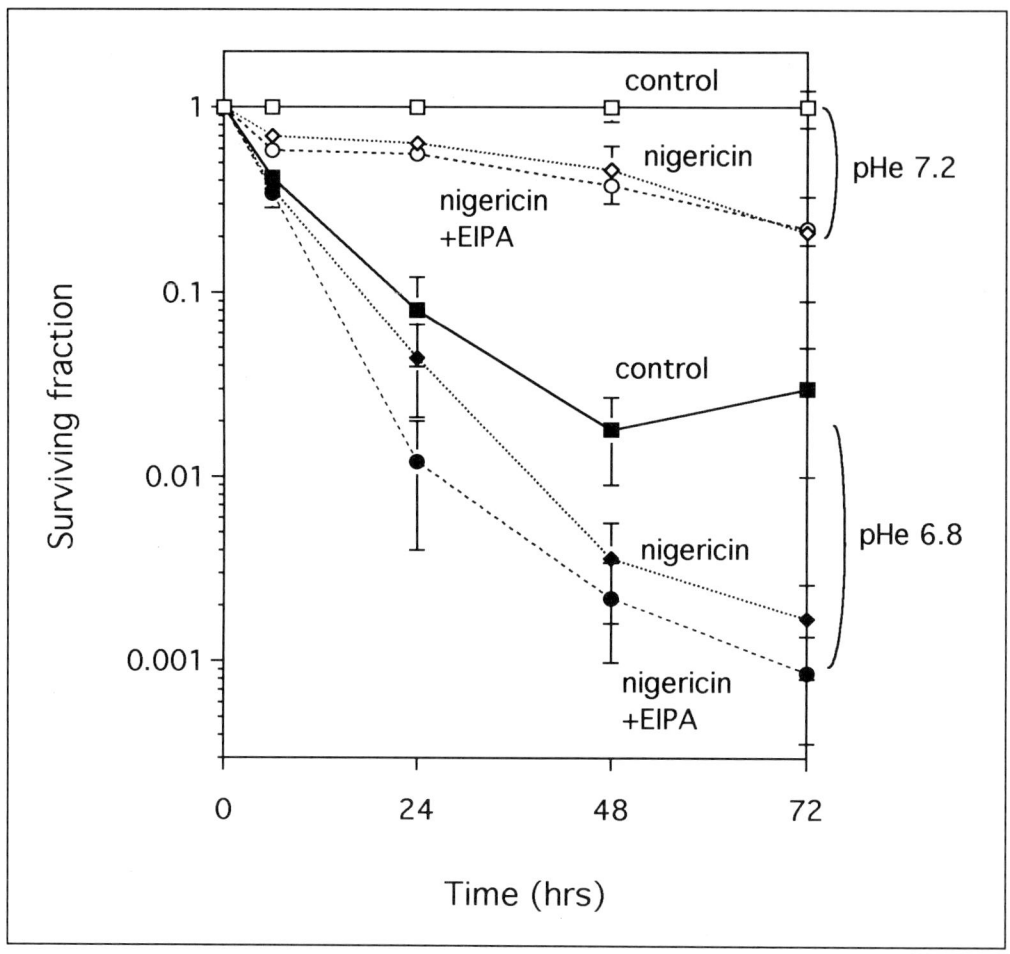

Fig. 14.8. Survival of KHT cells during 72 hour exposure to nigericin 0.033 µM with or without EIPA 0.5 µM at pH_e 7.2 or 6.8 in vitro.

with amiloride. In our laboratory, continuous infusion into mice was performed using micro osmotic pumps which have a capacity of 100µl and which pumped the fluid at a steady rate of 1 µl/hr for 72 hours. Pumps containing drugs were implanted into lightly anesthetized mice subcutaneously or intraperitoneally. Figure 14.10 shows the surviving fraction for KHT tumors treated by nigericin + hydralazine with or without DIDS and EIPA at drug doses that are close to the maximum levels tolerated by mice. Hydralazine has been shown to lower tumor pH_e by about 0.15 pH units (data not shown). The combination of these agents caused maximum cell killing with a surviving fraction per tumor of 2×10^{-3}, which is similar to the results obtained by bolus injection (Fig. 14.6).

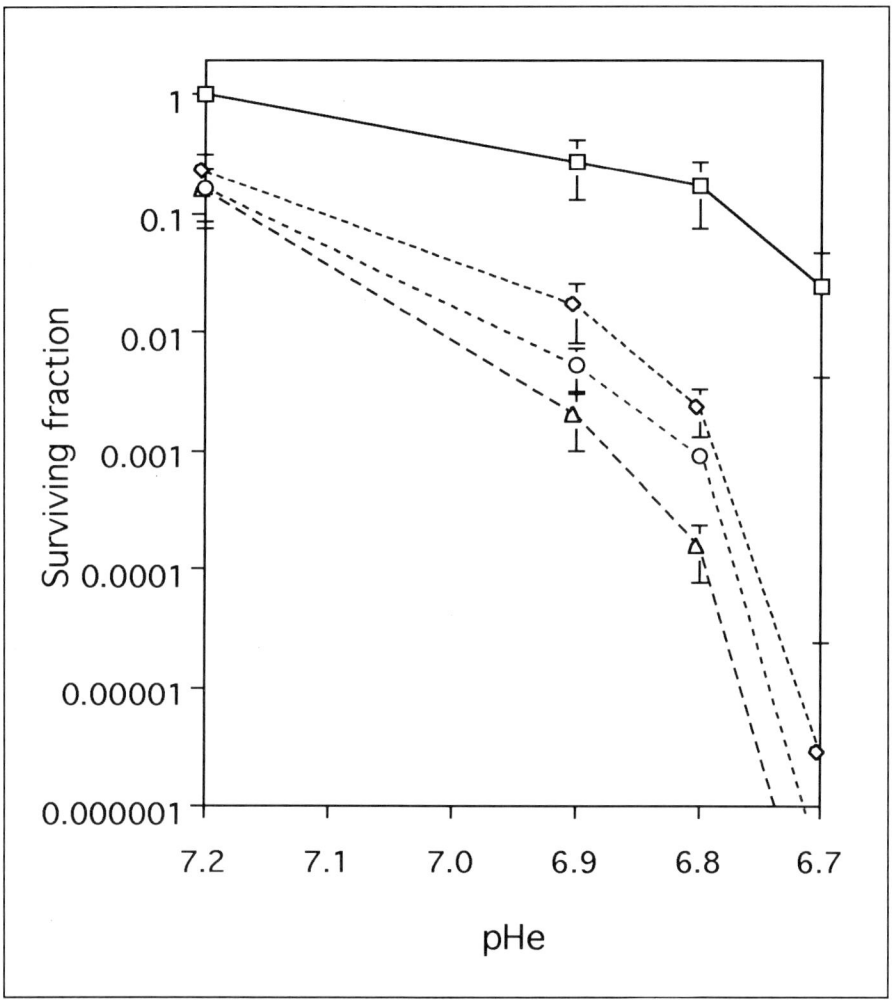

Fig. 14.9. Effect of pHe on survival of KHT cells exposed for 72 hours to various agents. ❑: Control, ◆▲: nigericin (0.033 μM), ○: nigericin with DIDS (50 μM), △: nigericin with DIDS and EIPA (5 μM).

Treatments leading to acidification of tumors include hydralazine as described above and administration of glucose.[36-39] Furthermore Jähde et al[50] and Kuin et al[41] have demonstrated that m-iodobenzylguanidine (MIBG), which is a mitochondrial inhibitor, can reduce the amount of glucose required for effective reduction of pH_e. Insulin or inorganic phosphate also have been reported to lower tumor pH_e.[50] These treatments are being investigated to determine whether they will augment therapeutic effects caused by chronic administration of nigericin, EIPA and DIDS.

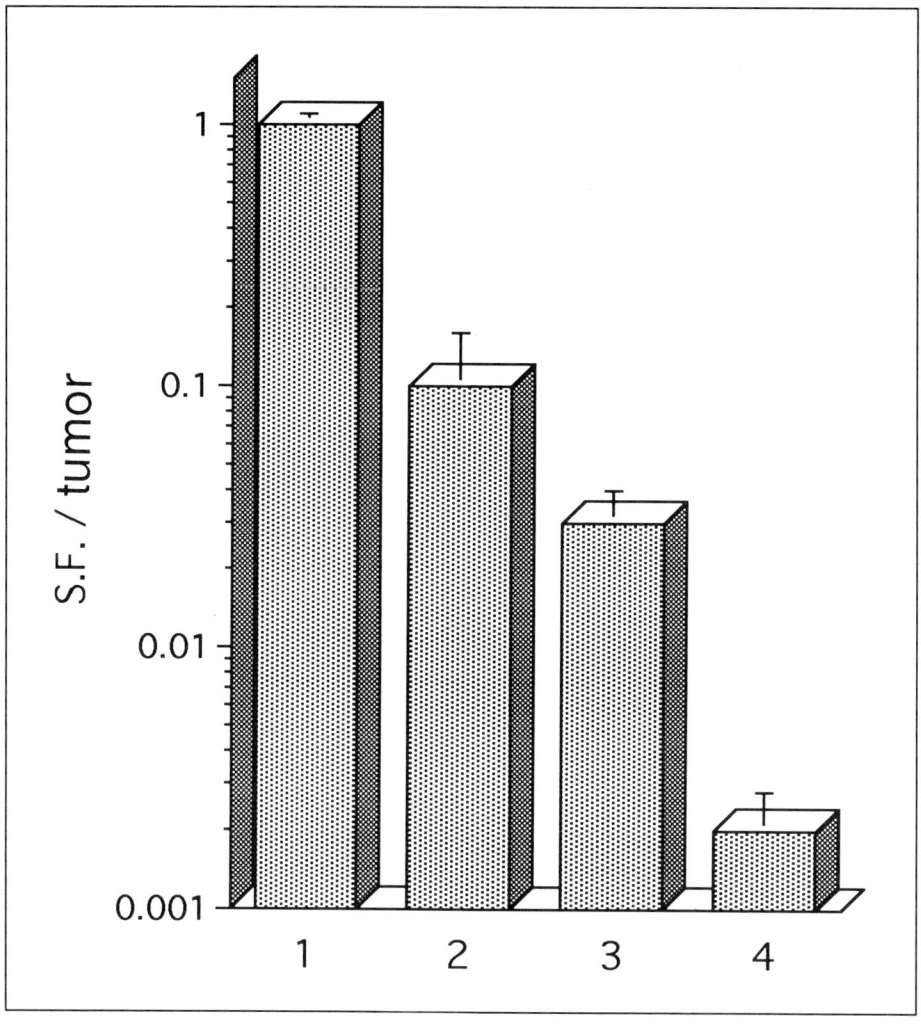

Fig. 14.10. Surviving fraction per tumor for the KHT tumors. The drugs were given by infusion from micro osmotic pumps. (1) control; (2) nigericin (2 mg/ml) + hydralazine (12.5 mg/ml); (3) nigericin + hydralazine + DIDS (10 mg/ml); (4) nigericin + hydralazine + DIDS + EIPA (3 mg/ml).

F. REGULATION OF pH_i AND HYPERTHERMIA

Heating cells in an environment at low pH_e enhances lethality, and pH_i appears to be the important determinant of thermosensitivity.[36,51-53] Also, hyperthermia has been demonstrated to lower tumor pHe, probably through a mechanism of vascular occlusion and inhibition of the clearance of metabolic acids. Combination therapy using agents such as analogs of amiloride which inhibit the regulation of pHi might therefore have potential to enhance the lethality of hyperthermia. Since Miyakoshi et al[54] reported that amiloride enhanced thermal killing of CHO cells, many articles have reported an enhancing effect of amiloride and its analogs on thermosensitivity.[54,55] Ruifrok and Konings[56] demonstrated that amiloride reduced the shoulder of the thermal survival curve and the thermotolerance of mouse fibroblasts. Kim et al,[55] and unpublished studies from our laboratory by Liu et al, have shown that heating cells at 43°C leads to a decrease in their pH_i, and that amiloride or its analogs can lead to a further reduction in pH_i and to additional cell killing. The measurements are made, however, on cells that have lost reproductive ability, so that it is unclear as to whether the decrease in pH_i is a cause or a result of heat-induced toxicity.

Recently, Song and his colleagues[53,57] demonstrated the importance of simultaneous inhibition of both the Na^+/H^+ exchanger and the Na^+-dependent HCO_3^-/Cl^- exchanger in enhancing thermosensitivity. They showed enhanced thermosensitivity of SCK cells and increased growth delay in vivo by using HMA or EIPA as an inhibitor of the Na^+/H^+ exchanger and DIDS or $R(+)$-[(5,6-dichloro-2,3,9,9a-tetrahydro-3-oxo-9a-propyl-1H-fluoren-7-yl)oxy] acetic acid as an inhibitor of the Na^+-dependent HCO_3^-/Cl^- exchanger. A combination of inhibitors of the regulation of pH_i might prove to be a clinically useful method for sensitizing tumor cells to hyperthermia.

G. REGULATION OF pH_I AND RADIATION

Tumors have been shown to be heterogeneous with respect to almost any biologic property that can be characterized.[58] Heterogeneity in the microenvironment may lead to hypoxic regions in tumors, and hypoxic cells are known to be resistant to irradiation, requiring about a 3-fold increase in dose as compared to aerobic cells to achieve the same level of cell killing. Tumor cells that survive treatment with radiation may be susceptible to agents which are toxic to hypoxic cells including drugs that require bioreduction for activity such as mitomycin C or tiripazamine.[59] Another exploitable property of hypoxic cells may be the low pH_e in their microenvironment. Therefore combined therapy with radiation and with drugs which inhibit regulation of pH_i might lead to a marked increase in cell killing of tumors as compared to radiation alone.

We have studied the use of nigericin and analogs of amiloride (with or without hydralazine) given immediately after irradiation of murine tumors.[33,34] In these experiments we found that cell killing was no greater than expected from knowledge of the toxic effects of radiation alone and of drugs alone. Failure to observe synergistic effects suggests that the average pHe of the subpopulation of tumor cells that survive radiation may not be lower than the average pHe in the tumor, particularly when hydralazine is also given to reduce blood flow. Previous reports have shown that cells which survive irradiation may demonstrate decreased sensitivity to acidic extracellular fluid perhaps because of increased repair of potentially lethal damage.[60,61] This effect might also influence the cytotoxicity of radiation used with nigericin and amiloride analogs.

H. REGULATION OF pH_i AND CHEMOTHERAPY

The activity and/or uptake into cells of many anticancer drugs depends on the pH gradient across the cell membrane. Drugs which are weak acids such as melphalan or chlorambucil will be concentrated in cells under conditions where $pH_e < pH_i$, and are more effective at low pH_e.[62-64] In contrast, drugs such as doxorubicin which is a weak base have decreased uptake and activity under acidic conditions.[65,66] Treatment with agents which reduce pH_i might have potential to enhance the uptake or cytotoxicity of anticancer drugs by changing pH_i and the pH gradient across the cell membrane. Consistent with this concept, Wood et al[67] reported that reduction of tumor pH_i by nigericin enhanced delay in growth of RIF-1 tumors induced by the alkylating agent melphalan.

One of the most important problems which limits the effectiveness of anticancer drugs is acquired resistance. Resistance to multiple drugs (MDR) may be caused by overexpression of a variety of membrane proteins such as P-glycoprotein or multidrug resistance protein.[68] Development of MDR in tumor cells has been correlated with an alkaline shift of pH_i.[69,70] It is probable that these shifts in pH_i of MDR cells are related to changes in the activity of membrane-based ion exchange mechanisms, including the Na^+/H^+ and Na^+-dependent HCO_3^-/Cl^- exchanger.[71] There are inconsistent data concerning the influence of expression of P-glycoprotein on exchange activity, and we are evaluating the activity of both exchangers in pairs of drug-resistant and drug-sensitive cells.

I. SUMMARY

Inhibitors of the regulation of intracellular pH might have therapeutic potential in tumors where the microenvironment is often acidic as compared to normal tissue. The therapeutic potential of inhibitors of the Na^+/H^+ exchanger in tumor-selective therapy is summarized below.

(1) Amiloride and its analogs which inhibit the activity of the $Na+/H^+$ exchanger do not have cytotoxicity when given alone. In order to obtain pH-dependent cell killing, it is necessary to combine them with agents such as nigericin or CCCP, which cause cellular acidification. Moreover, to enhance cell killing, both the Na^+/H^+ exchanger and the Na^+-dependent HCO_3^-/Cl^- exchanger should be inhibited.

(2) The mean pH_e in most tumors ($\sim pH_e$ 6.8-7.0), is higher than the pH at which acute treatments with nigericin and amiloride analogs lead to major cytotoxic effects. Strategies which lower tumor pH_e might enhance cell killing induced by nigericin and analogs of amiloride, especially when these agents are given by chronic infusion.

(3) Inhibition of the regulation of pH_i may have potential to enhance the lethality of hyperthermia.

(4) Tumor cells that survive treatment with radiation may be susceptible to drugs through mechanisms that relate indirectly to the hypoxic state. As yet, however, agents that cause cellular acidification and/or inhibit regulation of pH_i have provided only additive effects when used with radiation. These agents might stimulate the uptake and/or activity of anticancer drugs by modifying the pH gradient across the cell membrane.

Acknowledgments

Work described in this chapter was supported by research grants from the Medical Research Council and the National Cancer Institute of Canada.

References

1. Tannock IF, Rotin D. Acid pH in tumors and its potential for therapeutic exploitation. Cancer Res 1989; 49:4373-4384.
2. Tannock IF. The relation between cell proliferation and the vascular system in a transplanted mouse mammary tumor. Br J Cancer 1968; 22:258-273.
3. Wike-Hooley JL, Haveman J, Reinhold JS. The relevance of tumour pH to the treatment of malignant disease. Radiother Oncol 1984; 2:343-366.
4. Vaupel P, Kallinowsky F, Okunieff P. Blood flow, oxygen and nutrient supply, and metabolic microenvironment of human tumors: a review. Cancer Res 1989; 49:6449-6465.
5. Newell K, Franchi A, Pouysségur J et al. Studies with glycolysis-deficient cells suggest that production of lactic acid is not the only cause of tumor acidity. Proc Natl Acad Sci USA 1993; 90:1127-1131.
6. Van den Berg AP, Wike-Hooley JL, Van den Berg-Blok AE et al. Tumour pH in human mammary carcinoma. Eur J Cancer Clin Oncol 1982; 18:457-462.
7. Martin GR, Jain RK. Noninvasive measurement of interstitial pH profiles in normal and neoplastic tissue using fluorescence ratio imaging microscopy. Cancer Res 1994; 54:5670-5674.
8. Oberhaensli RD, Hilton-Jones D, Bone BJ et al. Biochemical investigation of human tumours in vivo with phosphorus-31 magnetic resonance spectroscopy. Lancet 1986; 2:8-11.
9. Daly PF, Cohen JS. Magnetic resonance spectroscopy of tumors and potential in vivo clinical applications: a review. Cancer Res 1989; 49:770-799.
10. Grinstein S, Rotin D, Mason MJ. Na^+/H^+ exchanger and growth factor-induced cytosolic pH changes. Role in cellular proliferation. Biochem Biophys Acta 1989; 988:73-97.
11. Cassel D, Scharf O, Rotman M et al. Characterization of Na^+-linked and Na^+-independent Cl^-/HCO_3^- exchange systems in Chinese hamster lung fibroblasts. J Biol Chem 1988; 263:6122-6127.

12. Cala PM. Volume regulation by *Amphiuma* red blood cells. The membrane potential and its implications regarding the nature of the ion-flux pathways. J Gen Physiol 1980; 76:683-708.
13. Aronson PS. Kinetic properties of the plasma membrane Na$^+$/H$^+$ exchanger. Annu Rev Physiol 1985; 47:545-560.
14. Kleyman R, Cragoe Jr EJ. Amiloride and its analogs as tools in the study of ion transport. J Membrane Biol 1988; 105:1-21.
15. Sardet C, Franchi A, Pouysségur J. Molecular cloning, primary structure, and expression of the human growth factor-activatable Na$^+$/H$^+$ antiporter. Cell 1989; 56:271-280.
16. Orlowski J, Kandasamy RA, Shull GE. Molecular cloning of putative members of the Na$^+$/H$^+$ exchanger gene family. J Biol Chem 1992; 267:9331-9339.
17. Tse C-M, Brant SR, Walker MS et al. Cloning and sequencing of a rabbit cDNA encoding an intestinal and kidney-specific Na$^+$/H$^+$ exchanger isoform (NHE3). J Biol Chem 1992; 267:9340-9346.
18. Tse C-M, Levine SA, Yun, CHC et al. Cloning and expression of a rabbit cDNA encoding a serum-activated ethylisopropylamiloride-resistant epithelial Na$^+$/H$^+$ exchanger isoform (NHE2). J Biol Chem 1993; 268:11917-11924.
19. Wang Z, Orlowski J, Shull GE. Primary structure and functional expression of a novel gastrointestinal isoform of the rat Na/H exchanger. J Biol Chem 1993; 268:11925-11928.
20. Rink TJ, Tsein RY, Pozzan T. Cytoplasmic pH and free Mg^{2+} in lymphocytes. J Cell Biol 1982; 95:189-196.
21. Schwartz MA, Cragoe Jr EJ, Lechence CP. pH regulation in spread cells and round cells. J Biol Chem 1990; 265:1327-1332.
22. Thomas JA, Buchsbaum RN, Zimniak A. Intracellular pH measurements in Ehrlich ascites tumor utilizing spectroscopic probes. Biochemistry 1979; 18:2210-2218.
23. Boron WF, Deweer P. Intracellular pH transients in squid giant axons caused by CO_2, NH_3, and metabolic inhibitors. J Gen Physiol 1976; 67:91-112.
24. Boyer MJ, Tannock IF. Regulation of intracellular pH in tumor cell lines: influence of microenvironmental conditions. Cancer Res 1992; 52:4441-4447.
25. Boyer MJ, Barnard M, Hedley DW et al. Regulation of intracellular pH in subpopulations of cells derived from spheroids and solid tumours. Br J Cancer 1993; 68:890-897.
26. Ganz MB, Boyarsky G, Sterzel RB et al. Arginine vasopressin enhances pH$_i$ regulation in the presence of HCO_3^- by stimulating three acid-base transport systems. Nature (Lond.) 1989; 337:648-651.
27. Pouysségur J, Sardet C, Franchi A et al. A specific mutation abolishing Na$^+$/H$^+$ antiport activity in hamster fibroblasts precludes growth at neutral and acidic pH. Proc Natl Acad Sci USA 1984; 81:4833-4837.
28. Agarwall N, Haggerty JG, Adelberg EA et al. Isolation and characterization of a Na$^+$/H$^+$ antiport-deficient mutant of LLC-PK1 cells. Am J Physiol 1986; 251:C825-C830.

29. Rotin D, Steele-Norwood D, Grinstein S et al. Requirement of the Na⁺/H⁺ exchanger for tumor growth. Cancer Res 1989; 49:205-210.
30. Maidorn RP, Cragoe Jr EJ, Tannock IF. Therapeutic potential of analogues of amiloride: inhibition of the regulation of intracellular pH as a possible mechanism of tumour selective therapy. Br J Cancer 1993; 67:297-303.
31. Rotin D, Wan P, Grinstein S et al. Cytotoxity of compounds that interfere with the regulation of intracellular pH: a potential new class of anticancer drugs. Cancer Res 1987; 47:1497-1504.
32. Newell K, Wood P, Stratford I et al. Effects of agents which inhibit the regulation of intracellular pH on murine solid tumours. Br J Cancer 1992; 66:311-317.
33. Luo J, Tannock IF. Inhibition of the regulation of intracellular pH: potential of 5-(N,N-hexamethylene) amiloride in tumour selective therapy. Br J Cancer 1994; 70:617-624.
34. Hasuda K, Lee C, Tannock IF. Anti-tumor activity of nigericin and 5-(N-ethyl-N-isopropyl) amiloride: an approach to therapy based on cellular acidification and the inhibition of regulation of intracellular pH. Oncol Res 1994; 6:259-268.
35. Newell KJ, Tannock IF. Reduction of intracellular pH as a possible mechanism for killing cells in acidic regions of solid tumors: effects of carbonylcyanide-3-chlorophenylhydrazone. Cancer Res. 1989;49: 4477-4482.
36. Jähde E, Rajewsky MF. Sensitization of clonogenic malignant cells to hyperthermia by glucose-mediated, tumor-selective pH reduction. J Cancer Res Clin Oncol 1982; 104:23-30.
37. Jähde E, Rajewsky MF. Tumor-selective modification of cellular microenvironment in vivo: effect of glucose infusion on the pH in normal and malignant rat tissues. Cancer Res 1982; 42:1505-1512.
38. Ward KJ, Jain RK. response of tumours to hyperglycemia: characterization, significance and role in hyperthermia. Int J Hyperthermia 1988; 4:223-250.
39. Reinhold HS, van den Berg-Blok AE, van den Berg AP. Dose-effect relationships for glucose-induced tumour acidification and erythrocyte flux. Eur J Cancer 1991; 27:1151-1154.
40. Volk T, Jähde E, Fortmeyer HP et al. pH in human tumour xenograft: effect of intravenous administration of glucose. Br J Cancer 1993; 68:492-500.
41. Kuin A, Smets L, Volk T et al. Reduction of intratumoral pH by the mitochondrial inhibitor m-iodobenzylguanidine (MIBG) and moderate hyperglycemia. Cancer Res 1994; 54:3785-3792.
42. Chaplin DL. Hydralazine-induced tumour hypoxia: a potential target for cancer chemotherapy. J Natl Cancer Inst 1989; 81:618-622.
43. Lin J-C, Song CW. Effects of hydralazine on the blood flow in RIF-1 tumors and normal tissues of mice. Radiat Res 1990; 124:171-177.
44. Yip MS, Coates PE, Thiessen JJ. High-performance liquid chromatographic analysis of amiloride in plasma and urine. J Chromatogr 1983; 307:343-350.

45. Shi RJ-Y, Benet LZ, Lin ET. High-performance liquid chromatographic assay of basic drugs in plasma and urine using a silica gel column and an aqueous mobile phase. J Chromatogr 1986; 377:399-404.
46. Somogyi A, Keal J, Bochner F. Sensitive high-performance liquid chromatographic assay for determination of amiloride in biologic fluids using an ion-pair extraction method. Ther Drug Monit 1988; 10:463-468.
47. Meng QC, Chen YF, Oparil S. High-performance liquid chromatographic determination of amiloride and its analogues in rat plasma. J Chromatogr 1990; 529:201-209.
48. Alliegro MA, Dyer KD, Cragoe Jr EJ et al. High-performance liquid chromatographic method for quantitating plasma levels of amiloride and its analogues. J Chromatogr 1992; 582:217-223.
49. Sparks RL, Pool TB, Smith NKR et al. Effects of amiloride on tumor growth and intracellular element contact of tumor cells in vitro. Cancer Res 1983; 43:73-77.
50. Jähde E, Volk T, Atema A et al. pH in human tumor xenografts and transplanted rat tumors: effect of insulin, inorganic phosphate, and m-iodobenzylguanidine. Cancer Res 1992; 52:6209-6215.
51. Chu GL, Dewey WC. The role of low intracellular or extracellular pH in sensitization to hyperthemia. Radiat Res 1988; 114:154-167.
52. Chu GL, Wang Z, Hyun WC et al. The role of intracellular pH and its variance in low pH sensitization of killing by hyperthemia. Radiat Res 1990; 122:288-293.
53. Song WC, Lyons JC, Griffin RJ et al. Increase in thermsensitivity of tumor cells by lowering intracellular pH. Cancer Res 1993; 53:1599-1601.
54. Miyakoshi J, Oda W, Hirata M, et al. Effects of amiloride on thermosensitivity of Chinese hamster cells under neutral and acidic pH. Cancer Res 1986; 46:1840-1843.
55. Kim GE, Lyons JC, Song CW. Effects of amiloride on intracellular pH and thermosensitivity. Int J Radiat Oncol Biol Phys 1991; 20:541-549.
56. Ruifrok ACC, Konings AWT. Effects of amiloride on hyperthermic cell killing of normal and thermotolerant mouse fibroblast LM cells. Int J Radiat Biol 1987; 52:385-392.
57. Lyons JC, Ross BD, Song CW. Enhancement of hyperthermia effect in vivo by amiloride and DIDS. Int J Radiat Oncol Biol Phys 1993; 25:95-103.
58. Heppner GH. Tumor heterogeneity. Cancer Res 1984; 44:2259-2265.
59. Kennedy KA, Rockwell S, Sartorelli AC. Preferential activation of mitomycin C to cytotoxic metabolites by hypoxic tumour cells. Cancer Res 1980; 40:2356-2360.
60. Holahan EV, Stuart PK, Dewey WC. Enhancement of survival of CHO cells by acidic pH after X irradiation. Radiat Res 1982; 89:433-435.
61. Hwang YC, Kim S-G, Evelhoch JL et al. Modulation of murine radiation-induced fibrosarcoma-1 tumor metabolism and blood flow in situ via glucose and mannitol administration monitored by ^{31}P and ^{2}H nuclear magnetic resonance spectroscopy. Cancer Res 1991; 51:3108-3118.

62. Tannock IF, Rotin D. Keynote address: mechanisms of interaction between radiation and drugs with potential for improvements in therapy. NCI Monographs 1988; 6:77-83.
63. Jähde E, Glüsenkamp K-H, Klünder I et al. Hydrogen ion-mediated enhancement of cytotoxitity of bis-chloroethylating drugs in rat mammary carcinoma cells in vitro. Cancer Res 1989; 49:2965-2972.
64. Karuri AR, Dobrowsky E, Tannock IF. Selective cellular acidification and toxicity of weak organic acids in an acidic microenvironment. Br J Cancer 1993; 68:1080-1087.
65. Born R, Eicholtz-Wirth H. Effect of different physiological conditions on the action of adriamycin on Chinese hamster cells in vitro. Br J Cancer 1981; 44:241-246.
66. Wike-Hooley JW, Van der Zee J, Van Rhoon GC et al. Human tumour pH changes following hyperthermia and radiation therapy. Eur J Cancer Clin Oncol 1984; 20:619-623.
67. Wood PJ, Sansom JM, Newell K et al. Reduction of tumour intracellular pH and enhancement of melphalan cytotoxicity by the ionophore nigericin. Int J Cancer 1995; 60:264-268.
68. Cole SPC, Bhardwaj G, Gerlach JH et al. Overexpression of a transporter gene in a multidrug-resistant human lung cancer cell line. Science 1992; 258:1650-1654.
69. Thiebaut F, Currier SJ, Whitaker J et al. Activity of the multidrug transporter results in alkalinization of the cytosol: measurement of cytosolic pH by microinjection of a pH-sensitive dye. J Histochem Cytochem 1990; 38:685-690.
70. Simon S, Roy D, Schindler M. Intracellular pH and the control of multidrug resistance. Proc Natl Acad Sci USA 1994; 91:1128-1132.
71. Boscoboinik D, Gupta RS, Epand RM. Investigation of the relationship between altered intracellular pH and multidrug resistance in mammalian cells. Br J Cancer 1990; 61:568-572.

QUESTIONNAIRE

Receive a FREE BOOK of your choice

Please help us out—Just answer the questions below, then select the book of your choice from the list on the back and return this card.

R.G. Landes Company publishes five book series: *Medical Intelligence Unit, Molecular Biology Intelligence Unit, Neuroscience Intelligence Unit, Tissue Engineering Intelligence Unit* and *Biotechnology Intelligence Unit*. We also publish comprehensive, shorter than book-length reports on well-circumscribed topics in molecular biology and medicine. The authors of our books and reports are acknowledged leaders in their fields and the topics are unique. Almost without exception, there are no other comprehensive publications on these topics.

Our goal is to publish material in important and rapidly changing areas of bioscience for sophisticated scientists. To achieve this goal, we have accelerated our publishing program to conform to the fast pace in which information grows in bioscience. Most of our books and reports are published within 90 to 120 days of receipt of the manuscript.

Please circle your response to the questions below.

1. We would like to sell our *books* to scientists and students at a deep discount. But we can only do this as part of a prepaid subscription program. The retail price range for our books is $59-$99. Would you pay $196 to select four *books* per year from any of our Intelligence Units–$49 per book–as part of a prepaid program?

                                                    Yes   No

2. We would like to sell our *reports* to scientists and students at a deep discount. But we can only do this as part of a prepaid subscription program. The retail price range for our reports is $39-$59. Would you pay $145 to select five *reports* per year–$29 per report–as part of a prepaid program?

                                                    Yes   No

3. Would you pay $39–the retail price range of our books is $59-$99–to receive any single book in our Intelligence Units if it is spiral bound, but in every other way identical to the more expensive hardcover version?

                                                    Yes   No

To receive your free book, please fill out the shipping information below, select your free book choice from the list on the back of this survey and mail this card to:

    R.G. Landes Company, 909 S. Pine Street, Georgetown, Texas 78626 U.S.A.

Your Name _____

Address _____

City_____ State/Province:_____

Country: _____ Postal Code:_____

My computer type is Macintosh_____ ; IBM-compatible _____ ; Other _____

Do you own ____ or plan to purchase ____ a CD-ROM drive?

Available Free Titles

*Please check three titles in order of preference.
Your request will be filled based on availability. Thank you.*

- ☐ Water Channels
  Alan Verkman,
  University of California-San Francisco

- ☐ The Na,K-ATPase:
  Structure-Function Relationship
  J.-D. Horisberger, University of Lausanne

- ☐ Intrathymic Development of T Cells
  J. Nikolic-Zugic,
  Memorial Sloan-Kettering Cancer Center

- ☐ Cyclic GMP
  Thomas Lincoln, University of Alabama

- ☐ Primordial VRM System and the Evolution
  of Vertebrate Immunity
  John Stewart, Institut Pasteur-Paris

- ☐ Thyroid Hormone Regulation
  of Gene Expression
  Graham R. Williams, University of Birmingham

- ☐ Mechanisms of Immunological Self Tolerance
  Guido Kroemer, CNRS Génétique Moléculaire et
  Biologie du Développement-Villejuif

- ☐ The Costimulatory Pathway
  for T Cell Responses
  Yang Liu, New York University

- ☐ Molecular Genetics of Drosophila Oogenesis
  Paul F. Lasko, McGill University

- ☐ Mechanism of Steroid Hormone Regulation
  of Gene Transcription
  M.-J. Tsai & Bert W. O'Malley, Baylor University

- ☐ Liver Gene Expression
  François Tronche & Moshe Yaniv,
  Institut Pasteur-Paris

- ☐ RNA Polymerase III Transcription
  R.J. White, University of Cambridge

- ☐ src Family of Tyrosine Kinases in Leukocytes
  Tomas Mustelin, La Jolla Institute

- ☐ MHC Antigens and NK Cells
  Rafael Solana & Jose Peña,
  University of Córdoba

- ☐ Kinetic Modeling of Gene Expression
  James L. Hargrove, University of Georgia

- ☐ PCR and the Analysis of the T Cell Receptor
  Repertoire
  Jorge Oksenberg, Michael Panzara & Lawrence
  Steinman, Stanford University

- ☐ Myointimal Hyperplasia
  Philip Dobrin, Loyola University

- ☐ Transgenic Mice as an In Vivo Model
  of Self-Reactivity
  David Ferrick & Lisa DiMolfetto-Landon,
  University of California-Davis and Pamela Ohashi,
  Ontario Cancer Institute

- ☐ Cytogenetics of Bone and Soft Tissue Tumors
  Avery A. Sandberg, Genetrix & Julia A. Bridge,
  University of Nebraska

- ☐ The Th1-Th2 Paradigm and Transplantation
  Robin Lowry, Emory University

- ☐ Phagocyte Production and Function Following
  Thermal Injury
  Verlyn Peterson & Daniel R. Ambruso,
  University of Colorado

- ☐ Human T Lymphocyte Activation Deficiencies
  José Regueiro, Carlos Rodríguez-Gallego
  and Antonio Arnaiz-Villena,
  Hospital 12 de Octubre-Madrid

- ☐ Monoclonal Antibody in Detection and
  Treatment of Colon Cancer
  Edward W. Martin, Jr., Ohio State University

- ☐ Enteric Physiology of the Transplanted Intestine
  Michael Sarr & Nadey S. Hakim, Mayo Clinic

- ☐ Artificial Chordae in Mitral Valve Surgery
  Claudio Zussa, S. Maria dei Battuti Hospital-Treviso

- ☐ Injury and Tumor Implantation
  Satya Murthy & Edward Scanlon,
  Northwestern University

- ☐ Support of the Acutely Failing Liver
  A.A. Demetriou, Cedars-Sinai

- ☐ Reactive Metabolites of Oxygen and Nitrogen
  in Biology and Medicine
  Matthew Grisham, Louisiana State-Shreveport

- ☐ Biology of Lung Cancer
  Adi Gazdar & Paul Carbone,
  Southwestern Medical Center

- ☐ Quantitative Measurement
  of Venous Incompetence
  Paul S. van Bemmelen, Southern Illinois University
  and John J. Bergan, Scripps Memorial Hospital

- ☐ Adhesion Molecules in Organ Transplants
  Gustav Steinhoff, University of Kiel

- ☐ Purging in Bone Marrow Transplantation
  Subhash C. Gulati,
  Memorial Sloan-Kettering Cancer Center

- ☐ Trauma 2000: Strategies for the New Millennium
  David J. Dries & Richard L. Gamelli,
  Loyola University

PART V

CHAPTER 15

The NHE Family of Na^+/H^+ Exchangers; Its Known and Putative Members, and What can be Learned by Comparing Them with Each Other

Otto Fröhlich

A. INTRODUCTION

The number of sodium-hydrogen exchangers (NHEs) that have been identified is still growing, which underscores the importance of this transporter family in cellular ionic homeostasis and other processes. The first part of this article is a compilation of the known members of the NHE family. It lists the cloned DNA sequences of Na^+/H^+ exchanger-related proteins that are currently available in the databases. The sequences are mainly from cDNA libraries, but there is also a growing number of genomic sequences. The list not only contains the NHE isoforms that have been demonstrated to mediate Na^+/H^+ exchange; it also includes DNA sequences which, when translated, yield peptide sequences with significant homologies to the demonstrated exchanger isoforms. The second part of this article is a brief discussion of what can be learned from a comparison of this list of related sequences.

The Na^+/H^+ Exchanger, edited by Larry Fliegel. © 1996 R.G. Landes Company.

B. THE MEMBERS OF THE NHE FAMILY OF Na^+/H^+ EXCHANGERS

The first cloned isoform of the Na^+/H^+ exchanger is the ubiquitous form that is now referred to as NHE1 (Sardet et al, 1989).[1] This isoform NHE1, is the one with the most representatives: from hamster, human, pig, rabbit and rat;[2-9] probably there are clones from more species whose sequences have not yet been published. NHE1 was also the first isoform to be isolated as a genomic clone.[10] Because of the great interest in the transcriptional regulation of Na^+/H^+ exchange, the sequence of the NHE1 promoter region is available for man, rabbit and mouse.[10-13] Altogether, four mammalian NHE isoforms, named NHE1 through NHE4 have been identified from cDNA libraries from several different species: NHE2 from rabbit and rat,[10-12] NHE3 from man, opossum, rabbit and rat,[9,13-15] and NHE4 from rat.[9] A fifth isoform, human NHE5, exists as a partial sequence from a genomic clone.[20] Genomic clones have also been isolated for the other isoforms and from other species (see refs. 20 and 21) but these have not yet been fully characterized and published. Finally, a human genomic clone has been reported which encodes a pseudogene that is closely related to NHE3;[21,22] this gene exists in the human genome but not that of rat.

Whether these five isoforms comprise the complete set of NHE isoforms in the mammalian plasma membrane remains to be seen. There are other cDNA clones from other nonmammalian species for which it is difficult to tell whether they are representatives of known mammalian isoforms or whether they constitute their own isoform. For example, the red blood cells of trout contain a form named β–NHE since, unlike NHE1, its activity is regulated via beta-adrenergic receptor pathways.[23] This transporter, although it differs significantly in portions of its C-terminal domain from that of mammalian NHE1, is otherwise closely related to NHE1, so close in fact that when its C-terminal regulatory domain was used in a chimeric construct to replace the native domain in NHE1, it imparted beta-adrenergic regulation onto the N-terminal, transport-mediating domain of NHE1.[24] Besides β–NHE, a cDNA clone has been reported from the green shore crab (*Carcinus maenas*),[25] but it is not clear to which extent this form represents any of the already known isoforms. All these sequences are listed in Table 15.1.

The efforts to sequence the entire genome of some model species has led to genomic sequences which, after appropriate splicing (where necessary) followed by translation, yield amino acid sequences with significant homologies to the mammalian Na^+/H^+ exchangers. Among these are three apparent NHE isoforms in the nematode *C. elegans*. Also, open reading frames with lower, but still significant peptide homology were found in yeast (*S. cerevisiae*) and in a bacterium (*E. coli*). These sequences are listed in Table 15.2. Note that these sequences show little relation to the Na^+/H^+ exchangers that have been characterized by genetic and functional means in several different yeast and

Table 15.1. The members of the NHE family

Isoform	Clone type	Species	Accession number	Comment	Ref.
NHE1	genomic	human	M63805	promoter	2
			M63806	"	2
			L25272	"	3
		mouse	L37525	"	4
		rabbit	U21015	"	5
	cDNA	human	J03163		1
			M81768		2
			S68616		3
		hamster	X68970		4
		pig	M89631		5
		rabbit	X56536	partial seq.	6
			X59935		7
			X61504		8
		rat	M85299		9
NHE2	cDNA	rabbit	L13733		10
		rat	L11236		11
			L11004	splice isoform	12
NHE3	cDNA	human	U28043		13
		opossum	L43522		14
		rabbit	M87007		15
		rat	M85300		9
NHE3B	genomic	human	U16020	pseudogene	22
NHE4	cDNA	rat	M85301		9
NHE5	genomic	human	U08607	partial seq.	20
β–NHE	cDNA	trout	M94581		23
NHE-?	cDNA	crab	U09274		25

In some cases, the same sequence is published more than once at Genbank or EMBL. In these cases only one accession number is given.

Table 15.2. Putative members of the NHE family, derived from genomic sequences

Species	Name	Accession number	Comment
C. elegans	NHE-cele-1	M23064	ref. 26 partial seq.
	NHE-cele-2	U21317	gene B0495.4
	NHE-cele-3	U28737	gene F14B8.1
S. cerevisiae	NHE-yeast	U33007	gene D9461.40
E. coli	NHE-eco	U00006	

bacterial species. In this discussion, those sequences are considered to belong to a different, perhaps distantly related, family of Na^+/H^+ exchangers (the nha gene family), and they are not included in this compilation.

Finally, large-scale screening efforts to identify randomly selected cDNA clones from different libraries have yielded clones that are referred to as expressed sequence tags or ESTs. These clones typically contain only small inserts with partial sequences; they might still be useful templates to produce hybridization probes for library screening. There are now dozens of human ESTs encoding partial sequences, and this number is growing rapidly. The great majority of ESTs are of NHE1, but there are also a few ESTs of other isoforms. Recently, ESTs for mouse and rat have also been reported. Because of the great number of ESTs, only EST clones representative for the different isoforms and the tissues of origin are included in Table 15.3.

C. A COMPARISON OF THE MEMBERS OF THE NHE FAMILY

What kind of information can be extracted from such a tabulation of related sequences? The answer to this question comes from a comparison of these sequences with each other. Obviously, aligning and comparing these sequences with each other provides the necessary evidence to group them into the family of Na^+/H^+ exchangers and to identify the isoform. Before we start a more detailed comparison, it is useful to define the different regions of the NHE protein according to whether they are homologous among isoforms or not. In the past, the exchanger proteins were divided into two portions, a N-terminal portion of approximately 500 amino acids and a C-terminal portion of

Table 15.3. Representative EST clones from different species and different isoforms

Species	Isoform	Accession number	Tissue of origin
human	NHE-1	T29869	thyroid
		H45159	adult brain
		H13059	placenta
		R73284	breast
	NHE-3	H10153	infant brain
		R10042	fetal liver/spleen
	NHE-4	T31631	embryo
rat	NHE-2	H33973	PC-12 cells
mouse	NHE-1	R75225	brain

about 300 amino acids (the precise length varying with the isoform). The N-terminal portion exhibits the wavy hydropathy pattern characteristic for membrane proteins;[27,28] it constitutes the membrane-resident portion that mediates Na^+/H^+ exchange. The C-terminal portion is mainly hydrophilic; it resides in the cytoplasm and is the portion through which the activity of the exchanger is regulated by phosphorylation or other mechanisms.[29]

In light of the sequence alignments discussed below, it will be useful to divide the N-terminal portion further into two domains, A and B, and to retain the C-terminal portion as domain C. Domain A comprises the first putative membrane-spanning segment and the first exoplasmic loop, and domain B comprises the remaining membrane-spanning segments. The rationale for this further division is that the sequence homology among different isoforms is confined to the B domain. There is essentially no similarity among the A domains of different isoforms, and the little similarity that may exist among the C domains is restricted to the beginning of this domain. It is noteworthy to point out that the A domains of the NHEs of mammals and of the other higher animals all possess what appears to be a peptide leader sequence.[30] If indeed they contain a leader sequence that is cleaved post-translationally, this would mean that the N-terminus of the mature Na^+/H^+ exchanger proteins contains only a portion of the A domain and resides in the exoplasmic space. The significance behind isoform-specific A domains is not known; one possible explanation is that this domain contains sorting signals which direct the peptide into different membrane compartments, such as the apical or basolateral side of an epithelial cell or the different regions of a neuronal cell. Of course, sorting signals could also reside in the C domain.

For categorizing a sequence as one of the NHE isoforms, the best approach is to examine the N- and C-terminal stretches which differ greatly among the different isoforms. If there is a significant similarity in the A and C domains with a known isoform, it is fairly safe to conclude that the peptide belongs to that isoform. With the B domain, the differences among isoforms are more gradual but still readily detected. The dendrogram of Figure 15.1 which shows the degree of relatedness was obtained using the full-length peptide sequences, but a similar result is also obtained when only the peptide stretch corresponding to domain B is used. According to Figure 15.1, NHE2 and NHE4 are more similar to each other than to the other forms, and so are NHE3 and NHE5. As one moves away from mammals, the evolutionary distance also increases, but it is quite clear that β–NHE from trout red cells is most closely related to mammalian NHE1. The NHE isoform isolated from crab might be most closely related to NHE5/NHE3, but when only the B domains were compared, NHE-crab exhibited similar distances to all mammalian isoforms (not shown).

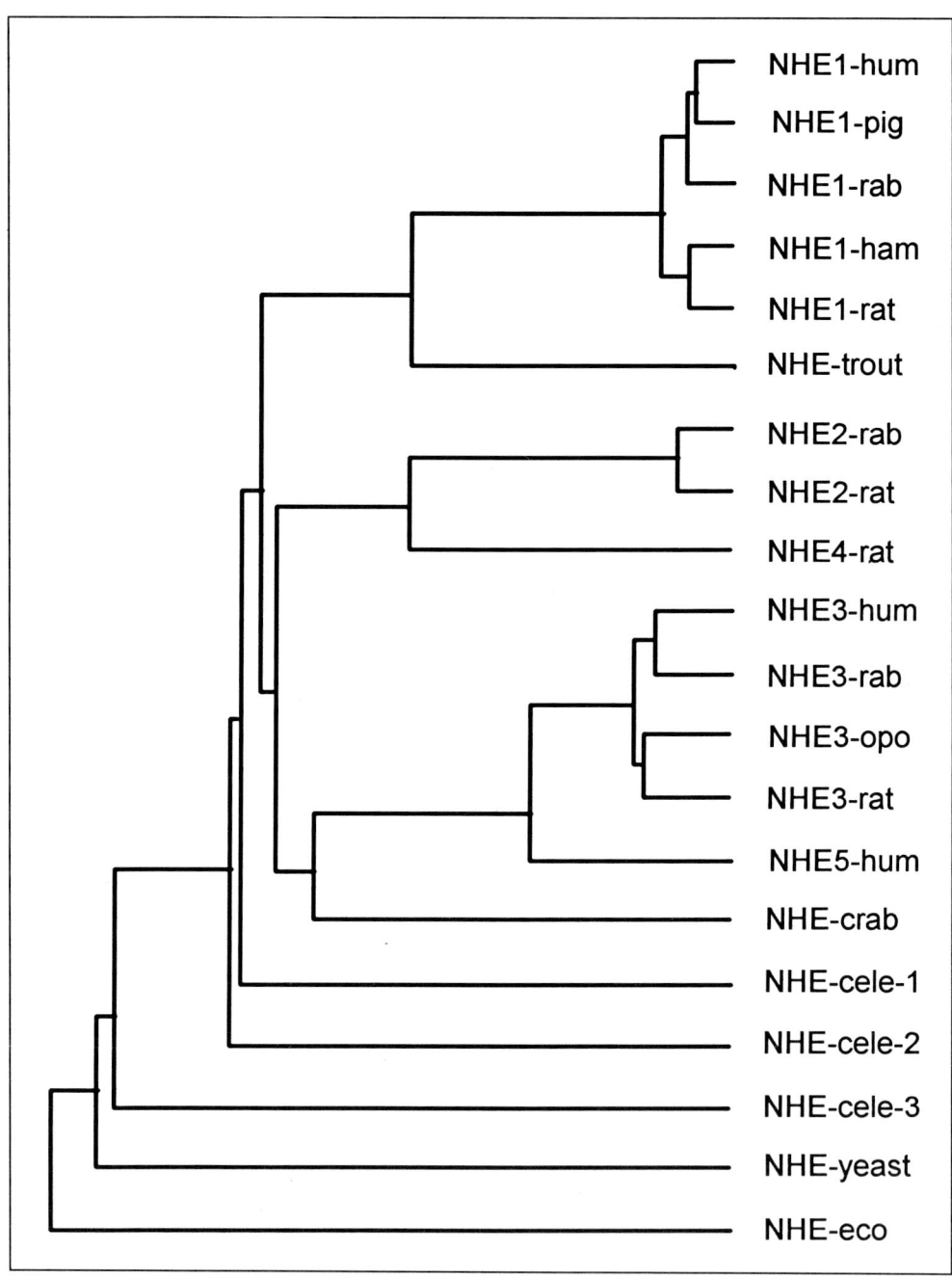

Fig. 15.1. Dendrogram depicting the relative similarities of the known and putative NHE proteins. This relationship was obtained using Pileup, a program of the GCG Genetics Computer Group suite of DNA analysis programs (Wisconsin Package, version 8.1).

As one examines the sequences from *C. elegans*, *S. cerevisiae* and *E. coli*, the distances become too large to permit an assignment to any of the mammalian isoforms. In fact, the distance is so great that one cannot exclude the possibility that these putative proteins are not Na^+/H^+ exchangers or at least not Na^+/H^+ exchangers located in the plasma membrane.

D. ARE ALL MEMBERS OF THE LIST TRUE Na^+/H^+ EXCHANGERS?

Despite the differences among isoforms, there are amino acid residues that are very well conserved. The conservation can be so strict that every isoform contains the same amino acid at its relative position in the peptide sequence (Fig. 15.2). The importance of this striking degree of conservation for Na^+/H^+ exchange is not clear on a mechanistic level. One could surmise that some sites or groups of residues need to be conserved to confer the substrate selectivity for Na^+ and H^+ ions. In other regions of the exchanger protein, strict conservation might be necessary to provide the structural elements and NHE-specific backbone that are essential for maintaining transport function in general. These are questions that will require considerable efforts of mutational structure-function analysis. Whatever the functional significance, these highly conserved stretches constitute fingerprints that are useful in classifying unknown proteins as candidate Na^+/H^+ exchangers.

The reason for including the putative gene products of Table 15.2 in the NHE family is that the translated peptide sequences contain these fingerprints. One could argue that when a gene is mutated over the course of time, its gene product might lose its original function and acquire a different one, while still maintaining a weak similarity with the original sequence. However, if that is the case one would expect that the amino acids that remain unchanged over time are not necessarily the same as those that are conserved when a new isoform with the same function evolves. The fact that the regions which are highly conserved among the mammalian NHEs are also largely conserved in the gene products of *C. elegans*, *S. cerevisiae* and *E. coli* therefore argues in favor of the notion that they still mediate Na^+/H^+ exchange. At least, one would expect them to perform a similar function. They might not necessarily mediate Na^+/H^+ exchange across the plasma membrane like the mammalian NHEs. Instead, they could, for example, mediate cation-cation or cation-proton exchange across an intracellular organellar membrane.

It might be significant in this context that the yeast sequence and two of the *C. elegans* sequences, NHE-cele-2 and NHE-cele-3, do not contain an obvious consensus sequence for a signal cleavage site. (No information is available for NHE-cele-1 since it represents a N-terminally truncated sequence.) For comparison, all mammalian NHE isoforms as well as those from trout, crab and *E. coli* appear to possess a signal

```
           171
nhe1-rat   VFFLFLLPPI ILDA-GYFLP LRQFTENLGT ILIFAVVGTL WNAFFLGGLL YAVCLVGGEQ IN-NIGLLDT LLFGSIISAV
nhe2-rat   V....Y.... V.......M. ..RP...... .FWY...... ..SIGI.LS. FGI.QIEAFG LS-D.T..QN ..........
nhe3-rat   L..FY..... V.......M. N.L.FG.... ..LY..I..I ..ATT.LS.. G.FLS.LMG E.LK.....F .....L.A..
nhe4-rat   IY..Y..... V.ES......M. T.P.F.I.S. ..WW.GL.A. I...GI.LS. .FI.QIKAFG LG-D.N..QN ........L.
nhe5-hum   T......... V.S.......M. S.L.FD...A ..TY...... ....TT.AA. WGLQQA.LVA PRVQA....F ........L.
nhe-trout  L...C..... .......... .......... ...I.P..V. ......V... ....L.Q.ESVG LS-GVD..AC .......V..
nhe-crab   V...Y..... .......V..M. N.L.FD..FT ..V.FD..FT ......VF.ITI ....IS.T.LFG --LD.PM.HM F.S.L.....
nhe-cele-1 T..Y...... .FGSS....M. N.A.F..FDS V.VFS.F.TI ....LTI.ITM ....LLMAQYDLFT --MSFTTFEI .V.SAL....
nhe-cele-2 V.M.Y.....L VF.......M. A.Q.FD.F.S ..CF.MI.TS ....T.AI.GS. ....W.IS.T.LFS V--ETP.MHM .VAAD.....
nhe-cele-3 M.MI.....A. VN.....LSMQ KKE.TIA.PK ..LF....TV VH.LLLAATM ....GIEPI.. FSFKP.F.HL FV.ATL....
nhe-yeast  Y.NV...... ..NS....E.N QVN.FN.MLS ..IF.IP.TF IS.VVI.II. ....IWTFLGLES I--D.SFA.A MSV.ATL..T
nhe-eco..  L.LVLFI..L LFAD-.WKT. T.E.L.HGRE .FGL.LALVV VTVVGI.F.I ...W.VPG--- I----P.IPA FALAAVL.PT
consensus  **.****##   **.*         .*         #          **.*              *                     * ***

           251
nhe1-rat   DPVAVLAVFE EIHINELLHI LVFGESLLND AVTVVLYHL- FEEFAS---Y EYVGISDIFL G--FLSFFVV SLGGVFVGVV
nhe2-rat   .......... N.V..Q.Y.. .......... .......N.- ..KS.CQM-- KTIQTV.V.A ..IAN..... GI...LI.IL
nhe3-rat   .......... .V.V.V.F.. I......... .......NV- ..S.VTLGG. DAVTGV.CVK ..IV...... ...TL....I
nhe4-rat   .......... .ARV..Q.YM MI...A.... .GIS.....NI- LIA.TKMHKF .DIEAV..LA ..CAR.VI.. GC....F.II
nhe5-hum   .......... .V.V.TVF.. I......... .......... .......... .......... .......... ..........
nhe-trout  .......... .......... .V........ .......... .......N.- ...SK---V GTVTVL.V.. ...VVC.... ....L..AI
nhe-crab   .......... .MQVE.V.F. .......... .......G.. .......... ..G.SELGE- ANIMAV..AS ..VA..LL.. A...TAI.II
nhe-cele-1 .......... .....I.... ..V..F.F.. N...A.F... .......G.. ..QC-SK.LIGS- NLSVL.YAT ..G....... A...AA..II
nhe-cele-2 .......... .....V.F.. .......V.F A......... .......G.A.. RM-.LT.SEIGT- .NLIT..YIN ..GV..L..AF.IGI.LL
nhe-cele-3 ...S..ST.A .LDVKKS.F. .I....C... .......... .ISI.MFETC INVIKTSDAS DEMLTITSVV SCVVQ.MI. CGF.LLC.A.
nhe-yeast  ...TI.SI.N AYKVDPK.YT II........ .ISI.MFETC QKFHGQPATF SS.FEGAGL- ........LMT FSVSLLI..L'
nhe-eco    .A..LSGIVG .GR.PKKIMG ILQ..A.M.. SGL.SLKFA VAVAMG---- TMIFTVGGAT ----VE.MK. AI..ILA.F.
consensus  #*#*        ***.#.##     ###.#.##                  #**           *         *   **        #
```

Fig. 15.2. Alignment of all NHE peptide sequences over a stretch of 160 amino acids. The numbering given corresponds to the amino acid sequence of NHE1-rat, which is also given in the first lane. A period marks where the other isoforms possess the same amino acid as NHE1. A dash denotes a gap in the sequence. The bottom row shows the locations with the highest degree of conservation: # means that all sequences examined contain the same amino acid; * means that at most two of the sequences deviate from the consensus sequence.

cleavage site after the first putative membrane-spanning (or TM, for transmembrane) segment. The absence of a signal cleavage site may be even more significant in the light that when the putative proteins of Table 15.2 are aligned with the mammalian NHE sequences, they appear to possess no A domain at all. Instead, their first TM segment starts at the position equivalent to the second TM of the mammalian NHEs. Finally, an interesting result is obtained by examining the yeast sequence by the PSORT program, a suite of algorithms developed by K. Nakai which is available through the PSORT server via the Internet.[31] According to this examination, the putative yeast NHE peptide is rated as a possible protein of the inner mitochondrial membrane. This finding is consistent with the observation that the mitochondrial membrane contains cation-hydrogen exchangers which have not yet been identified on the molecular level.[32]

E. ESTABLISHING A CONSENSUS HYDROPATHY PROFILE OF NHE PROTEINS

The standard analysis for estimating where the TM segments are located on the amino acid sequence of a membrane protein is to examine the hydropathy profile for alternating stretches containing primarily nonpolar or polar amino acids.[27] Unfortunately, despite many refinements, this type of analysis is still relatively unreliable in predicting the precise location of a TM segment. However, if one assumes that the general topology of a membrane protein is the same among all its isoforms, it might be useful to examine the NHE proteins collectively. Figure 15.3 shows the result of a TM analysis by the PSORT server for the B domain of all NHE isoforms (omitting the first TM segment that is part of the leader sequence and shows no homology among the different isoforms). Not surprisingly, the predicted TM segments of the different mammalian NHE members are not precisely in register with each other. More surprisingly, however, there is also uncertainty in just being able to predict the presence of a TM segment. For example, the algorithm did not acknowledge the eighth TM segment of NHE2 and postulated two TM segments at this location for NHE4. More apparent discrepancies are found for the nonmammalian proteins. This could mean that these proteins have a different topological arrangement in the membrane or that the predictive power of the analysis is rather limited.

However, when all sequences are viewed together, a consensus TM pattern emerges. This pattern predicts 11 TM segments for the B domain of the mammalian NHE members. Thus it appears that the collective TM analysis of all members of a membrane protein family brings us closer to the real picture of the protein's membrane topology. It is understood that any systematic deviation from the hydropathy profile rules, mainly due to conserved "atypical," charged or polar, residues in the middle of a membrane-spanning segment, is not eliminated.

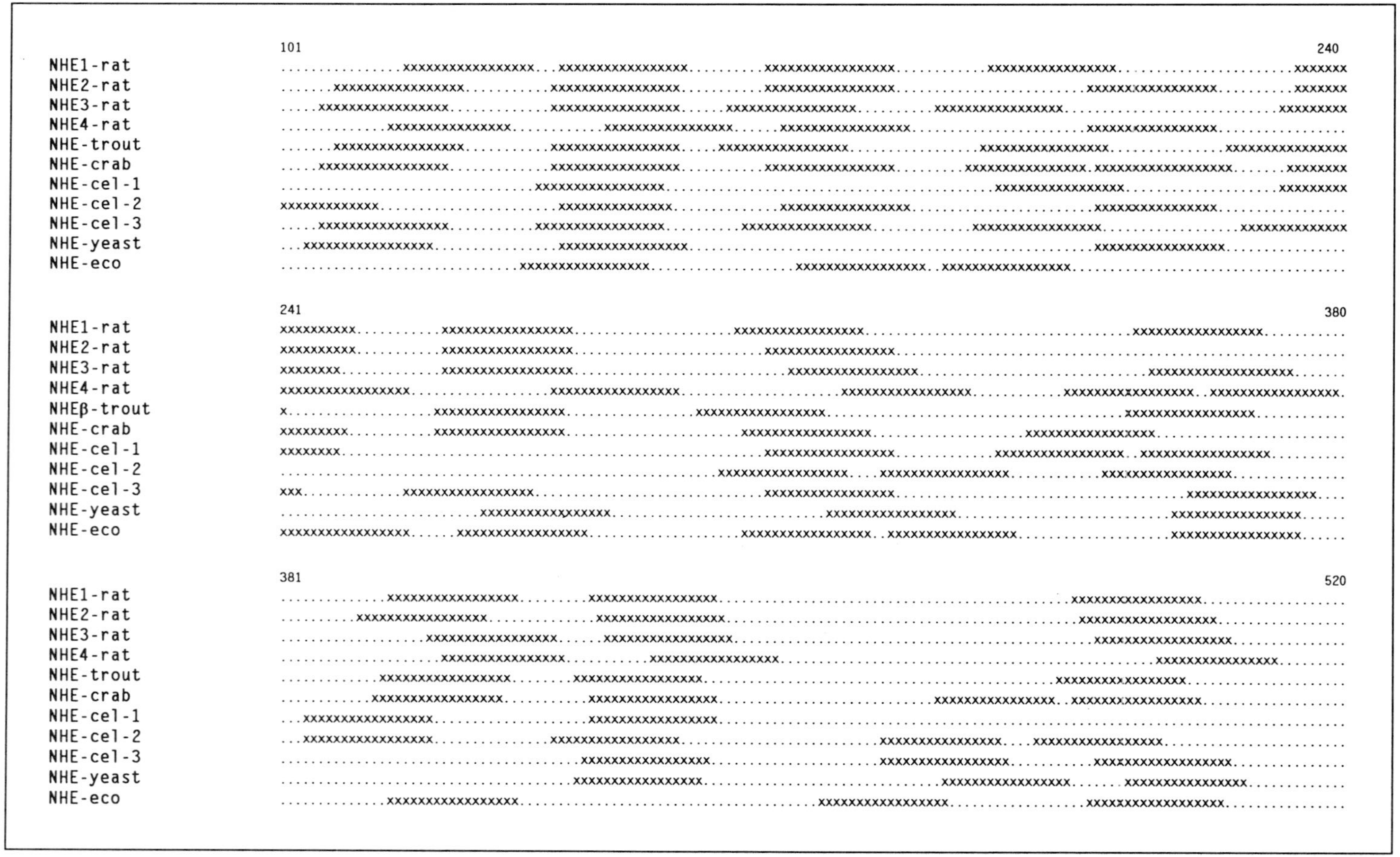

Fig. 15.3. Predicted location of the TM segments as predicted by the PSORT suite of algorithms.[31] The strings of x's mark the stretch of 17 residues that is most likely to traverse the membrane. The sequences have been shifted relative to each other in order to align the second TM stretch which is predicted to be virtually the same for all isoforms.

Nevertheless, this method has reduced some of the noise and might serve as a guide for future mutational and labeling studies of the topology of the Na^+/H^+ exchanger protein.

F. SUMMARY

In conclusion, the family of NHE proteins is a growing one. There are at least five mammalian isoform, in addition to isoforms in trout and crab. These isoforms are very highly conserved in several stretches, which permits one to establish a fingerprint sequence that can be used to identify new members of the NHE family. Furthermore, the databases contain several DNA sequences which code for putative members of the NHE family in *C. elegans*, yeast and *E. coli*. Even though their homology with the mammalian exchangers is much weaker than what is found among the mammalian isoforms, they can be tentatively included in the NHE family on the basis of the fingerprint sequences. When all NHE forms are aligned and collectively analyzed for their hydropathy profile, one arrives at a consensus profile which might become a useful working hypothesis for experimental verification.

REFERENCES

1. Sardet C, Franchi A, Pouysségur J. Molecular cloning, primary structure, and expression of the human growth factor-activatable Na^+/H^+ antiporter. Cell 1989; 56:271-280.
2. Miller RT, Counillon L, Pages G et al. Structure of the 5'-flanking regulatory region and gene for the human growth factor-activatable Na^+/H^+ exchanger NHE1. J Biol Chem 1991; 266:10813-10819.
3. Kolyada AY, Lebedeva TV, Johns CA et al. Proximal regulatory elements and nuclear activities required for transcription of the human Na^+/H^+ exchanger (NHE1) gene. Biochim Biophys Acta 1994; 1217:54-64.
4. Dyck JRB, Silva NL, Fliegel L. Activation of the Na^+/H^+ exchanger gene by the transcription factor AP-2. J Biol Chem 1995; 270:1375-1381.
5. Blaurock MC, Reboucas NA, Kusnezov JL et al. Phylogenetically conserved sequences in the promoter of the rabbit sodium-hydrogen exchanger isoform 1 gene (NHE1/SLC9A1). Biochim Biophys Acta 1995; 1262: 159-163.
6. Takaichi K, Wang D, Balkovetz DF et al. Cloning, sequencing and expression of Na^+/H^+ antiporter cDNAs from human tissues. Am J Physiol 1992; 262:C1069-C1076.
7. Fliegel L, Dyck JRB, Wang H et al. Cloning and analysis of the human myocardial Na^+/H^+ exchanger. Mol Cell Biochem 1993; 125:137-143.
8. Counillon L, Pouysségur J. Nucleotide sequence of the Chinese hamster Na^+/H^+ exchanger NHE1. Biochim Biophys Acta 1993; 1172:343-345.
9. Reilly RF, Hildebrandt F, Biemesderfer D et al. cDNA cloning and immunolocalization of a Na^+-H^+ exchanger in LLC-PK1 renal epithelial cells. Am J Physiol 1991; 261:F1088-F1094.
10. Fliegel L, Sardet C, Pouysségur J et al. Identification of the protein and cDNA of the cardiac Na^+/H^+ exchanger. FEBS Lett 1991; 279:25-29.

11. Tse C-M, Ma AI, Yang VW et al. Molecular cloning and expression of a cDNA encoding the rabbit ileal villlus cell basolateral membrane Na^+/H^+ exchanger. EMBO J 1991; 10:1957-1967.
12. Hildebrandt F, Pizzonia JH, Reilly RF et al. Cloning, sequence, and tissue distribution of a rabbit renal Na^+/H^+ exchanger transcript. Biochim Biophys Acta 1991; 1129:105-108.
13. Orlowski J, Kandasamy RA, Shull GE. Molecular cloning of putative members of the Na^+/H^+ exchanger gene family. cDNA cloning, deduced amino acid sequence, and mRNA tissue expression of the rat Na^+/H^+ exchanger NHE1 and two structurally related proteins. J Biol Chem 1992; 267:9331-9339.
14. Tse CM, Levine SA, Yun CC et al. Cloning and expression of a rabbit cDNA encoding a serum-activated ethylisopropylamiloride-resistant epithelial Na^+/H^+ exchanger isoform (NHE2). J Biol Chem 1993; 268:11917-11924.
15. Wang Z, Orlowski J, Shull GE. Primary structure and functional expression of a novel gastrointestinal isoform of the rat Na^+/H^+ exchanger. J Biol Chem 1993; 268:11925-11928.
16. Collins JF, Honda T, Knobel S et al. Molecular cloning, sequencing, tissue distribution and functional expression of a sodium/hydrogen ion exchanger (NHE2). Proc Natl Acad Sci USA 1993; 90:3938-3942.
17. Brant SR, Yun CH, Donowitz M et al. Cloning, tissue distribution, and functional analysis of the human Na^+/H^+ exchanger isoform, NHE3. Am J Physiol 1995; 269:C198-C206.
18. Amemiya M, Yamaji Y, Cano A et al. Acid incubation increases NHE3 mRNA abundance in OKP cells. Am J Physiol 1995; 269:C126-C133.
19. Tse C-M, Brant SR, Walker MS et al. Cloning and sequencing of a rabbit cDNA encoding an intestinal and kidney-specific Na^+/H^+ exchanger isoform (NHE3). J Biol Chem 1992; 267:9340-9346.
20. Klanke CA, Su YR, Callen DF et al. Molecular cloning and physical and genetic mapping of a novel human Na^+/H^+ exchanger (NHE5/SLC9A5) to chromosome 16q22.1. Genomics 1995; 25:615-622.
21. Szpirer C, Szpirer J, Riviere M et al. Chromosomal assignment of four genes encoding Na^+/H^+ exchanger isoforms in human and rat. Mammalian Genome 1994; 5:153-159.
22. Brant SR, Bernstein M, Wasmuth JJ et al. Physical and genetic mapping of a human apical epithelial Na^+/H^+ exchanger (NHE3) isoform to chromosome 5p15.3. Genomics 1993; 15:668-672.
23. Borgese F, Sardet C, Cappadoro M et al. Cloning and expression of a cAMP-activated Na^+/H^+ exchanger: evidence that the cytoplasmic domain mediates hormonal regulation. Proc Natl Acad Sci USA 1992; 89:6765-6769.
24. Borgese F, Malapert M, Fievet B et al. The cytoplasmic domain of the Na^+/H^+ exchangers (NHEs) dictates the nature of the hormonal response: behavior of a chimeric human NHE1/trout beta NHE antiporter. Proc Natl Acad Sci USA 1994; 91:5431-5435.

25. Towle DW, Wu W. Positive identification and amplification of Na$^+$/H$^+$ antipoter cDNA prepared from crab (Carcinus Maenas) gill mRNA. Bull Mt Desert Biol Lab 1994; 33:122-123.
26. Marra MA, Prasad SS, Baillie DL. Molecular analysis of two genes between let-653 and let-56 in the unc-22(IV) region of Caenorhabditis elegans. Mol Gen Genet 1993; 236:289-298.
27. Engelman D, Steitz T, Goldman A. Identifying nonpolar transbilayer helices in amino acid sequences of membrane proteins. Annu Rev Biophys Biophys Chem 1986; 15:321-353.
28. Fliegel L, Fröhlich O. The Na$^+$/H$^+$ exhanger: an update on structure, regulation and cardiac physiology. Cardiovasc Res 1993; 296:273-285.
29. Wakabayashi S, Fafournoux P, Sardet C et al. The Na$^+$/H$^+$ antiporter cytoplasmic domain mediates growth factor signals and controls "H$^+$-sensing." Proc Natl Acad Sci USA 1992; 89:2424-2428.
30. von Heijne G. Patterns of amino acids near signal-sequence sites. Eur J Biochem 1983; 133:118-127.
31. The PSORT server can be reached by pointing the WWW browser at psort@ nibb.ac.jp.
32. Garlid KD. Mitochondrial cation transport: a progress report. J Bioenerg Biomembr 1994; 26:537-542.

INDEX

Page numbers in italics denote figures (f) or tables (t).

A

Aalkjaer C, 49
Acidosis, 10, 12, 34, 37, 50, 91, 96, 134, 164, 173, 187, 188, 191, 192, 203, 204, 207
Adrenosine triphosphate (ATP), 104, 106, 108, 137, 173
-ase (ATPase), 27, 257, 260
Aicken CC, 49
Alanine, 57
Alberta Heritage Foundation for Medical Research, 223
Alpern RJ, 32
Alpha (α)-intercalated cells, 27, $29t$
Amiloride(s), 3, 4, 6, 11, 12, 23-25, 27, 47, 49, 54, 55, 69, 70, 81-83, 102, 107, 112, 132, 133, 138, 140, 151, 176, 188, 192, 195, 198, $200t$, $201t$, 203-205, 218, 220, 231, 232, 237, 238, 254, 257, 265, 269, 270, 273, 276, 278, 279
analogs, 11-13, 70, 104, 112, 132, 138, 198, 218, 232, 265, 270, 271, 273, 278-280
methylisobutyl, $195-199f$
Anderson SE, 203
Angioplasty, 207
restenosis after, 11
Angiotensin II, 8, 29, $30t$, 49, 51, $52f$, 54, 59, 151, 153, 163, 176, 194
receptor, 58
Antiporter(s), 6, 91, 94, 192, 194, 195, 201, 206, 207, 218, 244, 252, 254-256, $258t$, $259t$, 269, 271
Ca^{2+}/H^+, 243
Na^+/H^+, 1, 2, 3, 11, 72, 187, 188, 222, 237, 238, $240f$, 242, 243, 263, 273
basic characteristics, 3
molecular dissection of bacterial, 227-244
molecular nature and properties of, 229-243
antiporters of *E. coli*, 229-235
function as a pH sensor and titrator, 232-234
NhaB's pH dependence, 234, 235
sensitivity to inhibitors, 231, 232
stoichiometry, 229-231
antiporters of other bacteria, 235-243
common denominator search, 237-239
extreme alkaliphiles, 242
homologous/heterologous expression, 242, 243
of vibrio genus, 235-237
tetracycline-resistance locus, 239-242
other Na^+ excretion machinery, 243, 244
regulation of nhaA expression, 228, 229
role in Na^+ homeostasis, 227, 228
stimulation, 8

Na^+/K^+, 1
nonmammalian, 4
role of Na^+/H^+ isoforms in cell volume regulation, 101-116
clinical significance and pathophysiology, 102
in regulatory volume increase, 102-104
isoforms of NHE, 111-114
cloning, 111
functional and pharmacological properties, 112-114
tissue and cellular distribution, 111
mechanisms of activation, 104-111
responisveness to osmolarity, 114-116
sod2, 256, 257, 260
Apical membrane, 26, 72, 81, 111, 112, 129, 149
amiloride resistance, 81, $82t$
Na^+/H^+ exchanger, 22, 23, 25, 27, 29-31, 34, 36, 112, 150
protein, 24
AP-1, 10, 30, 31, 35, 36, 56, 74, $84t$, 85, 92-94, 96, 104, $105f$, 128, 134-137
site, 95
AP-2, 10, 56, 95, 98, 128, 137
site(s), 94-96
Aronson PS, 132
ARPA, 261
Arrhythmia(s), 11, 171-173, 175-179, 195
Arrythmogenesis, 13, 171, 173, 174, 176, 178, 179, 198
Asimakis GK, 205
Atherosclerosis, 49
Atrial natriuretic peptide (ANP), 153
Avkiran M, $200t$

B

Banuelos MA, 260
Barclay Foundation, David and Frederick, 179
Basolateral membrane, 72-74, 81, 85, 112, 129, 149
Na^+/H^+ activation, 24-28, 150
$Na^+/H^+CO_3/CO_3$ symporter, 29, 34
Berk BC, 55, 59
Bertrand B, 114, 115
Beta (β)-intercalated cells, 27, $29t$
Beta (β)-NHE, 31, 77, 80, 112, 135, 152, 288-290
Bianchini L, 115
Bicarbonate, 50, 115, 149, 269
absorption, 33, 34, 115, 150
Blaurock NC, 96
Blood vessel wall, *see* Vascular smooth muscle cell(s).
Bobik A, 54, 55
Bond JM, $200t$
Boyer MJ, 275

Brewster JL, 109
British Heart Foundation, 179
Brooks WW, 178
Bugge E, 201t

C
Cabado A, 136
Caco-2, 73, 74, 84t, 85
Cala PM, 103
Calcineurin
  gene disruption strains, 257
Calcium
  cytosolic, 32, 107, 192
  effect on NHE isoforms, 113t
  excess, 11
  extracellular, 54, 205, 219f
  intracellular, 51, 54, 57, 59, 107, 114, 153, 154, 158, 175, 192-194, 198, 204, 205, 207, 223
  ionophores, 8, 59, 107, 153
Calmodulin (CaM), 2t, 8, 9f, 32, 52f, 75, 76f, 84, 114, 115, 151, 154, 155f, 158, 160, 162f
  antagonists, 59
  kinase, 30, 32, 33, 57, 59, 107, 114, 130, 154, 158, 164, 188
cAMP, 30, 31, 33, 73, 74, 77, 84, 85, 92, 106, 107, 112, 135, 136, 151, 161, 163
  analogs, 135, 136, 152
  response element binding site, 94, 128
Canessa M, 61
Cano A, 163
Cardiac
  disorder possible treatments, 13
  hypertrophy, 11
Cardiodepressant actions/effects, 176, 204
Cardiomyocytes
  as exchange pathway, 216
Cardioprotective actions/effects, 195, 198, 201, 203, 204, 206, 207, 218
Carr P, 49
cDNA, 4, 11, 94, 111, 124-126, 138, 139, 161, 287, 288, 290
Cellular alkalinization, 51, 52, 54, 57, 59
cGMP, 73, 153
Chemotactic factors, 3, 8
Chemotherapy, 264
Chimeric exchangers, 75-81, 157f
  proteins, 158
Chou PY, 154
Cimetidine, 70, 133t
Cl^-/HCO_3^- exchange, 3, 24, 27, 48, 82, 104, 216, 217f
Clonidine, 70, 133t, 232

Cloning, 1, 3, 4, 48, 57
  expressed sequence tags (ESTs), 290
  of a cytoplasmic cofactor, 31
  of antiporter(s), 227, 235, 242, 243
  of cDNA, 111, 139, 287, 290
  of NHE gene family, 21, 125, 287
  of NHEs, 73-81, 111, 288
  of NHE 1, 69, 111, 124, 288
  of NHE1 promoter, 92-95
  of NHE3, 20, 125, 288
  of NHE4, 111, 288
CLUSTAL W alignment algorithm, 127f
Colchicine, 35
Collins JF, 124
Contractile recovery during reperfusion, 196f, 197f
  time to restoration of sustained, 202f
Corr PB, 178
Coughlin SR, 177
Cragoe Jr EJ, 270
C-terminal
  domain, 70, 75, 77, 139, 150, 288
    phosphorylation of, 8, 291
  region, 2t, 5, 130, 134, 135
Cycloheximide, 35
Cytoplasmic domain, 10, 31, 124, 136, 160, 162f, 164
  anatomy, 150-152
  hydrophilic, 48
  model of NHE1, 9f
  of NHE3, 31, 136
Cytosensor microphysiometer, 257-259f

D
Dascalu A, 107
Deletion mutants, 157f, 160, 227-229, 232-236, 243
Dennis SC, 174, 200t
Dexamethasone, 37
Diabetes, 34t
Diabetic cardiomyopathy, 221
Diacylclycerol, 8, 188, 191f, 194
  analogs, 193
Dibrov P, 238
4,4-diisothiocyanstilbene 2,2-disulfonic acid (DIDS), 265, 267f, 271, 273-279
Donowitz M, 81, 83
Dopamine, 29, 30t
Duan J, 192, 198, 200t, 207
Duff HJ, 198, 200t
de Toit EF, 200t, 201
Dyck J, 13

E
Elder E, 61
Endocytosis to regulate NHE, 3, 31, 32
Endothelin, 30t, 33, 51, 54, 176, 188, 189f, 194
Engelman DM, 5f, 70

Epidermal growth factor (EGF), 3, 8, 51, 52, 59, 80, 85, 104, 152, 153
Escherichia coli, 2t, 139, 227, 229-239, 242-244, 288, 289t, 293, 297
　antiporter, 72
$ET_{A/B}$ receptors, 33
Ethyl-isopropyl amiloride (EIPA), 24, 26, 35, 50, 51, 51, 55, 59, 70, 133t, 154, 155, 157f, 175, 200t, 201t, 203, 205, 206, 265, 267f, 270-279
Eukaryotic isoform
　first gene isolated, 1
　Na^+/H^+ exchange, 22
Extracellular signal-related kinase(s) (ERK), *see* Mitogen-activated protein kinase(s).
Extracellular sodium, 1, 51

F
Faes FC, 201t
Fasman GD, 154
Fertilization, 3, 8
Fetal bovine serum (FBS), 74t, 76f-78f, 80
Fibroblast growth factor (FGF), 55, 73-77, 81, 153
Fliegel L, 13, 57, 96, 98, 223, 261
Focal adhesions
　NHE1 in, 164
Foster CD, 49
Free oxygen radical(s), 172, 173, 178, 179

G
Ganz MB, 163, 269
Gap, 238f
GCG Genetics Computer Group, 292f
Geck P, 103
Genistein, 85
Gerchman Y, 139
Gilchrist JSC, 223
Glucocorticoids, 10, 34, 36, 37, 83, 137
　receptor, 128
　response element, 93f
Glu 262, 5
　Ile, 6
Glycosylation, 138
Goldman A, 70
Goldstein JA, 177
Goss G, 108, 134
G-protein(s), 60, 137, 153, 161-164, 191f
　-coupled receptors, 57
Grinstein S, 52, 104, 107, 108, 134, 136, 268
Grotyohann, LW, 204
Growth factors, 2t, 3, 8

H
H^+
　-ATPases, 3, 22, 27, 28, 251
　cytoplasmic, 3
　extracellular, 173, 175, 215
　intracelllular, 58, 72, 73, 112, 139, 155, 158, 161, 162f, 164, 191, 215, 219, 223
　secretion, 36, 115, 129
　sensor, 72, 81, 137, 139
　transport, 27, 28, 35, 137
Haddock P, 179
Hahnenberger, K, 261
Haist J, 207
Han J, 109
Harmaline, 70, 133t, 231, 232
Hata K, 200t
Haworth R, 13, 98
HCO_3^-, 22, 23, 25-27, 49, 54, 265, 266
　absorption, 33-35, 129
　/Cl^- exchanger, 264-266, 268f, 269, 271, 273, 275, 279, 280
　-free media, 12, 103, 104
　transport, 32, 37
Hearse DJ, 171, 179
Heart and Stroke Foundation
　of Canada, 13, 98, 207
　of Manitoba, 223
Heat-shock protein 70 (hsp70), 2t, 10, 134
Hemmingsen S, 261
Hendrikx M, 200t
Hexamethyl amiloride, 11
High osmolarity glycerol (HOG), 109, 252
Ho J, 207
Hoescht, 198
HOE642, 175, 176, 188, 198, 200t
HOE694, 11, 50, 70, 112, 133t, 175, 176, 188, 195, 198, 200t, 201t
Homeostasis
　acid-base, 29
　cellular, 123, 150, 207, 263, 287
　extracellular fluid volume, 26, 29, 150
　intracellular
　　Ca^{2+}, 178, 179
　　ionic, 173, 174f
　water, 26, 116
Hopfer U, 21
Hori M, 200t
Hormone-receptor coupling, 52f
Howard RL, 56
Huang CL, 59
Hydrogen, *see* H^+.
Hyperosmolarity, 33, 57, 73, 74, 115, 137
　effect on NHE isoforms, 113t
Hyperosmotic challenge, 8, 102, 103
Hyperplastic agents, 51, 52, 54, 60
Hypertension, 48, 49, 52f, 59, 60
　studies, 47, 55t
Hyperthermia, 278, 279, 281
Hypertonicity, 22, 26
Hypertrophic agents, 51, 52, 54, 60
Hypertrophy, 51, 178, 215, 222, 223
Hypoxia, 102, 177, 203, 218-221, 264, 279. *See also* Ischemia.

I

Ibuki C, 179, 200t
Ikeda T, 161
Ileal brush border membranes, 32, 77, 83
Insulin, 3, 153, 221, 277
Intestinal NHE studies, see Na+/H+ exchanger(s).
Ionomycin, 157f, 158
Ischemia, 10, 11, 13, 102, 133, 171-173, 176, 177, 187, 191-194, 196f-199f, 201, 203-205, 207, 218-221. See also Hypoxia.
Ischemic cardiomyocyte, 193
Israel Academy of Sciences, 244
Itoh T, 108

J

Jähde E, 277
Jia Z, 252

K

K
  secretion, 27
K_i
  amiloride for NHE1/2, 70
  apical membrane for amiloride, 81
  basolateral membrane for amiloride, 82
Kaplan SH, 201t
Kapus A, 136
Karmazyn M, 195-197f, 199f-202f, 204
Khandoudi N, 200t, 207
Kidney
  acid excretion, 21
  Na+/H+ exchange in, see Na+/H+ exchanger(s).
Kihara Y, 178
Killing effects
  cell, 271, 273, 275, 279, 280
  of tumors, 12
Kim GE, 279
Kinne R, 21
Kitakaze M, 200t
Klein HH, 201t
Kleyman R, 270
Knickelbein RG, 81
Kolyada AY, 56, 94
Konings AWT, 279
Kranzhofer R, 50
Krapf R, 128
Krulwich T, 242
Kuin A, 277
Kusuhara M, 61
Kyte and Doolittle parameters, 6, 70

L

Lactate, 204
Ladilov YV, 201t
Lazdunski M, 173, 174
LeBlanc G, 234
Lemasters JJ, 204
Leu167, 6, 138
Leukotriene, 134, 206
Levine SA, 136
Little PJ, 59
Liu, 279
LLC-
  PK cells, 24, 35
  PK1 cells, 32, 34, 35, 74, 95
LLK-PK cells, 35, 113t
Loop of Henle, 115
Lucchesi PA, 56, 61
Lucena N, 13
Lukacs G, 136

M

MACAW, 238f, 240f
MacVector™, 93f
Mahrer, 83
Maidorn RP, 270
Manning AS, 171
Mathur S, 207
Matsuda S, 200t
MCT cells, 34-36, 92, 96, 113t
Medical Research Council of Canada, 13, 207, 223
Meng H-P, 200t
Message(s), 12, 72, 73, 111, 221, 223
Messenger ribonucleic acid (mRNA), 10, 11, 24, 26-28, 34-37, 47, 51, 55, 56, 60, 83, 91, 96, 98, 111, 123, 128-130, 137, 161, 178, 221-223
Methylisobutyl amiloride (MIA), 195-202f
Methylprednisolone, 83
Michaelis-Menten
  behavior, 112
  kinetics, 132
Miller RT, 92
Mitchell P, 1, 2t, 227
Mitochondrial exchanger, 123
Mitogen-activated protein (MAP) kinase (also MEK), 57-59, 108-110f, 223, 252
  -kinase (MEKK), 109
Mitogenesis, 12
Mitsuka M, 50, 51, 56, 61
Miyakoshi J, 278
Mochizuki S, 200t
Moe OW, 32
Moffat MP, 192, 196f-202f, 207
Molecular Devices Corporation, 257
Molecular Evolutionary Genetics Analysis (MEGA) program, 127f
Morgan JP, 178
Moyle J, 227
Murer H, 21
Murphy E, 200t

Muslin A, 61
Myers ML, 200t, 207
Myocardial preconditioning, 205, 206
Myocardium
 Na+H+ exchanger in, 10
Myocytes, 11, 91, 96, 175, 179, 188, 191, 192, 198
Myosin light chain kinase (MLCK), 107-111, 137

N

Na^+
  absorption, 30, 83
  extracellular, 54, 219f, 228, 229
  tolerance and export from yeast, 251-261
    bidirectional exchange on sod2, 256, 257
    fungal sodium/proton antiporter, 252-256
    heterologous expression of sod2 sodium/proton antiporter, 260
    PMR2/ENA1-type sodium export pumps, 257-259
    ZSOD2 sodium/proton antiporter, 256
  transport, 30, 31, 137
Na^+-ATPase(s), 251, 252, 256, 257
NaCl
  absorption, 23-27, 29, 33, 82-84t
  transport, 4, 21, 26, 29
Na^+/H^+ antiporter. See Antiporter.
$NaHCO_3$
  absorption, 22-24, 26, 33
  symport, 187, 191
  transport, 21, 28, 29, 215, 217f
Na^+/H^+ exchange inhibition's protective effects in ischemic reperfused heart, 200, 201
Na^+/H^+ exchanger(s)
  biochemistry and molecular biology of, 1-13
    future prospects, 12, 13
    history, 1-3
      first cloning, 1
    importance in muscle, 10, 11
    importance in tumor cells, 11, 12
    pharmacology of mammalian, 4
    physiology of, 3, 4
    regulation of expression of NHE1, 10
    role of cytoplasmic region, 8-10
    structure of transmembrane region of NHE1, 4-8
    transport capacity, 11
  evidence for role in cell growth, 223t
  in mammalian kidney, 21-37
    acute regulation of renal Na^+/H^+ exchanger, 28-33
      by Ca^{2+} calmodulin and tyrosine kinases, 32, 33
      in proximal tubule transport and apical membrane, 29, 30
      of NHE3 by
        cAMP-dependent protein kinase, 30-32
        protein kinase C, 32
    chronic regulation of renal Na^+/H^+ exchanger, 33-37
      chronic metabolic acidosis, 34-36
      glucocorticoids, 36, 37
    in nephron, 22-28
      cortical collecting duct, 27, 28
      distal convoluted tubule, 26, 27
      in descending limb, 25
      in proximal tubule, 22-25
        apical membrane, 23, 24
        basolateral membrane, 24, 25
        NaCl absorption, 23
        $NaHCO_3$ absorption, 22, 23
        NH_4^+ secretion, 23
      in thick ascending limb, 25
        apical membrane, 26
        basolateral membrane, 26
        HCO_3^- absorption, 25
        NaCl absorption, 26
      medullary collecting duct, 28
    in plasma membrane, see Plasma membrane.
  mediating myocardial ischemic and reperfusion injury, 187- 207
    exchange and injury, 193-207
      a non-Ca^{2+}-dependent component of reperfusion injury, 204, 205
      experimental evidence for NHE involvement, 195
      NHE activation in platelets and neutrophils, 206, 207
      NHE in lactate-induced depression of recovery, 204
      NHE's role in myocardial preconditioning, 205, 206
      nonpharmacological protection: NHE inhibition, 203, 204
      pharmacological studies, 195-198
      protective effects of NHE inhibitors, 198-203
      theories for NHE involvement, 193-195
    future directions, 207
    intracellular pH regulatory mechanisms, 191
    regulation of Na^+ and Ca^{2+} concentrations, 191-193
  molecular studies of intestinal epithelial, 69-85
    cellular distribution of, 72, 73
    members of the NHE gene family, 69-72
    regulation of, 73-85
      cloned exchangers, 73-81
        chimeric NHE, 75-81
        role of C-terminus, 75
        short-term regulation, 73, 74
      intact intestinal tissue studies, 81-85
NHE family, 287-297
  comparison of members, 290-293
  consensus hydropathy profile, 295-297
  members, 288-290
  true NHEs, 293-295
quaternary structure, 139
rate in PS120 fibroblasts, 74t
regulation in vascular smooth muscle, 47-62
  in blood vessel wall, 48, 49
  in hypertension, 59-61
  in vivo: tone and growth, 49, 50

in VSMC growth, 51-55
    activation by hyperplastic/hypertrophic stimuli, 52-54
    growth response dependency on extracellular Na^+, 54
    inhibitors blocking growth, 54, 55
    VSMC proliferation, 54
in VSMC migration and inflammation, 51
of function, 55-59
    by kinases and phosphates, 56-59
        PKC-dependent pathways, 58, 59
        PKC independent pathways, 59
    by mRNA and protein expression, 55, 56
regulatory cytoplasmic domain of, 149-165
    anatomy of cytoplasmic domain, 150-152
    calcium/calmodulin, 153-160
    G-proteins, 161-164
    phosphorylation, 152, 153
    regulatory cofactors, 160, 161
Na^+/H^+ exchanger inhibitor(s)
    therapeutic potential in tumor selective therapy, 263-281
        acidic microenvironment of tumors, 263, 264
        chemotherapy and pH_i regulation, 280
        hyperthermia and pH_i regulation, 278, 279
        intracellular pH regulation, 264-268
        potential therapeutic application, 269, 278
            acute administration in vivo, 271-273
            prolonged administration, 273-278
            short-term effects in vitro, 270, 271
        radiation and pH_i regulation, 279, 280
        role in tumor cell proliferation and survival, 268, 269
Nakai K, 295
Na^+/K^+-ATPase, 11, 25, 30, 175, 176, 191f, 192, 194, 251
$Na^+/K^+/2Cl^-$ cotransporter, 102, 103, 108
Na^+/K^+ pump, 173-176, 178, 179
Neely JR, 195, 200t, 204
Neointima, 49, 50
Nephron(s), 22, 26, 28
    juxtamedullary, 24, 25, 128
Neutrophils
    NHE in, 206, 207
Neve KA, 163
Newell K, 271
NH_4
    secretion, 23
    trapping in lumen, 25
NhaA, 2t, 3, 72, 227, 229-239
NhaB, 3, 227, 229-238
NhaC, 238
NhaR, 227-229
NHE
    inhibition, 188, 193, 195-201, 203, 204, 206, 207
    isoforms'
        characteristics, 126t
        distribution in kidney, 29t
        evolutionary relationship of eukaryotic members, 127f
        pharmacological properties, 133t

NHE1, 2t, 4-6, 8, 10, 24-29, 31, 34-37, 47, 48, 50, 57, 69, 70, 72, 74, 75, 77, 80, 81, 83, 85, 108, 111-115, 124-126, 128, 130,134-139, 149-155, 157f, 158, 160, 161, 164, 165, 187, 188, 220, 222, 223, 232, 239, 255f, 265, 288-290, 294f
    candidate membrane-spanning segments of, 6t
    distribution, 72
    expression, 10, 11, 56, 96, 97f, 129, 132, 178
    gene, 10, 56, 91, 92, 94-96, 98, 125, 128
        regulation, 91, 94, 97f
    kinases, 61, 153, 154
    phosphorylation, 60, 80
    promoter
        regulation of expression, 91-98
            cloning and analysis of, 92-95
            comparison of mouse, human and rabbit, 95, 96
            future directions, 96-98
    regulation model, 162f
    topology, 5
NHE2, 2t, 4, 24, 28, 29t, 48, 111-115, 124-126, 132, 135-138, 152, 165, 265, 288-290, 295
    distribution, 72, 73, 129
    epithelial, 69, 70, 72, 74, 77, 80-85
    gene, 128
NHE3, 2t, 4, 24-27, 29, 34-37, 48, 111-115, 124-126, 130, 132, 136-139, 150-152, 157f-161, 163, 165, 265, 288-290
    acute regulation systems, 30-33
    C-terminus, 76f, 81
    distribution, 72, 73, 129
    epithelial, 69, 70, 72, 74, 75, 77, 80-85
    gene, 124
    phosphorylation, 31, 32, 81
NHE4, 2t, 4, 24, 27, 29t, 48, 72, 80, 111, 113t, 115, 124-126, 138, 153, 255f, 265, 288-290, 295
    distribution, 72, 73, 130
    expression, 129
    gene, 128
NHE5, 2t, 4, 111, 124-126, 138, 153, 288-290
    expression, 130
    gene, 128
Nigericin, 12, 192, 230, 270-275, 277-280
Nippon Medical School, 179
Nonpolar cells, 22
Norepinephrine, 49, 177
Normoxic cells, 191f. See also Ischemia.
N-terminal, 124, 125, 139, 290, 291
    domain, 6, 70, 72, 75, 150
    hydrophobic domain, 5, 130

O

Okadaic acid, 58, 73-76f, 80, 151, 152
OKP cells, 33, 35-37
Oligomerization, 2t, 10
Opie LH, 200t, 201
Opossum kidney (OK) cells, 24, 74, 113t, 114, 136

Orlowski J, 69, 85, 112, 115, 130, 134, 136
Osmolytes, 101, 103, 252
Osmotic shrinking, 8
Oubain, 192

P

Padan E, 3, 229, 238, 244
Parathyroid hormone (PTH), 30-32, 153
Parker JC, 103
pH
  cytoplasmic, 3, 8, 12, 30, 47, 153, 156f
    sensor, 57, 58
  external, 12
  extracellular (pH$_e$), 12, 156f, 173, 174, 264, 265, 271-275, 277-280
  homeostasis, 2, 3, 112, 241. *See also* Homeostasis.
  intracellular (pH$_i$), 1, 10, 12, 49, 51, 52, 54, 78f, 80-82, 98, 103, 104, 112, 123, 132, 139, 149-151, 154-156f, 161, 163, 164, 187-189f, 191-193, 204-207, 216, 217f, 263, 264
    regulation, 1, 3, 10, 11, 48, 129, 242, 264-268
  low, 34
  maintenance domain, 160, 161
  paradox, 204
  regulation, 4, 28
    in cells, 3
  regulatory proteins, 3
  -sensor, 8, 151, 232, 234, 239
Phe, 6, 138
Phorbol esters, 8, 10, 32, 52, 56, 58, 104, 106, 133-136, 152, 177, 188, 193
Phorbol myristate acetate (PMA), 32, 58, 73, 75-77, 79f-81, 114,135, 151, 158, 161, 163, 188, 189f
Phosphorylation, 152, 153, 161, 188
Pierce GN, 176, 200t, 201, 223
Pileup, 238f, 240f, 292f
Pinner E, 238
Pitts RF, 21
Piwnica-Worm D, 192
pK, 233
Plasma membrane Na$^+$/H$^+$ exchanger
  characteristics of, 123-140
    functional characteristics, 132, 133
      biochemical properties, 132
      pharmacological properties, 132, 133
    genomic organization, 128
    membrane topology, 130, 131
    molecular heterogeneity, 124-127
      chromosomal mapping, 126, 127
      isolation of multiple isoforms, 124, 125
    regulation of activity, 133-137
      acute, 134-137
      chronic, 137
    structural components of, 137-139
      determinants of amiloride sensitivity, 138, 139
      glycosylation, 138
      H$^+$-modifier site, 139
      quaternary structure, 139
    tissue expression, 128-130
Platelet-derived growth factor (PDGF), 8, 10, 51, 52, 55, 59, 153, 161
Platelets
  NHE in, 206, 207
Pogwizd SM, 178
Polarized epithelia, 22, 72, 111, 129, 132, 133, 149
Postischemic ventricular recovery
  NHE in lactate-induced depression, 204
Pouysségur J, 48, 57, 61, 104, 138, 165, 232, 269
Primary response genes, 35
Prinzmetals' angina, 171
Prokaryotic Na$^+$/H$^+$ exchange, 22
Protein kinase
  A (PKA), 30, 31, 33, 58, 74, 81, 106, 107, 113t-115, 130, 135, 136, 152, 153, 160, 161, 163
    inhibitors, 35
  C (PKC), 8, 10, 30, 32, 35, 36, 54, 57-59, 74, 84, 85, 96, 106, 113t-115, 130, 135-137, 151-154, 163, 177, 188, 191f, 193, 194
  G, 153
Proximal tubule, 22-27, 29t, 32-36, 58, 74, 115, 128, 129, 136, 149
  functions, 22, 23
PSORT program, 295, 296f

Q

Quaternary structure of exchangers, 139, 150

R

Rabbit S$_2$ proximal tubule cells (RKPC-2), 10, 24, 113t, 129
Radiation of tumors, 264, 279-281
Rao GN, 56, 61
Regulatory
  cofactors, 160, 161
  volume
    decrease (RVD), 101, 103, 104
    increase (RVI), 102-106
Renal
  cortical brush border vesicles, 30-32, 138
  tubular Na$^+$/H$^+$ exchange, 21, 29
Reperfusion, 10, 13, 187, 192, 193, 196f-198, 207
  anoxic, 172
  injury, 133, 198, 204-206, 218
  nonpharmacological protection, 203, 204
  of myocardium, 11, 171-179, 195
Restenosis, 11
Resting tension of heart, 199f
Retenoic acid, 91, 95, 98
Rothman, 229
Rotin D, 104, 271
Ruifrok ACC, 279
Ryanodine, 178

S

Sack S, 200t
Saint Thomas' Hospital Heart Research Trust (STRUTH), 179
Sarcolemmal Na$^+$/H$^+$ exchanger
   expression and activity in health and disease, 215-223
      intracellular hydrogen, 215, 216
      kinetics, 216-218
      response to disease, 218-223
         diabetic cardiomyopathy, 221
         ischemia and hypoxia, 218-221
         normal and abnormal cardiac growth, 221-223
   role in arrhythmogenesis, 171-179
      calcium, common arrythmogenic mediator, 178, 179
      free oxygen radical involvement, 172, 173
      Na$^+$/H$^+$ exchangers' key role, 173-178
         arrythmogenic role evidence, 174, 175
         chronic regulation impact, 178
         interaction with Na$^+$/K$^+$ pump, 175, 176
         mechanism of activation, 173
         neurohormonal modulation, 176, 177
         selectivity of pharmacological inhibitors, 176
Sardet C, 5, 56, 69, 104, 111, 123, 124, 152, 288
Schmitt, 153
Scholz W, 200t
Schuldiner S, 229, 238
Shilo, Moshe, Center for Biogeochemistry, 244
Shimada Y, 179
Shrode LD, 106
Shull G, 69, 130
Siczkowski M, 60
Signaling pathways of NHE activation, 109, 111, 133, 134, 137, 151, 161, 165
   schematic, 110f
Singh D, 13
Sodium, see Na$^+$.
sod2, 252-260
Solcoz M, 128
Song CW, 279
Sparks RL, 275
Spontaneously hypertensive rats (SHR), 54, 56, 60
SP-1 site(s), 94, 128
Steenbergen C, 205
Steitz TA, 70
Stress-activated protein kinase(s) (SAPK), 109, 163
Studies demonstrating Na$^+$/H$^+$ inhibitions' protective effects in ischemic reperfused heart, 200, 201
Szpirer C, 126

T

Takeda M, 108
Takewaki S, 56
Tani M, 195, 200t
Taubman MB, 51
Tetracycline, 239, 241
Thiry-Vella loops, 83
Threonine, 108
   kinase, 130
Thrombin, 8, 51, 54, 59, 104, 107, 152, 157f-159f, 164, 177, 206
   alpha (α), 80, 161
   receptor, 57
Thromboxane A$_2$, 206
Thyroid hormone, 10, 34t, 93f, 134
Topology of
   membrane, 130, 131f
   Na$^+$/H$^+$ exchanger protein, 12
   NHE isoforms, 30, 70, 71f
TopPredII, 5f, 6t
Transmembrane
   domain(s), 5, 6, 252
      model of NHE1, 7f
   helices, 5, 70, 72
   passage, 5
   region(s), 4, 48, 138, 139
   spanning segments (TMS), 124, 130, 138, 151, 152, 160, 229, 239, 295, 296f
Tse C-M, 72, 73, 81
Tseng H, 61
Tumors
   acidic microenvironment, 263, 264, 269, 271
   Na$^+$/H$^+$ exchanger in, 11, 12, 263-281
   selective therapy, 13, 263-281
Tyrosine, 108
   kinase, 30, 33, 35, 36, 52f, 57, 85, 115, 137
      inhibitors, 33, 35
      receptors, 153
   phosphorylation, 33, 36

V

Vairo G, 55
Valinomycin, 230, 231
Vallega G, 61
Vascular
   smooth muscle, 10, 11
      and Na$^+$/H$^+$ exchanger regulation, 47-62
   cell growth, 11
   mitogens, 52, 54, 56, 58, 60
Vasopressin, 3, 49
Ventricular
   fibrillation, 171, 172, 174-179
   tachycardia, 171, 172, 174, 178
Villareal ML, 153

W

Wakabayashi S, 57, 59, 104, 139, 160
Wang D, 138
Wang H, 13
Wang Z, 125
Ward CA, 201*t*, 207
Watson J, 207
Weinman EJ, 161
Weiss RG, 200*t*
Wike-Hooley JL, 264
Williams B, 56
Winkel GK, 57, 134
Winston DC, 175
Wistar-Kyoto (WKY) rats, 54, 56, 60
Wood PJ, 280

Y

Yamada M, 172
Yang W, 13, 98
Yasutake M, 200*t*
Yeast, 251-261, 295
Yeo EJ, 83
Yip, 72, 73, 81
Young PG, 252, 261
Ytrehus K, 201*t*
Yu f, 136

Z

zsod2, 256

Molecular Biology Intelligence Unit
Available and Upcoming Titles

- ☐ Organellar Proton-ATPases
  Nathan Nelson, Roche Institute of Molecular Biology
- ☐ Interleukin-10
  Jan DeVries and René de Waal Malefyt, DNAX
- ☐ Collagen Gene Regulation in the Myocardium
  M. Eghbali-Webb, Yale University
- ☐ DNA and Nucleoprotein Structure In Vivo
  Hanspeter Saluz and Karin Wiebauer, HK Institut-Jena and GenZentrum-Martinsried/Munich
- ☐ G Protein-Coupled Receptors
  Tiina Iismaa, Trevor Biden, John Shine, Garvan Institute-Sydney
- ☐ Viroceptors, Virokines and Related Immune Modulators Encoded by DNA Viruses
  Grant McFadden, University of Alberta
- ☐ Bispecific Antibodies
  Michael W. Fanger, Dartmouth Medical School
- ☐ Drosophila Retrotransposons
  Irina Arkhipova, Harvard University and Nataliya V. Lyubomirskaya, Engelhardt Institute of Molecular Biology-Moscow
- ☐ The Molecular Clock in Mammals
  Simon Easteal, Chris Collet, David Betty, Australian National University and CSIRO Division of Wildlife and Ecology
- ☐ Wound Repair, Regeneration and Artificial Tissues
  David L. Stocum, Indiana University-Purdue University
- ☐ Pre-mRNA Processing
  Angus I. Lamond, European Molecular Biology Laboratory
- ☐ Intermediate Filament Structure
  David A.D. Parry and Peter M. Steinert, Massey University-New Zealand and National Institutes of Health
- ☐ Fetuin
  K.M. Dziegielewska and W.M. Brown, University of Tasmania
- ☐ Drosophila Genome Map: A Practical Guide
  Daniel Hartl and Elena R. Lozovskaya, Harvard University
- ☐ Mammalian Sex Chromosomes and Sex-Determining Genes
  Jennifer A. Marshall-Graves and Andrew Sinclair, La Trobe University-Melbourne and Royal Children's Hospital-Melbourne
- ☐ Regulation of Gene Expression in *E. coli*
  E.C.C. Lin, Harvard University
- ☐ Muscarinic Acetylcholine Receptors
  Jürgen Wess, National Institutes of Health
- ☐ Regulation of Glucokinase in Liver Metabolism
  Maria Luz Cardenas, CNRS-Laboratoire de Chimie Bactérienne-Marseille
- ☐ Transcriptional Regulation of Interferon-γ
  Ganes C. Sen and Richard Ransohoff, Cleveland Clinic
- ☐ Fourier Transform Infrared Spectroscopy and Protein Structure
  P.I. Haris and D. Chapman, Royal Free Hospital-London
- ☐ Bone Formation and Repair: Cellular and Molecular Basis
  Vicki Rosen and R. Scott Thies, Genetics Institute, Inc.-Cambridge
- ☐ Mechanisms of DNA Repair
  Jean-Michel Vos, University of North Carolina
- ☐ Short Interspersed Elements: Complex Potential and Impact on the Host Genome
  Richard J. Maraia, National Institutes of Health
- ☐ Artificial Intelligence for Predicting Secondary Structure of Proteins
  Xiru Zhang, Thinking Machines Corp-Cambridge
- ☐ Growth Hormone, Prolactin and IGF-I as Lymphohemopoietic Cytokines
  Elisabeth Hooghe-Peters and Robert Hooghe, Free University-Brussels
- ☐ Human Hematopoiesis in SCID Mice
  Maria-Grazia Roncarolo, Reiko Namikawa and Bruno Péault DNA Research Institute
- ☐ Membrane Proteases in Tissue Remodeling
  Wen-Tien Chen, Georgetown University
- ☐ Annexins
  Barbara Seaton, Boston University
- ☐ Retrotransposon Gene Therapy
  Clague P. Hodgson, Creighton University
- ☐ Polyamine Metabolism
  Robert Casero Jr, Johns Hopkins University
- ☐ Phosphatases in Cell Metabolism and Signal Transduction
  Michael W. Crowder and John Vincent, Pennsylvania State University
- ☐ Antifreeze Proteins: Properties and Functions
  Boris Rubinsky, University of California-Berkeley
- ☐ Intramolecular Chaperones and Protein Folding
  Ujwal Shinde, UMDNJ
- ☐ Thrombospondin
  Jack Lawler and Jo Adams, Harvard University
- ☐ Structure of Actin and Actin-Binding Proteins
  Andreas Bremer, Duke University
- ☐ Glucocorticoid Receptors in Leukemia Cells
  Bahiru Gametchu, Medical College of Wisconsin
- ☐ Signal Transduction Mechanisms in Cancer
  Hans Grunicke, University of Innsbruck
- ☐ Intracellular Protein Trafficking Defects in Human Disease
  Nelson Yew, Genzyme Corporation
- ☐ apoJ/Clusterin
  Judith A.K. Harmony, University of Cincinnati
- ☐ Phospholipid Transfer Proteins
  Vytas Bankaitis, University of Alabama
- ☐ Localized RNAs
  Howard Lipshitz, California Institute of Technology
- ☐ Modular Exchange Principles in Proteins
  Laszlo Patthy, Institute of Enzymology-Budapest
- ☐ Molecular Biology of Cardiac Development
  Paul Barton, National Heart and Lung Institute-London
- ☐ RANTES, *Alan M. Krensky, Stanford University*
- ☐ New Aspects of V(D)J Recombination
  Stacy Ferguson and Craig Thompson, University of Chicago

Neuroscience Intelligence Unit

Available and Upcoming Titles

- Neurodegenerative Diseases and Mitochondrial Metabolism
  M. Flint Beal, Harvard University

- Molecular and Cellular Mechanisms of Neostriatum
  Marjorie A. Ariano and D. James Surmeier,
  Chicago Medical School

- Ca^{2+} Regulation By Ca^{2+}-Binding Proteins in Neurodegenerative Disorders
  Claus W. Heizmann and Katharina Braun,
  University of Zurich, Federal Institute for Neurobiology, Magdeburg

- Measuring Movement and Locomotion: From Invertebrates to Humans
  Klaus-Peter Ossenkopp, Martin Kavaliers and Paul Sanberg, University of Western Ontario and University of South Florida

- Triple Repeats in Inherited Neurologic Disease
  Henry Epstein, University of Texas-Houston

- Cholecystokinin and Anxiety
  Jacques Bradwejn, McGill University

- Neurofilament Structure and Function
  Gerry Shaw, University of Florida

- Molecular and Functional Biology of Neurotropic Factors
  Karoly Nikolics, Genentech

- Prion-related Encephalopathies: Molecular Mechanisms
  Gianluigi Forloni, Istituto di Ricerche Farmacologiche "Mario Negri"-Milan

- Neurotoxins and Ion Channels
  Alan Harvey, A.J. Anderson and E.G. Rowan,
  University of Strathclyde

- Analysis and Modeling of the Mammalian Cortex
  Malcolm P. Young, University of Oxford

- Free Radical Metabolism and Brain Dysfunction
  Irène Ceballos-Picot, Hôpital Necker-Paris

- Molecular Mechanisms of the Action of Benzodiazepines
  Adam Doble and Ian L. Martin,
  Rhône-Poulenc Rorer and University of Alberta

- Neurodevelopmental Hypothesis of Schizophrenia
  John L. Waddington and Peter Buckley,
  Royal College of Surgeons-Ireland

- Synaptic Plasticity in the Retina
  H.J. Wagner, Mustafa Djamgoz and Reto Weiler,
  University of Tübingen

- Non-classical Properties of Acetylcholine
  Margaret Appleyard, Royal Free Hospital-London

- Molecular Mechanisms of Segmental Patterning in the Vertebrate Nervous System
  David G. Wilkinson,
  National Institute of Medical Research-UK

- Molecular Character of Memory in the Prefrontal Cortex
  Fraser Wilson, Yale University